# Methods in Neurosciences

Volume 14

## Paradigms for the Study of Behavior

# Methods in Neurosciences

Edited by

## P. Michael Conn

*Department of Pharmacology*
*The University of Iowa*
*College of Medicine*
*Iowa City, Iowa*

Volume 14

# Paradigms for the Study of Behavior

**ACADEMIC PRESS, INC.**
*A Division of Harcourt Brace & Company*
San Diego   New York   Boston   London   Sydney   Tokyo   Toronto

This book is printed on acid-free paper. ∞

Academic Press, Inc.
1250 Sixth Avenue, San Diego, California 92101

*United Kingdom Edition published by*
Academic Press Limited
24–28 Oval Road, London NW1 7DX

International Standard Serial Number: 1043-9471

International Standard Book Number: 0-12-185277-6 (Hardcover) (alk. paper)

PRINTED IN THE UNITED STATES OF AMERICA
93  94  95  96  97  98    EB    9  8  7  6  5  4  3  2  1

# Table of Contents

## Section VI  Other Behavior

# Contributors to Volume 14

*Article numbers are in parentheses following the names of contributors. Affiliations listed are current.*

J. H. ABRAINI (16), Biologie des Hautes Pressions, Faculte de Medicine Nord, 13326 Marseille Cedex 15, France

ENRICO ALLEVA (7), Section of Behavioral Pathophysiology, Laboratorio di Fisiopatologia di Organo e di Sistema, Instituto Superiore di Sanita, 1-00161 Rome, Italy

DAVID E. ANDERSON (26), Laboratory of Behavioral Sciences, National Institute on Aging, National Institutes of Health, Gerontology Research Center, Baltimore, Maryland 21224

JULIO M. AZCURRA (10), Departamento de Ciencias Biológicas, Facultad de Ciencias Exactas y Naturales, Universidad de Buenos Aires, Ciudad Universitaria, 1428 Buenos Aires, Argentina

J. LEE BEVERLY (20), Department of Surgery, University Hospital, State University of New York, Health Science Center, Syracuse, New York 13210

RICHARD J. BODNAR (18), Department of Psychology and Neuropsychology, Queens College, City University of New York, Flushing, New York 11367

GEORG ANDREES BÖHME (14), Rhône-Poulenc Rorer, Centre de Recherche de Vitry-Alfortville, 94403 Vitry-sur-Seine Cedex, France

JUAN L. BRUSÉS (10), Department of Physiology and Neurobiology, The University of Connecticut, Storrs, Connecticut 06269

THOMAS W. CASTONGUAY (19), Department of Human Nutrition and Food Systems, University of Maryland, College Park, Maryland 20742

MARIA ESTER CELIS (24), Department of Pharmacology, Facultad de Ciencias Quimicas, 5016 Cordoba, Argentina

PAUL F. CHAPMAN (28), Department of Psychology and Graduate Program in Neuroscience, University of Minnesota, Minneapolis, Minnesota 55455

JAMES A. CHERRY (1), Cold Spring Harbor Laboratory, Beckman Neuroscience Center, Cold Spring Harbor, New York 11724

JOHN T. CLARK (3, 4), Department of Physiology, Meharry Medical College, Nashville, Tennessee 37208

MARGARET G. CUTLER (6), Department of Biological Sciences, Glasgow Polytechnic, Glasgow G4 OBA, Scotland, United Kingdom

GABRIELA DÍAZ-VÉLIZ (11), Departamento Preclínicas, Facultad de Medicina, Division Ciencias Médicas Oriente, Universidad de Chile, Santiago 9, Chile

DEAN DLUZEN (5), Department of Anatomy, Northeastern Ohio Universities, College of Medicine, Rootstown, Ohio 44272

SAMUEL F. DWORKIN (21), Department of Oral Medicine, University of Washington, School of Dentistry, Seattle, Washington 98195

BERNARD T. ENGEL (26), Laboratory of Behavioral Science, National Institute on Aging, National Institutes of Health, Gerontology Research Center, Baltimore, Maryland 21224

KEVIN C. F. FONE (12), Department of Physiology and Pharmacology, Medical School, Queen's Medical Centre, Nottingham University, Nottingham NG7 2UH, England

JOHN R. GLEASON (20), Department of Statistics and Measurement, College of Arts and Sciences, Syracuse University, Syracuse, New York 13244

CHRISTOPHER J. GORDON (17), Neurotoxicology Division, Health Effects Research Laboratory, U.S. Environmental Protection Agency, Research Triangle Park, North Carolina 27711

JAANUS HARRO (23), Laboratory of Psychopharmacology, Institute of General and Molecular Pathology, Tartu University, Verski 34, EE-2400 Tartu, Estonia

KENJI ICHIHARA (13), Department of Chemical Pharmacology, Faculty of Pharmaceutical Science, Meijo University, Nagoya 468, Japan, and Research Laboratory, Maruko Pharmaceutical Co., Ltd., Kasugai, 486, Japan

THEODORE J. KALOGERIS (19), Department of Physiology and Biophysics, Louisiana State University School of Medicine, Shreveport, Louisiana 71130

ERNEST D. KEMBLE (8), Division of Social Sciences, University of Minnesota–Morris, Morris, Minnesota 56267

PHYLLIS E. MANN (22), Department of Comparative Medicine, Tufts University School of Veterinary Medicine, North Grafton, Massachusetts 01536

PETER MARLER (29), Department of Zoology, Section of Animal Physiology, Animal Communication Laboratory, University of California, Davis, Davis, California 95616

MICHAEL M. MEGUID (20), Department of Surgery, University Hospital, State University of New York, Health Science Center, Syracuse, New York 13210

SERGIO MORA (11), Departamento Preclínicas, Facultad de Medicina, División Ciencias Médicas Oriente, Universidad de Chile, Santiago 9, Chile

TOSHITAKA NABESHIMA (13), Department of Neuropsychopharmacology and Hospital Pharmacy, Nagoya University School of Medicine, Showa-ku, Nagoya 466, Japan

DOUGLAS A. NELSON (29), Department of Zoology, Section of Animal Physiology, Animal Communication Laboratory, University of California, Davis, Davis, California 95616

M. J. PARKES (15), Department of Physiology, Royal Free Hospital School of Medicine, London NW3 2PF, England

EDWARD O. PRICE (2), Department of Animal Sciences, University of California, Davis, Davis, California 95616

VICTOR D. RAMIREZ (5), Department of Physiology and Biophysics, University of Illinois, Urbana, Illinois 61801

ROBERTO REFINETTI (17), Department of Psychology, College of William and Mary, Williamsburg, Virginia 23187

MICHEL REIBAUD (14), Rhône-Poulenc Rorer, Centre de Recherche de Vitry-Alfortville, 94403 Vitry-sur-Seine Cedex, France

J. C. ROSTAIN (16), Biologie des Hautes Pressions, Faculte de Medicine Nord, CNRS-URA 1330, 13326 Marseille Cedex 15, France

ZOLTAN SARNYAI (9), Alcohol and Drug Abuse Research Center, McLean Hospital–Harvard Medical School, Belmont, Massachusetts 02178

LONNIE L. SEARS (28), Department of Pediatrics, University of Louisville, Louisville, Kentucky 40292

LENNART SVENSSON (25), Department of Pharmacology, University of Göteborg, S-413 90 Göteborg, Sweden

MARK I. TALAN (26), Laboratory of Behavioral Sciences, National Institute on Aging, National Institutes of Health, Gerontology Research Center, Baltimore, Maryland 21224

C. Tomei (16), Biologie des Hautes Pressions, Faculte de Medicine Nord, 13326 Marseille Cedex 15, France

Enrique Torre (24), Department of Neurosciences, University of Virginia, School of Medicine, Charlottesville, Virginia 22908

T. R. Vidyasagar (27), Center for Visual Science, and Division of Neurosciences, John Curtin School of Medical Research, Canberra, ACT 2601, Australia

Leanne Wilson (21), Department of Oral Medicine, University of Washington, School of Dentistry, Seattle, Washington 98195

Zhong-Jin Yang (20), Department of Surgery, University Hospital, State University of New York, Health Science Center, Syracuse, New York 13210

# Preface

Behavior is the culmination of the gene expression, ligand–receptor interactions, and all the biochemistry that makes an organism precisely that! The measurement of behaviors is, owing to its intrinsic complexity, exceedingly difficult, and approaches differ from one research group to another.

This volume is dedicated to detailed descriptions of the means for measuring a wide range of behaviors. It describes sexual, reproductive, and maternal behaviors, social interactions, aggression, drug- and environment-induced behaviors, amnesia, maze tests, feeding behavior, illness behavior, exploratory behavior, grooming behavior, cardiovascular responses, analysis of memory, and classical conditioning.

The goal of this volume—and of others in this series—is to provide in one source a view of the contemporary techniques significant to a particular branch of neurosciences, information which will prove invaluable not only to the experienced researcher but to the student as well. Of necessity some archival material will be included, but the authors have been encouraged to present information that has not yet been published, to compare (in a way not found in other publications) different approaches to similar problems, and to provide tables that direct the reader, in a systematic fashion, to earlier literature and as an efficient means to summarize data. Flow diagrams and summary charts will guide the reader through the processes described.

The nature of this series permits the presentation of methods in fine detail, revealing "tricks" and shortcuts that frequently do not appear in the literature owing to space limitations. Lengthy operating instructions for common equipment will not be included except in cases of unusual application. The contributors have been given wide latitude in nomenclature and usage since they are best able to make judgments consistent with current changes.

I wish to express my appreciation to Mrs. Sue Birely for assisting in the organization and maintenance of records and to the staff of Academic Press for their efficient coordination of production. Appreciation is also expressed to the contributors, particularly for meeting their deadlines for the prompt and timely publication of this volume.

P. Michael Conn

# Methods in Neurosciences

# Section I

## Sexual and Reproductive Behavior

# [1] Measurement of Sexual Behavior: Controls for Variables

James A. Cherry

Sexual behavior in many animals can be readily observed under standard laboratory conditions. Although often considered to be a complex form of behavior, mating typically consists of a series of stereotyped and well-characterized events that are relatively conserved between a wide range of species. Thus, with some patience and a reasonably small budget, the study of sexual behavior is within the realm of virtually any neuroscientist. This chapter outlines basic considerations for setting up the proper conditions for observing sexual behavior and then discusses possible methods of scoring and analysis. Finally, a standard procedure for observing behavior in the laboratory is described. To generalize for the largest audience, discussion focuses primarily on rodent sexual behavior.

## Factors Influencing Sexual Behavior and Its Measurement

It is a simple matter to place adult male and female rats together and be relatively assured that behaviors of a sexual nature will occur. In fact, mating can proceed in many species in the laboratory under most conditions without apparent distraction. However, the ease with which one observes such behavior can obscure the fact that a number of important factors may contribute to both the quality and quantity of behaviors that are ultimately displayed. The strain, age, and sexual experience of animals all affect sexual performance and must be standardized. Husbandry practices that include photoperiod and housing or social environment may be equally important. Conditions of testing, including the size and complexity of the test arena, presence of odors, hormonal condition of subjects, and the time of day, will also affect performance. Fortunately, in most cases there are no "correct" conditions or factors that must be used. What is important, however, is maintaining tight control over whatever conditions are decided on. This is the best way to reduce the variability inherent in behavioral data as well as to prevent undesired influences from confounding the experimental treatments.

*Methods in Neurosciences, Volume 14*

## Species/Strain

The question of what species or strain to examine is usually dictated by an investigator's general past usage. However, in cases where this choice can be made, there are several factors that need to be considered. For laboratory-bred rodents, strains can differ greatly with respect to reliable performance during staged matings, as well as in absolute behavioral measures such as mount frequency and ejaculation latency (1–3). If pregnancy and measures of fertility are part of the experimental design, one should be aware that fecundity is well known to vary between different strains of rats and mice.

A thorough understanding of the mating cycle of a species will also be required in order to run controlled behavioral studies in the laboratory. For most standard laboratory rodents—rats, mice, hamsters—ovulation and periods of sexual receptivity occur spontaneously in regular 4- or 5-day cycles. However, for "reflex" ovulating species, including voles, rabbits, cats, and ferrets, ovulation is induced by sensory and mechanical stimulation provided by the male. In ferrets, as with many mammalian species, onset of reproductive activity occurs in response to increasing day length (see below). Unlike the spontaneously ovulating females that exhibit reoccurring cycles of receptivity, female ferrets, once in estrus, may remain sexually receptive for weeks if mating with a male does not occur. These species differences thus require modified procedures for the preparation and maintenance of reproductively active animals (see section on Stimulus Animals).

## Age

Rodents mature quickly. Hamsters, rats, and mice are capable of reproducing anywhere from 7 to 10 weeks of age if allowed to mature naturally, but they can display sexual behaviors much earlier if provided with exogenous steroids (4, 5). Peak sexual performance in rodents can usually be observed for at least 6 months, and of course much longer for species with long life spans. Conversely, decline in reproductive output, for a variety of known and unknown reasons, occurs with age (6, 7).

## Sexual Experience

Copulation is usually not as polished in first-time maters at it is after experience has been acquired. In hamsters, the behavior of sexually inexperienced males is often characterized by incorrect orientation toward the female, resulting in ectopic mounts of the head, for example (8). To the extent that it has been studied, sexually mature female rodents seem to exhibit the appropriate sexual behaviors on their first exposure to sexually active males. Unless the ontogeny of behavior is of interest, it is important to give males a sufficient amount of sexual experience with receptive females prior to testing in order to bring all males "up to speed." Inexperienced male rodents generally require exposure to sexually receptive females at least two or three times on separate days for periods of 10 min or more before consistent measures are obtained in behavioral tests. This practice reduces the variability in measurements made between repeated trials and as well provides a controlled background of sexual history for all male subjects. For certain experiments, pretest screening can be used to help eliminate males that fail to perform to a criterion level of sexual behavior, such as two ejaculations in 10 min. Although screening can ensure that only sexually capable males will be included in an experiment, one must recognize that the selected group will not be a random sample and therefore will no longer represent a population value. In such cases, care must be taken not to overly generalize when interpreting the results.

## Photoperiod

Many rodents are photoperiodic; that is, they rely on light cues to govern seasonal as well as daily activities, including reproductive behavior. The coupling of reproductive rhythms to photoperiod varies from species to species. In hamsters, for example, less than exactly 12.5 hr of light per day results in gonadal regression in males (9), accompanied by consequent reduction in sexual behavior (10). For most species, 14–16 hr of light per day is sufficient to ensure reproductive activity.

For females in particular, control of the daily light/dark cycle can be important for establishing consistent periods of estrus, or sexual heat. In rats, for example, constant light—such as might occur if the lights are inadvertently left on in the colony room—disrupts the natural estrous cycle, causing females to enter a state of persistent estrus and sexual receptivity (11). Although these effects may be much more acute for females, constant lighting over time can also affect aspects of sexual behavior in males (12).

In addition, sexual behavior is known to vary at different periods of time during the day relative to the photoperiod. As owners of pet hamsters or gerbils know, many rodents are most active at night, during the dark. In fact, controlled experiments indicate that, for hamsters, males initiate copulation more rapidly and ejaculate sooner during the dark phase than in the light phase of the daily cycle (13). Although systematic observations of reproductive behavior at specific time points throughout the day have not been carried out for most animals, this example emphasizes the importance of testing animals at the same time each day. For nocturnally active animals, it is often helpful to set up reversed light/dark cycles in which lights are scheduled to turn off in the morning at the beginning of the working day, or at any other time that is convenient for observers to arrange behavioral observations.

## Housing Environment

After birth, litters are usually left with their mother (or a foster mother, if appropriate) until weaning at around 3 weeks of age for most rodents. Several choices for housing animals can be made at this time. Sexes may be separated, animals from different litters may be integrated, and/or individuals may be isolated in single cages. As mentioned earlier, there is no correct choice in most instances, but all animals in the same experiment should be given the same treatment. For example, the behavior of isolated individuals may differ in any number of ways from the behavior of group-housed animals. Similarly, other variations in rearing conditions have long been known to affect later, adult expression of behavior (14).

It is likely that juveniles or adults ordered directly from an animal supplier will have been raised under grouped conditions. For nonrodent species, or just to be sure, a call to the supplier about husbandry practices can be informative. Finally, of course, all animals in the same experiment should be kept in the same room, or, if necessary, in an identical room that receives the same care and is subject to the same environmental conditions (temperature, photoperiod, food/water).

## Test Apparatus

A perhaps surprisingly critical issue is in what environment the animals are to be tested. Animals often behave very differently in their home residence than in the cage of another. Because this may be of concern, animals are usually observed in a "neutral" arena. The arena can be virtually any container or enclosure that the observer chooses, so long as it permits unimpeded viewing and can be easily wiped clean between tests. The simplest type of arena is a glass aquarium; usually a 10 gallon (38 liter) container is of sufficient size for most rodents. For novices, it may be helpful to raise the aquarium on blocks and place a mirror beneath the glass floor at a 45° angle. The entire apparatus may be arranged at eye level to enable visualization of copulation from below, a perspective that can facilitate the distinction between mounts without intromissions from actual intromissions.

Most studies of sexual behavior have been conducted using an apparatus such as that just described. There is no question, however, that the sexual behavior of rodents in the wild occurs under different circumstances, and some investigators have attempted to simulate more accurately these conditions. To examine rat sociosexual behaviors under a more natural context while maintaining the controlled conditions of the laboratory, McClintock and Adler (15) constructed a "seminatural" environment. This was nothing more than a large wooden box containing numerous objects that added to the complexity of the surroundings. Observation of sexual behavior in these enclosures enabled the authors to reveal aspects of feminine sexual behavior not usually exhibited in the more confined aquarium setup.

However, even the simple aquarium can be modified to create a different kind of and, it might be argued, more natural environment in which to observe so-called pacing behavior during mating in rats. In this arrangement, a partition having a hole small enough for only the female to fit through is placed into the aquarium, dividing it into two compartments. Females, by virtue of being smaller than males, are free to travel in and out of both compartments through the small hole in the partition. Males, however, are too large to fit through the hole and are confined to only one compartment. Similar conditions have been arranged by other investigators through the use of tethers or harnesses fitted to males, effectively restricting the mobility of males to one side of the test apparatus (16). Under such circumstances, females can "pace" the number of contacts that she receives from the male. Not surprisingly, measurements of sexual behavior, such as the interval between intromissions and duration of estrus, differ considerably when females, and not males, are permitted to regulate sexual interactions (17).

In addition to the size of the arena and the presence of barriers to limit access between participants, the shape of the test apparatus may also be varied to study specific types of behaviors. For example, Steel (18) has used donut-shaped, concentric runways for the study of sociosexual interactions between male and female hamsters.

## Stimulus Animals

Typically, one is interested in examining the performance of individual subjects (as opposed to pairs), requiring that partners be provided by the investigator to induce or stimulate sexual responses in the experimental animal. (The "experimental" animal or subject refers to the individual that received an experimental treatment and whose behavior is the topic of interest.) Because the behavior of the experimental subject can depend critically on the behavioral state of the partner in a sexual pairing, the partner, or "stimulus animal," can have a profound impact on the outcome of a test. It is therefore essential that sufficient numbers of stimulus males and females will be available once behavioral observations begin.

Unless measures of pregnancy and fertility are of interest, gonadally intact animals are not desirable for use as stimulus animals for several reasons. First, the investigator has less control over maintaining the reproductive competency of gonadally intact animals compared to gonadectomized subjects treated with exogenous hormones. And, of course, intact females in natural estrus can provide only a limited period of sexual receptivity due to the likelihood that pregnancy or pseudopregnancy may ensue following interactions with males. In addition, the behavior of gonadally intact animals—particularly males—will often be much less reliable from day to day than the behavior of hormone-treated, gonadectomized subjects. There is also a tendency for variability in the behavior of different individuals to be greater in gonadally intact subjects compared to hormone-treated castrates.

Generally, the most efficient strategy for generating sexually active stimulus animals is to create and maintain as part of the colony a group of gonadectomized males and females that can be used repeatedly for behavioral testing. These animals can be kept in a state of sexual readiness by the administration of exogenous, replacement hormones. This is accomplished either through daily injection or by subcutaneous implantation of capsules or pellets that tonically release small quantities of the hormone desired. A large literature over the past several decades has described the various effects and effectiveness of replacement regimens of different androgens and estrogens in restoring copulatory behavior in vertebrates. In short, sexual activity in most castrated male vertebrates can be maintained or restored quite well with testosterone, given either as implants or as daily injections after dissolving in an oil vehicle. In the latter case, the propionated form of testosterone is most widely used. For spontaneously ovulating rodents, sexual receptivity in females can be induced by giving injections of 0.5 to 10 $\mu$g of estradiol benzoate 24 and/or 48 hr before testing followed by 250 to 500 $\mu$g of progesterone 4 hr before testing. This regimen of hormone replacement for rats, hamsters, mice, and guinea pigs will effectively generate highly receptive females, although variations in the hormone doses and injection schedules may produce equivalent or better results (19). For reflex ovulating females, progesterone is not required for the display of sexual receptivity, and estradiol alone, given either tonically or applied in pulses, is sufficient to induce receptive behaviors.

Because one is usually interested in evaluating the sexual performance of subjects under optimal conditions, the stimulus should be an animal that is prepared by the investigator to display high levels of sexual activity. However, the quality of the stimulus used can be varied depending on the experimental question being asked. Subtle differences between the behavioral capacities of animals subjected to different experimental treatments may be better revealed by testing under suboptimal conditions (i.e., with a less-than-willing partner). In such cases, the hormone replacement regimen for the stimulus can be adjusted accordingly.

Although there are studies indicating that measurements of sexual behavior obtained following appropriate hormone replacement of gonadectomized animals may not be identical to those from intact animals (20, 21), the ability to control hormone dosage is an advantage that often outweighs this problem. After using a number of animals, one also recognizes eventually that the same dose of hormone given to individual animals can produce quite different levels of response. This observation has led to studies of the "dud–stud" phenomenon, in which certain males fail to exhibit the expected level of sexual performance even after high doses of testosterone are given (22, 23). Such males are obviously not suitable as stimulus animals for examining sexual behavior in female subjects and should not be used. This problem illustrates the importance of constantly inspecting and evaluating the performance of stimulus subjects, and immediately replacing inadequate animals.

It is also best not to overuse stimulus animals, as they can become sexually exhausted and less likely to provide optimal sexual stimulation. Ideally, stimulus females would be used only once per session, as just-mated females may be perceived differently by experimental males than fresh, unmated females. Also, females may become less receptive to subsequent males, especially if tests are prolonged. For stimulus males, repetitive use may cause them to tire if used for too may trials with females on the same day. This may be more of a problem if males are intact and being used to inseminate experimental females. Sperm becomes depleted in males after multiple ejaculations have been delivered (24, 25), and a week or more may be required before sperm counts return to normal.

## Measurement of Sexual Behavior

### Operational Definitions

Before formal collection of data is initiated, it is critical that a sufficient amount of preparative time be invested observing behaviors in the identical context in which actual measurements will be made. There are at least four important reasons for this premeasurement period. First, it is necessary to decide on what behaviors will be measured and, second, what the operational definitions for each behavior will be. Such definitions consist of a precise description of each behavior being measured in order that a consistent definition for the occurrence of a particular behavior will always be applied within and between tests. These definitions may be trivial, such as to describe the occurrence of simple acts like single vocalizations. For certain behaviors, however, the transition from one to the next may be particularly subtle, requiring strict and consistently applied definitions. As an example, one can imagine a scenario during rat mating where in a matter of seconds a female might exhibit "approach to male" (female moves to within one body length of male beginning from a distance of at least two body lengths), followed by a "crawl over" (female climbs over male from one side to another) and a "dart" [a run of several steps terminated by a crouching posture; definitions modified from Edwards and Pfeifle (26)]. A more thorough discussion of operational definitions and data sampling techniques can be found elsewhere (27, 28).

It is also important that these definitions be formulated well in advance of the start of an experiment, then tested for relevance and accuracy while observing animals in the context in which the actual experiment will occur. Slight changes in the testing apparatus or odor environment may alter the expression of some behaviors, while adding or eliminating others. Finally, the need to practice real-time collection of data is imperative, especially for neophytes. There is no quicker waste of data than to discover toward the middle or end of an experiment that the criteria for defining particular behaviors have not been applied consistently during each test, resulting in imprecise and faulty measurements. The usual reason this occurs is that not enough time was spent sharpening one's behavioral eye during the preobservation period. Time invested in practice observations is rarely wasted.

## What to Measure

The decision of what behavioral variables need to be measured is a matter of informed judgment by the investigator. A number of measurements describing rodent sexual behavior have been presented in the literature, and they may be as obvious or obscure as required. For example, if an investigator has reason to believe that scent glands located on a specific region of the body provide specific information to the subject, sniffing duration or frequency may be measured separately for relevant areas.

In some cases, certain absolute measurements can be used in turn to make inferences about other aspects of behavior that are difficult to measure. For example, certain temporal variables, such as the latency to a first mount attempt or duration of anogenital investigation, can indicate something about the motivational state of the subject. However, the point has accurately been made that latencies to mount or intromit, rather than reflecting arousal or motivation, may instead represent impairments in performance ability or changes in erectile function (29). Thus, the variables selected for scoring need to be chosen with some insight, and their interpretation should be approached with appropriate caution.

The conceptual division of rodent sexual behavior into a distinct series of events, from precopulatory to postejaculatory, provides a framework for determining what behaviors to measure during a mating test. In general, most measures will fall into one of three categories: frequency, duration, or latency. Occasionally, as with the determination of female receptivity in rats, the intensity of a particular behavior—lordosis—can also be scored. The drawback to measuring intensity during the course of testing is that without the use of specific equipment to provide actual measurements of a given parameter, intensity becomes a subjective interpretation of the observer. Vocalization intensity, for example, can be measured with proper sound equipment, or, if such accuracy is not required, vocalizations can be scored as either loud, intermediate, or soft. However, there is no simple instrument available to measure lordosis in female rats. This has resulted in some observers computing a lordosis rating to measure receptivity, using a four-point scale based on a judgment of the intensity of each lordotic response (30).

In addition to the latency, frequency, and total duration of any given behavior, the time or interval between specific events can also be informative. Two interval measures commonly computed for rodent sexual behavior are the interintromission and postejaculatory intervals (III and PEI, respectively). The PEI, which is simply the time between an ejaculation and the resumption of mating, in particular has long been of interest to psychologists because it is thought to encompass a period in which dynamic changes in sexual arousal of the male occur (31). This measure has also been used to compare species differences in copulatory patterns (32). For a more extensive analysis of the theory and interpretation of various measures of rat copulatory behavior, see Sachs and Barfield (33).

## How to Score

There are many ways to collect data. The simplest and oldest way is to use a timer, pencil, and paper. Technology first entered with the advent of motorized chart recorders that could be used in conjunction with a switchboard or keyboard to store the occurrence of behavioral events as pen deflections on endless rolls of paper. These instruments made an improvement in the amount of data that could be collected during a single session, but at a price of countless hours spent converting the frequency and length of pen deflections into an interpretable format. The dust thickened on such units when dedicated computerized data recorders became available. The early instruments generally suffered from being virtually unprogrammable, limited in the number of events that could easily be scored, and relatively expensive, yet the immediate availability of event frequency and duration provided by such units at the touch of a button was a convenience that justified their use. Today, it would behoove most researchers to invest in a portable laptop computer. They are reasonably affordable, can store weeks' worth of data on a single hard disk, and may be programmed to suit the needs of an individual investigator. Furthermore, a number of programs written especially for the quantification of sexual behavior have been described in the literature and may be available from the respective authors on request.

Certain behaviors that are performed quickly or are otherwise difficult to view in real time may be accessible through the use of video recording. Purchase of video equipment places the effort to measure sexual behavior onto a different level of expense, yet, for certain applications, recording of test sessions may be necessary. Analysis of behaviors that might occur infrequently over a period of many hours is made possible by time-lapse videotaping, in which many hours of real-time behavior can be distilled into a much shorter period for review.

The first extensive studies of rat sexual behavior were accomplished using standard motion picture analysis (34, 35). More recently, video analysis was used to help measure the decline in female-directed behaviors by males following castration. Recording the 30 min interactions on tape enabled the observers to play back entire tests and accurately quantify rapid and subtle behaviors such as approach, pursuit, and climbing over (36). In fact, behavioral sessions can be taped routinely, whether or not a complete behavioral analysis will be undertaken. A record of each test can be useful to review in case more information is desired or in the event of lost data, and it is simple enough to record over and reuse videotapes if the information is not needed. Of course, unless certain data are difficult to obtain otherwise, videotaping test sessions for later review is no substitute for recording behavioral observations in real time. The length of time required to analyze behavior from a videotape can be quite forbidding depending on what measures are being scored.

Finally, there are specific gadgets that have been developed by different investigators for fairly esoteric applications. As one example, a design placing male and female rats in series within a relatively simple electric circuit has been used to measure the duration of sexual contacts during mating (37). Wet contact between the male and female (such as that which would occur during penile insertion) causes the circuit to be completed, and a pen deflection on an event recorder, or direct reading by a computer, marks the incident.

Use of this simple device has shown, for example, that in rats intromission duration is shorter than ejaculation duration, and that intromission duration is affected by the pacing of sexual contacts by the female (17).

## Procedure

After having controlled for the various factors itemized in the first section (Factors Influencing Sexual Behavior and Its Measurement), determined precisely what and how specific components of sexual behavior will be measured, and convinced oneself that he/she is a skilled observer of behavior, data collecting can begin. The following description outlines a typical behavioral session, with attention focused on small details that this author has found important.

Rat, mouse, and hamster females in natural estrus begin to become receptive in the early evening. Whereas gonadectomized females will not be subject to the same hormonal constraints, the most consistent behaviors will be observed with a hormone replacement regimen that mimics the natural hormonal profile (i.e., estradiol and progesterone treatments will be scheduled such that females will be receptive when the lights go out). Sexual behavior of rodents is therefore usually observed early during the dark phase of the light/dark cycle, using a small lamp with a dim (15–40 W) red light to provide sufficient illumination. Even for species in which copulation may occur throughout the day, such as in some vole species, observation during the dark phase under dim lighting may help to render the observer slightly less conspicuous.

The setup for observing sexual behavior in rodents is quite simple. The test chamber is placed in a quiet room in a location that can be easily viewed from the observer's position. It is a good idea to keep all animals other than the pair being tested in a separate room or at least a good distance from the testing chamber, as odors and/or vocal signals produced by other animals may interfere with or influence the behavior of the experimental pair.

It is important to allow all experimental subjects a period of habituation to the testing arena. This practice minimizes anxiety as well as the amount of exploration that animals make once being placed into a novel environment. Depending on one's time limitations, this period can vary anywhere from 5–10 min to overnight. Stimulus animals generally do not require habituation. They are typically handled frequently and thus experience less anxiety from handling. Furthermore, by design they are fully primed by exogenous hormones to exhibit sexual behavior and are usually less distracted by nonrelevant cues than are the experimental subjects. They are also familiar with the test situation from past experiences.

After the experimental animal has habituated to the testing environment, the stimulus animal is introduced, and scoring of behavior may begin. Some observers prefer not to begin formally collecting data until the occurrence of a particular behavior, such as sniffing or investigation by the experimental subject. In this way, tests are then standardized on the basis of a common, introductory event, rather than on a strict time basis.

Once observations for a pair of animals have been completed, cleaning the apparatus between tests with lightly soapy water or ethanol and water may be recommended if different animals will be used in each test. Not only may buildup of odors interfere with the behaviors of subsequent subjects, but the accumulation of urine and feces can quickly soil arena walls and thus make viewing difficult. This procedure also helps to provide a consistent environment for each new pair of animals.

In my view, sexual behavior can be one of the easiest forms of behavior to observe, quantify, and evaluate accurately in the laboratory. With some practice and attention to the factors outlined above, the study of sexual behavior can be effectively managed by any responsible technician.

# References

1. T. E. McGill, *Behaviour* **19,** 341 (1962).
2. T. E. McGill, *in* "Reproductive Behavior and Evolution" (J. S. Rosenblatt and B. R. Komisaruk, eds.), p. 73. Plenum, New York, 1977.
3. C. L. Moore and L. Wong, *Physiol. Behav.* **51,** 569 (1992).
4. M. J. Baum, *J. Comp. Physiol. Psychol.* **78,** 356 (1972).
5. M. Diamond, M. Mast, and R. Yanagimachi, *Horm. Behav.* **5,** 129 (1974).
6. M. H. R. Huber, F. H. Bronson, and C. Desjardins, *Biol. Reprod.* **23,** 305 (1980).
7. J. P. Toner and N. T. Adler, *Physiol. Behav.* **35,** 113 (1985).
8. L. L. Miller, J. M. Whitsett, J. G. Vandenbergh, and D. R. Colby, *J. Comp. Physiol. Psychol.* **91,** 245 (1977).
9. J. A. Elliot, *Fed. Proc.* **35,** 2339 (1976).
10. L. P. Morin and I. Zucker, *J. Endocrinol.* **77,** 249 (1978).
11. S. M. Schwartz, *Neurosci. Biobehav. Rev.* **6,** 391 (1982).
12. B. D. Fantie, R. E. Brown, and W. H. Moger, *J. Reprod. Fertil.* **72,** 435 (1984).
13. G. A. Eskes, *Brain Res.* **293,** 127 (1984).
14. V. H. Denenberg and R. W. Bell, *Science* **131,** 227 (1960).
15. M. K. McClintock and N. T. Adler, *Behaviour* **67,** 67 (1978).
16. M. S. Krieger, D. Orr, and T. Perper, *Behav. Biol.* **18,** 379 (1976).
17. M. S. Erskine, E. Kornberg, and J. A. Cherry, *Physiol. Behav.* **45,** 33 (1989).
18. E. Steel, *Horm. Behav.* **15,** 141 (1981).
19. A. S. Clark and E. J. Roy, *Physiol. Behav.* **30,** 561 (1983).
20. M. N. Perkins, M. S. Perkins, and B. D. Bunnell, *Behaviour* **89,** 29 (1984).
21. D. F. Hardy and J. F. DeBold, *Horm. Behav.* **2,** 287 (1971).

22. F. A. Beach, *Psychosom. Med.* **4,** 173 (1942).
23. A. S. Clark and E. J. Roy, *Horm. Behav.* **19,** 227 (1985).
24. D. Austin and D. A. Dewsbury, *Physiol. Behav.* **37,** 627 (1986).
25. U. W. Huck and R. D. Lisk, *J. Comp. Physiol. Psychol.* **99,** 98 (1985).
26. D. A. Edwards and J. K. Pfeifle, *Physiol. Behav.* **30,** 437 (1983).
27. P. N. Lehner, "Handbook of Ethological Methods." Garland STPM Press, New York, 1979.
28. P. Martin and P. Bateson, "Measuring Behavior." Cambridge Univ. Press, New York, 1986.
29. B. J. Everitt, *in* "Behavioral Biology: Neuroendocrine Axis" (T. Archer and S. Hansen, eds.), pp. 111–122. Lawrence Erlbaum, Hillsdale, New Jersey, 1991.
30. D. B. Hardy and J. F. DeBold, *Physiol. Behav.* **7,** 643 (1971).
31. F. A. Beach and A. M. Holz-Tucker, *J. Comp. Physiol. Psychol.* **42,** 433 (1949).
32. D. A. Dewsbury, *Science* **190,** 947 (1975).
33. B. D. Sachs and R. J. Barfield, *in* "Advances in the Study of Behavior" (J. S. Rosenblatt, R. A. Hinde, E. Shaw, and C. Beer, eds.), Vol. 7, p. 91. Academic Press, New York, 1976.
34. C. Diakow, *J. Comp. Physiol. Psychol.* **88,** 704 (1975).
35. D. W. Pfaff and C. Lewis, *Horm. Behav.* **5,** 317 (1974).
36. J. B. Mitchell and J. Stewart, *Brain Res.* **491,** 116 (1989).
37. J. T. Peirce and R. L. Nuttall, *J. Comp. Physiol. Psychol.* **54,** 585 (1961).

# [2] Practical Considerations in the Measurement of Sexual Behavior

Edward O. Price

## Introduction

This chapter summarizes those variables, techniques, and conditions that are of general importance in the measurement of sexual behavior in mammals. Emphasis is placed on rodents and ungulates since the author is most familiar with those groups.

Most mammalian species exhibit polygamous or promiscuous mating systems (1). Sexual behaviors consist of precopulatory activities (including courtship) and events associated with copulation (2). There is no substitute for a thorough understanding of the sexual repertoire of the species studied in terms of the sexual behaviors exhibited and the sequence in which they normally occur. Most sexual behaviors are relatively discrete and short-lived. Hence, frequency of events is more important than duration. In certain studies, the timing of sexual behaviors can be of critical importance.

Sexual behavior in the female mammal is ultimately controlled by physiological events associated with her estrous cycle. Her attractiveness to males (attractivity), her interest in males (proceptivity), and her willingness to remain immobile when mounted (receptivity) signal her estrous condition (3). Males are less cyclic, frequently testing females for willingness to interact sexually. Through experience, males learn the cues (some very subtle) associated with female receptivity.

The motor patterns associated with the sexual behavior of male and female mammals are basically innate. Certain aspects of male behavior, such as mount orientation and ability to attain intromissions, may be refined through experience or learning (4–6). Dewsbury (7) has proposed that the copulatory behaviors of different species of male mammals can be distinguished based on the presence (or absence) of four events or criteria:

1. Intravaginal thrusting: Some species (e.g., laboratory rats, cats, cattle, sheep, goats) exhibit a single, deep intravaginal thrust during ejaculation, whereas other species (e.g., most primates) display repetitive thrusting.
2. Multiple intromissions: Some species (e.g., laboratory rats) never ejaculate on their first intromission, whereas other species (e.g., cattle, sheep, goats) are capable of ejaculating on their first penile insertion.

*Methods in Neurosciences, Volume 14*

3. Multiple ejaculations: Most species ejaculate multiple times when exposed to receptive females, whereas a few species (e.g., guinea pigs) cease copulating following their first ejaculation.
4. Locking: In a few species (e.g., dogs, stumptail macaques), a "lock" or mechanical tie occurs between the penis and vagina during copulation which makes separation difficult or impossible.

As each of these four criteria has two alternatives, male mammals have 16 possible patterns of copulatory behavior.

## Copulatory Behaviors of Mammals

The basic copulatory behaviors measured during investigations of the sexual behavior of male mammals are (1) mounts without intromission, (2) mounts with intromission (but no ejaculation), and (3) mounts with ejaculation. These are often referred to as mounts (M), intromissions (I), and ejaculations (E). Ejaculations are typically followed by a period of sexual inactivity (refractory period); hence, the series of mounts and intromissions culminating in an ejaculation are referred to as an ejaculatory series. In some cases (e.g., frequently in cattle, sheep, and goats), the only event in the series is a mount with ejaculation.

From the frequency and timing of the above three variables, we can construct additional measures of sexual performance:

1. Mount latency (ML): time from exposure to a female to the first mount (with or without intromission)
2. Intromission latency (IL): time from exposure to a female to the first intromission (with or without an ejaculation)
3. Ejaculation latency (EL): time from exposure to a female to the first ejaculation
4. Mount frequency (MF): the number of mounts in an ejaculatory series
5. Intromission frequency (IF): the number of intromissions in an ejaculatory series
6. Mean interintromission interval (MIII): the mean time interval separating the intromissions of a series, including the time from the last intromission to the ejaculation
7. Postejaculatory interval (PEI): time from an ejaculation to the first mount (or intromission) of the next ejaculatory series; the PEI typically increases in length with successive ejaculations
8. Ejaculation frequency (EF): the number of ejaculations preceding satiety or end of the test period

9. Ejaculation rate: the number of ejaculations attained per unit of time; the ejaculation rate is sometimes referred to as "serving capacity" (8)

## Precopulatory Behaviors

Precopulatory behaviors are more varied and complex in some mammalian species than others. There are also sex differences. Female laboratory rats often solicit males by approaching them, engaging in physical contact (e.g., nudging, grooming, crawling over), and then running or darting away (9). Female control of the timing of copulation in this species through solicitation behaviors may be a function of enclosure size and complexity (9). Female sheep actively seek out rams early in their estrous period, and, when anogenital inspection by the male ensues, she may waggle her tail and look back at him while standing immobile (10).

Precopulatory behaviors are not as prominent in the male rat as in some other species. Male sheep often engage an intense inspection of the anogenital region of the female prior to mounting. The flehmen or lip curl response may be exhibited in response to olfactory stimulation (11). Ewe receptivity is tested when the ram propels its foreleg (often stiffened) toward the ewe, frequently striking her hind legs or inguinal region. These leg kicks may be interspersed with nudging in which the ram moves forward and parallel to the ewe with head lowered and cocked sideways while uttering a low-frequency, grumbling vocalization (10). Prior to mounting, the ram may exhibit mount intention movements in which his head is raised over the ewe's rump and his weight is shifted rearward as if he is about to mount. Sometimes one front foot is lifted off the ground in the process. Mount intentions in the bull sometimes include head throws in which the head and neck are rapidly moved in the direction of the female with little or no movement of the body (12). The bull may terminate a mount intention by resting its chin on the rump of the cow.

## Individual Variation in Sexual Performance

Individual males will differ in the frequency with which they exhibit various sexual behaviors. Table I lists the frequencies of sexual behaviors of three 18-month-old Hereford bulls when individually exposed to a single female for 40 min. Bull A engaged in a relatively large number of precopulatory behaviors, and, although he mounted 10 times, he had no successful copulations. Bull B exhibited fewer precopulatory behaviors and only one ejaculation in 23 mounts. Bull C attained a relatively large number of ejaculations

TABLE I   Frequencies of Sexual Behaviors of Bulls[a]

|  | Bull | | |
| Variable | A | B | C |
| --- | --- | --- | --- |
| Flehmen responses | 3 | 9 | 1 |
| Head throws | 59 | 14 | 12 |
| Mount intentions | 60 | 24 | 1 |
| Mount attempts | 1 | 1 | 3 |
| Mounts without ejaculation | 10 | 22 | 10 |
| Ejaculations | 0 | 1 | 5 |
| Time with female (min) | 34.0 | 40.0 | 24.9 |

[a] Three 18-month-old Hereford bulls were individually exposed to a single restrained, nonestrous female for 40 min.

with 15 mounts and exhibited very few precopulatory behaviors. How, then, does one characterize the overall sexual performance of these three individuals?

## Subjective Scoring Systems

Some investigators have resorted to subjective or composite scoring systems to obtain a single measure of overall sexual performance (13, 14). Such systems are normally based on estimated values or "weightings" of the behaviors the animal is expected to exhibit. The major problem with subjective scoring systems is in assigning quantitative values to the behaviors observed and in using those scores to rank or compare individuals or groups of animals. Table II illustrates the values assigned to the sexual performance of male pigs when exposed to an estrous sow for 15 min. Everyone would agree that a boar who attains an ejaculation during the test period has exhibited better sexual performance than one that shows only "repeated false mounts" (inappropriate orientation). But is the performance of the former worth twice as much (score of 10 versus 5, respectively, in Table II)? In Table I, Bull A exhibited 119 head throws and mount intentions but attained no ejaculations (10 mounts). Under the same circumstances, Bull C exhibited only 13 head throws and mount intentions, 10 mounts without ejaculation (same as Bull A), but attained 5 ejaculations. How do these bulls compare in terms of their overall sexual performance, and could a subjective scoring system be devised (for bulls) that would accurately reflect these differences? I think it is clear that subjective scoring systems based on differential weight-

TABLE II     Subjective Scoring System Used to Rate the
Sexual Performance of Male Pigs[a]

| Score | Behavior |
|-------|----------|
| 0 | No sexual interest |
| 1 | Some sexual interest with sniffing of uroanal region |
| 2 | Sexual interest with nose-to-nose contact and flank contact |
| 3 | Great deal of sexual interest |
| 4 | One false mount |
| 5 | Repeated false mounts |
| 6 | One correct mount |
| 7 | Repeated correct mounts |
| 8 | Penis extension |
| 9 | Intromission |
| 10 | Ejaculation |

[a] Ref. 14.

ing of variables are only as valid as one can justify the relative weightings assigned.

The matter becomes more complicated when one considers that failure to perform a behavior at one level (e.g., ejaculation) can result in either an increase or decrease in the frequency of some other behavior at a different level. For example, mounting frequency increases manyfold if males are unable to attain intromissions because of a physical defect, lack of experience, or if the experimenter artificially prevents vaginal penetration by covering the perineum of the female (15). It is clear that one must consider each variable separately when interpreting the results of sexual performance tests.

## Mating Efficiency

Measures of mating efficiency may be useful for certain types of research and when sustained breeding of males is required (e.g., livestock industry). Definitions of mating efficiency include the number of mounts or intromissions per ejaculation, the number of ejaculations attained prior to reaching sexual satiety, the number or proportion of females impregnated during a designated time interval, and the number or proportion of females in a group that received copulations (16). Persons collecting semen from males for artificial insemination may consider a male efficient if it exhibits a relatively short ejaculation latency (17). The term "efficiency" is obviously subject to one's point of view and should be used sparingly as a substitute for more objective measures of sexual performance.

## Selecting Subjects

Choice of subjects is almost always a concern in investigations dealing with sexual behavior. Certain species may be selected because of the nature of their sexual behaviors, availability, size, and information base. Strain or breed differences in sexual behavior are well known (18, 19). Rearing conditions (i.e., physical and/or biological environments) may also affect sexual performance and may interact with genotype (20). Using subjects as their own controls in experimental manipulations is recommended, whenever possible.

## Handling Subjects

Handling of animal subjects and the presence of observers may inhibit sexual behavior in some individuals or groups of animals. Regular handling or contact with humans prior to testing will normally reduce or eliminate these concerns. In some instances, handling is believed to improve sexual performance (21). Techniques should be developed to handle animals safely and humanely. Transport cages are sometimes preferred over physical restraint. Wire-mesh gloves can prevent serious bites by small mammals.

## Preparation of Stimulus Females

Female sexual performance is dependent on her estrous state, which also affects male sexual behavior. Dewsbury (22) discusses the effects of mode of estrus on the sexual behaviors of both female and male rodents. Female rodents may attain estrus through natural cycles, postpartum estrus, or male-induced estrus (e.g., in response to male odors), or they may be induced to exhibit estrus by hormone administration. The latter procedure, which may be used with intact or ovariectomized females, typically involves the injection of an estrogen (e.g., estradiol benzoate) 2–3 days before testing and progesterone 4–6 hr before testing (22). Estrus can be induced in laboratory rats with 0.1 mg intramuscular (i.m.) injections of estradiol benzoate and 1.0 mg injections of progesterone (23). Ovariectomized ewes can be induced to exhibit estrus by 25 mg i.m. injections of progesterone 5 and 3 days prior to testing and 200 $\mu$g of estradiol benzoate 24 hr prior to testing (24). An alternative procedure for inducing (and synchronizing) estrus in intact ewes is to insert vaginal sponges containing progestin 12–14 days before testing, with removal of the sponges 32–34 hr prior to testing (25). Note that the order of administration of estrogen and progesterone is reversed in rodents and ungulates.

In some rodents, males mating with naturally cycling females attain more ejaculations than those mating with hormone-induced estrous females (22). Males mating with females in postpartum estrus have higher intromission frequencies than when the females are cycling naturally or are induced to exhibit estrus. On the other hand, ejaculation rates are similar for rams exposed to both confined, intact ewes cycling naturally or synchronized with progestin sponges and hormone-induced ovariectomized ewes (25). Confinement may reduce variation in female behaviors related to mode of estrus.

## Presentation of Stimulus Females

Immobility of the female when mounted is important to the mating success of male mammals. When mounted, sexually receptive female rodents exhibit "lordosis" in which the perineum and head of the female are elevated while her four limbs are extended from the initial crouching position (26). Receptivity of ewes is minimally affected by the stimulus of copulation. Receptivity in the female laboratory rat declines progressively with successive copulations (27). This phenomenon is even more pronounced in female cattle, and female bison become unreceptive after only a single successful copulation (28). Such species differences can affect the timing of behavioral events and may determine the number of females required to test male sexual performance.

Physical restraint of females to simulate immobility reduces the number of females needed when assessing male sexual performance. Bulls of the *Bos taurus* breeds readily copulate with both estrous and nonestrous cows that are restrained (8); immobility is more important than estrous condition. However, *Bos indicus* bulls do not respond well to restrained females (29). Ejaculation frequencies are slightly greater for rams when exposed to unrestrained rather than restrained estrous ewes (29), but the differences are typically nonsignificant. Restrained nonestrous female ungulates (e.g., cattle, sheep, goats) will sometimes exhibit evasive behaviors when mounted and, thus, can discourage some males from mounting; such females should not be used for testing. Mineral oil can be used to lubricate the vaginas of nonestrous females although, after the first copulation, semen provides a natural lubricant. An analgesic/tranquilizer may also be administered to restrained females to minimize the stress of handling and restraint. Of course, animal welfare must be considered when females are restrained.

   Avoidance behaviors of unrestrained nonreceptive females can entice males to follow or chase, often to the extent that receptive females are ignored (30). Consequently, nonreceptive females should be removed from the group when assessing the sexual performance of males.

## Dealing with Aggression

   Males of some species (e.g., male cattle, sheep, and goats) often become aggressive when competing for access to females (31). Aggression and/or the fear of more dominant conspecifics can inhibit sexual behavior (32). Although sexual performance may be improved, in some instances, by the presence of other males, competition is generally detrimental to sexual performance, particularly when accompanied by serious aggression. Males quickly learn their vulnerability to attack when they are mounting females, so the slightest approach by another male may inhibit or terminate copulation. If it is necessary or desirable to test two or more males in the same arena, problems with aggression can be partially alleviated by using additional stimulus females to reduce competition. With ungulates, one can reduce aggression among the test subjects by restraining stimulus females and spacing them far enough apart to discourage interaction. Multiple test arenas can be used to test more than one male at a time under conditions which preclude direct competition. A visual barrier between test arenas is usually helpful. It should be noted that both wild and domestic male Norway rats are not aggressive toward one another in multimale tests of sexual behavior, even when pursuing the same estrous female (23).
   Females may also compete with one another for the attention of males. Estrous ewes will frequently block access of the male to other females in the group. While female competition may influence mate choice by males, it does not appear to have a major effect on male performance.

## Physical Features of the Test Environment

   Sexual behaviors can be influenced by physical features of the test environment. Relatively small enclosures may limit the range of behaviors observed, particularly for females (9). Relatively large enclosures may make identification and observation of animals more difficult. Circular arenas eliminate corners which often interfere with rear-oriented mounting. With rodents, clear plexiglass cylinders may facilitate copulation by the male (33) and improve observer accuracy. Ventral viewing of copulation in smaller mammals can be achieved by constructing test arenas with clear plastic floors.

Observing the animals ventrally can assist the observer in differentiating between mounts with and without intromissions and ejaculations (34). Outdoor test enclosures used with large mammals should at least have solid sidewalls to block extraneous visual stimulation (e.g., conspecific interactions, human activity). Outdoor arenas should be free of vegetation (e.g., weeds, grass) which can encourage feeding responses in low libido males and, thus, reduce sexual performance. Mounting activity is usually greater on a dirt or rough substrate (35).

Photoperiod is an important consideration when testing nocturnal species and seasonal breeders. For example, the sexual performance of male rats is usually greatest during the last part of the dark phase of their light/dark cycle (33, 36). For this reason, investigators frequently reverse the light/dark cycle of rodent facilities so tests can be conducted during normal working hours. Dim lighting or red lighting may be used to illuminate the test arena during testing. However, evidence indicates no particular advantage of dim over normal room lighting (37).

The sexual performance of cattle, pigs, and most rodents is similar year-round. Sheep and goats, however, are "short-day" breeders and should be observed during late summer or fall. Other species (e.g., horses) are sexually active during periods of increasing day length. Mating activity can be stimulated at other times of the year by artificially exposing animals to increasing or decreasing day length (38). Hormone treatment can also be used to stimulate sexual activity during the nonbreeding season.

Most sexual behavior studies with rodents are conducted under temperature-controlled conditions. However, very high temperatures (>35°C) and heavy rainfall can inhibit the sexual performance of large domestic ungulates tested outdoors (39).

## Data Collection Techniques

Computerized event recorders and software programs developed specifically for behavioral research (40) have greatly increased the efficiency of data collection and analysis. Recording of behavioral events on videotape has largely replaced the movie camera. Time-lapse video recorders have increased precision in analyzing the motor patterns associated with sexual behavior (41). Most sexual behaviors are relatively discrete and short-lived and, consequently, should be recorded using the "all occurrences" sampling technique (42).

Various devices have been developed to determine whether females are mounted in the absence of the observer. Harnesses are commercially available for rams and bulls which leave a colored crayon or ink mark on the rump or back of females when they have been mounted (43, 44). Crayon or paint marks can be placed on the rump of females which are obliterated by repeated mounting (45). Pressure-sensitive mount detectors are commercially available for cattle which activate (i.e., change color) when the animal is mounted (46). Activity meters have been developed to detect the approximately 3-fold increase in locomotor activity that accompanies estrus in female ungulates (47, 48). Other estrus-detection techniques used with livestock include measurements of declines in progesterone in blood or milk, increased blood serum levels of luteinizing hormone, increased vaginal temperature, and increased electrical conductivity in the mucous lining of the vagina (49). Dogs and rats have been trained to detect estrus-related odors from cattle (50, 51).

Intromissions have been verified using electronic circuitry attached to both males and females that records the reduced resistance associated with genital contact (52). Substances can be placed in the vagina of females that color the penis if intromission occurs (53). Accelerometers have been attached to the backs of males to record thrusting responses (54).

## Animal Identification

Rapid identification of individuals is important in multianimal tests since sexual behaviors are generally of short duration and often occur in rapid succession. Nyanzol dye which permanently blackens the hair of mammals provides identifying markings for a month or more, depending on the rate of shedding. It is best to anesthetize rodents when the dye is applied and until the hair dries to prevent ingestion of the chemicals via grooming. Fast-drying white curb paint on metal branding irons has been used to place identifying numbers on large animals (e.g., black Angus cattle), and latex wool paint is routinely used to mark sheep. Freeze marking ("branding") may be used with rodents and larger animals to achieve distinguishing white pelage markings (55, 56). Again, animal welfare considerations apply.

## Test Length

Testing is typically terminated when the male has reached a predetermined criterion for sexual satiety or when a preset time interval has elapsed. The criterion for satiety in the laboratory rat is typically 30 min without an ejaculation (36). Male rats average about 7 ejaculations before reaching this criterion (57), whereas domestic rams average about 5.5 ejaculations with a 20-min satiety criterion (58). Depending on the questions being asked, one might decide to terminate testing after a preset time period has elapsed rather than test each male to satiety. The former approach may be sufficient when the primary objective is to rank a relatively large number of individuals with respect to ejaculation rate (8). Other techniques have been devised to streamline testing large numbers of animals. For example, with domestic rams it has been found that the time from the first ejaculation to the first mount interaction after the second ejaculation correlates highly ($r = -0.91$) with total number of ejaculations attained over a much longer period (25).

Time-limited tests create problems of interpretation when some of the subjects do not initiate sexual responses until well into the test period (e.g., if subjects or animals are sexually inexperienced or not used to being handled). Under such conditions, one might choose to record the latency for the animal's first sexual response and start the test period at that point. Another disadvantage of time-based tests is that the test period sometimes ends just before an ejaculation, thus underestimating the ejaculation rate of the male and raising questions about including any mounts or intromissions that may have been exhibited in the aborted ejaculatory series. If it is likely that the frequency, timing, and/or sequence of behavioral events will differ within and between animals over successive ejaculatory series, one might want to terminate testing after a predetermined number of ejaculations have been attained. A measure of ejaculation rate could still be obtained by recording the total time that had elapsed when the criterion was reached.

## Testing Young Animals

Sexual behavior tests administered at about the time of puberty (for the species) will frequently underestimate sexual performance. Early maturing individuals will likely exhibit adult levels of sexual performance, but late-maturing individuals will often perform poorly or be sexually inactive and should be retested at a later date. Pretest exposure to estrous females (i.e., sexual experience) at puberty can enhance the sexual performance of young males to levels commonly exhibited by sexually naive individuals exposed to sexually receptive females for the first time at a later stage of development (59). Aggressive females can inhibit the sexual behaviors of young males. In some species (e.g., ungulates), older females may be too large for younger (i.e., smaller) males to attain intromissions. As a rule of thumb, young males should be tested with females the same size (particularly height) as themselves. Older males should be tested with larger mature females.

## Sexual Stimulation

The sexual performance of males of some species is enhanced by pretest sexual stimulation. For example, the ejaculation rate of male cattle and goats is increased by about 75% if they are allowed to observe the mounting behavior of other males before exposure to estrous females (60, 61). The same treatment has no effect on male sheep (62). In male laboratory rats, pretest handling and brief electric shock have been shown to improve sexual performance (21, 63).

The introduction of novel families will accelerate sexual recovery or forestall sexual satiation (17, 64). Hence, ejaculation rate is increased by exposing males to several females rather than a single stimulus individual. This also facilitates mate choice. Finally, changing the environmental context in which the stimulus animals are presented (i.e., introducing environmental novelty) or restraining males in the presence of females can also enhance sexual performance (17).

## Measuring Libido

Sexual performance is based on the animal's libido (sexual motivation), as well as its ability to execute the motor patterns constituting its sexual behaviors. Occasionally, the research question(s) being addressed requires measurement of the animal's libido, per se. Individuals with relatively good libido will not always be good performers due to such factors as lack of experience, genital abnormalities, or locomotor problems. Hence, libido does not always correlate well with sexual performance, and separate tests measuring libido may be warranted.

One approach has been to measure the propensity of an individual to remain in proximity to a sexual stimulus which is restrained or confined to prevent direct contact with the test subject (65). "Obstacle" tests have been used with males to determine how fast or long he is willing to work or how much aversive stimulation he is willing to·endure to gain access to female conspecifics (66, 67). The task may involve traversing a runway or simple maze or performing an operant-conditioned task such as pressing a lever to gain access to a female. The older literature includes accounts of measuring the amount of electric shock a male rat would endure on an electrified grid floor to reach a female on the other side of the apparatus.

Another approach is to allow males to access and mount females but to prevent intromissions (and ejaculations) by surgically denervating or deviating the penis (68), by applying a topical anesthetic to the penis (69), or by covering the perineum of the female with tape (70) or cloth (68). By precluding copulation, one eliminates the variation related to intromission frequency as well as the temporary loss of libido following each ejaculation. The latter point is particularly important when one considers that refractory periods increase with successive ejaculations and that different males ejaculate at different rates. Perineum covers reduce the frequency of anogenital sniffs, but the frequency of other precopulatory behaviors and mounts increase significantly (68). It should be noted that some males may become aggressive toward females after repeated mounting without intromission.

## Mate Choice and Sexual Orientation

Mate choice and sexual orientation can be measured by frequencies of approaches and sexual behaviors directed toward two or more appropriate stimulus animals. Time spent in proximity to available stimulus animals may also reflect preferences when the latter are restrained and/or confined to prevent direct physical contact with the test subject (71). T- or Y-shaped mazes are sometimes used to measure approach frequency and proximity time (72–74). Simultaneous choice tests are usually more informative than tests conducted sequentially, since choices are made under identical conditions. Repeated testing is desirable to determine the repeatability of choices made and so that the relative positions of the stimulus animals can be varied to control for possible location or position preferences. The information provided by preference tests is of questionable value when the test subjects exhibit low libido (i.e., the subjects are reluctant to engage in sustained contact with one or more of the stimulus animals). Differences in familiarity or experience with the stimulus animals (or animals similar in appearance and/or behavior) can affect the preferences of a subject both on initial exposure and after repeated contact. Mate preferences and sexual orientation may change over time. Short-term tests are more subject to the effects of handling and stimulus novelty than longer exposure periods.

Direct exposure to the stimulus animals tend to sustain the libido of the subject and provides more direct measures of preference. Tilbrook and Lindsay (75) exposed rams to a small group of estrous ewes and, after each 5-min test segment, removed the female that had received the greatest attention from the ram. Restraint of the stimulus animals eliminates individual differences in aggressiveness (i.e., competitiveness) which can bias the choices of the test subjects. Restraint can also be used to eliminate much of the behavioral feedback provided by the stimulus animals, if doing so is desirable for the question(s) being asked. Restraint or confinement is necessary when nonestrous females or males are used as sexual stimuli since evasive and aggressive behaviors vary greatly with different individuals and may inhibit the test subjects. Price *et al.* (76) covered the perinea of restrained male and female stimulus animals used to test the sexual orientation of male sheep. Doing so precluded development of a potential bias resulting from copulations with female (but not male) stimulus animals.

# References

1. J. F. Wittenberger, "Animal Social Behavior." Duxbury, Boston, Massachusetts, 1981.
2. B. D. Sachs and R. L. Meisel, in "The Physiology of Reproduction" (E. Knobil and J. D. Neill, eds.), p. 1393. Raven, New York, 1988.
3. F. A. Beach, *Horm. Behav.* **7**, 105 (1976).
4. D. A. Dewsbury, *Anim. Behav.* **17**, 217 (1969).
5. J. S. Rosenblatt, in "Sex and Behavior" (F. A. Beach, ed.), p. 416. Wiley, New York, 1965.
6. G. V. Silver and E. O. Price, *Appl. Anim. Behav. Sci.* **15**, 287 (1986).
7. D. A. Dewsbury, *Q. Rev. Biol.* **47**, 1 (1972).
8. M. A. de B. Blockey, *Appl. Anim. Ethol.* **7**, 307 (1981).
9. M. K. McClintock and N. T. Adler, *Behaviour* **58**, 67 (1978).
10. E. M. Banks, *Behaviour* **23**, 249 (1964).
11. B. L. Hart, in "Veterinary Clinics of North America: Food Animal Practice" (E. O. Price, ed.), Vol. 3, p. 463. Saunders, Philadelphia, Pennsylvania, 1987.
12. E. O. Price and S. J. R. Wallach, *J. Anim. Sci.* **69**, 1019 (1991).
13. P. J. Chenoweth, J. S. Brinks, and T. M. Nett, *Theriogenology* **12**, 223 (1979).
14. K. L. Ebenshade and B. H. Johnson, *Theriogenology* **27**, 581 (1987).
15. E. O. Price, H. Erhard, R. Borgwardt, and M. R. Dally, *J. Anim. Sci.* **70**, 3376 (1992).
16. E. O. Price, *J. Anim. Sci.* **61**, 62 (1985).
17. E. B. Hale and J. O. Almquist, *J. Dairy Sci.* **43** (Suppl.), 145 (1960).
18. T. E. McGill, in "Contributions to Behavior–Genetic Analysis: The Mouse as a Prototype" (G. Lindzey and D. D. Thiessen, eds.), p. 57. Appleton, New York, 1970.
19. J. D. Neely and O. W. Robison, *J. Anim. Sci.* **56**, 1033 (1983).
20. W. Riss, E. S. Valenstein, J. Sinks, and W. C. Young, *Endocrinology* (*Baltimore*) **57**, 139 (1955).
21. K. Larsson, *Behaviour* **20**, 110 (1963).
22. D. A. Dewsbury, *Neurosci. Biobehav. Rev.* **14**, 147 (1990).
23. E. O. Price, *Anim. Behav.* **28**, 657 (1980).
24. R. J. Kilgour and R. G. Whale, *Aust. J. Exp. Agric. Anim. Husb.* **20**, 5 (1980).
25. E. O. Price, R. Borgwardt, R. BonDurant, and M. R. Dally, unpublished data (1991).
26. D. W. Pfaff and C. Lewis, *Horm. Behav.* **5**, 317 (1974).
27. D. F. Hardy and J. F. DeBold, *J. Comp. Physiol. Psychol.* **78**, 400 (1972).
28. D. F. Lott, *Z. Tierpsychol.* **56**, 97 (1981).
29. E. O. Price, in "Veterinary Clinics of North America: Food Animal Practice" (E. O. Price, ed.), Vol. 3, p. 405. Saunders, Philadelphia, Pennsylvania, 1987.

30. J. J. Zenchak, L. S. Katz, E. O. Price, and S. J. R. Wallach, *J. Anim. Sci.* **66,** 2851 (1988).
31. E. O. Price and S. J. R. Wallach, *J. Anim. Sci.* **69,** 1034 (1991).
32. D. R. Lindsay, D. G. Dunsmore, J. D. Williams, and G. J. Syme, *Anim. Behav.* **24,** 818 (1976).
33. O. Mercier, J. Perraud, and J. Stadler, *Lab. Anim.* **21,** 125 (1987).
34. B. N. Bunnell, B. D. Boland, and D. A. Dewsbury, *Behaviour* **61,** 180 (1976).
35. L. D. Vailes and J. H. Britt, *J. Anim. Sci.* **68,** 2333 (1990).
36. D. A. Dewsbury, *in* "Endocrine Control of Sexual Behavior" (C. Beyer, ed.), p. 3. Raven, New York, 1979.
37. E. Hard and K. Larsson, *J. Comp. Physiol. Psychol.* **66,** 805 (1968).
38. R. H. BonDurant, B. J. Darien, C. J. Munro, G. H. Stabenfeldt, and P. Wang, *J. Reprod. Fertil.* **63,** 1 (1981).
39. C. G. Winfield, P. H. Hemsworth, D. B. Galloway, and A. W. Makin, *Aust. J. Exp. Agric. Anim. Husb.* **21,** 39 (1981).
40. L. P. J. J. Noldus, *Behav. Res. Methods, Instrum. Comput.* **23,** 415 (1991).
41. J. A. Metzler, E. O. Price, R. L. Kitchell, and R. H. BonDurant, *Physiol. Behav.* **43,** 207 (1988).
42. P. Martin and P. Bateson, "Measuring Behaviour: An Introductory Guide." Cambridge Univ. Press, New York, 1986.
43. C. V. Hulet, M. Shelton, J. R. Gallagher, and D. A. Price, *J. Anim. Sci.* **38,** 1210 (1974).
44. R. H. Foote, *J. Dairy Sci.* **58,** 248 (1975).
45. K. L. Macmillan, V. K. Taufa, D. R. Barnes, A. M. Day, and R. Henry, *Theriogenology* **30,** 1099 (1988).
46. W. F. Williams, D. R. Yver, and T. S. Gross, *J. Dairy Sci.* **64,** 1738 (1981).
47. C. A. Kiddy, *J. Dairy Sci.* **60,** 235 (1977).
48. A. S. Moore and S. L. Spahr, *J. Dairy Sci.* **74,** 3857 (1991).
49. J. W. Smith, S. L. Spahr, and H. B. Puckett, *J. Dairy Sci.* **72,** 693 (1989).
50. C. A. Kiddy, D. S. Mitchell, D. J. Bolt, and H. W. Hawk, *Biol. Reprod.* **19,** 389 (1978).
51. M. Dehnhard and R. Claus, *Theriogenology* **30,** 1127 (1988).
52. S. G. Carlsson and K. Larsson, *Scand. J. Psychol.* **3,** 189 (1962).
53. J. K. O'Hanlon, R. L. Meisel, and B. D. Sachs, *Behav. Neural. Biol.* **32,** 269 (1981).
54. C. Beyer, J. L. Contreras, G. Morali, and K. Larsson, *Physiol. Behav.* **27,** 727 (1981).
55. K. Ohwada, *Exp. Anim.* **40,** 395 (1991).
56. R. P. Lemenager, *in* "Handbook of Livestock Management Techniques" (R. A. Battaglia and V. B. Mayrose, eds.), p. 93. Burgess, Minneapolis, Minnesota, 1981.
57. F. W. Beach and L. Jordan, *Q. J. Exp. Psychol.* **8,** 121 (1956).
58. W. E. Pepelko and M. T. Clegg, *Anim. Behav.* **13,** 249 (1965).
59. G. W. Boyd, V. M. Healy, R. G. Mortimer, and J. R. Piotrowski, *Theriogenology* **36,** 1015 (1991).

60. D. R. Mader and E. O. Price, *J. Anim. Sci.* **59,** 294 (1984).
61. E. O. Price, V. M. Smith, and L. S. Katz, *Appl. Anim. Behav. Sci.* **13,** 83 (1984).
62. E. O. Price, S. J. R. Wallach, and M. R. Dally, *Appl. Anim. Behav. Sci.* **30,** 333 (1991).
63. R. J. Barfield and B. D. Sachs, *Science* **161,** 392 (1968).
64. K. Larsson, "Conditioning and Sexual Behavior in the Male Albino Rat." Almqvist and Wiksell, Stockholm, 1956.
65. E. O. Price and S. J. R. Wallach, *J. Anim. Sci.* **68,** 3572 (1990).
66. B. J. Meyerson and B. Bohus, *Pharmacol. Biochem. Behav.* **5,** 539 (1976).
67. D. Jowaisas, J. Taylor, D. A. Dewsbury, and E. F. Malagodi, *Psychon. Sci.* **25,** 287 (1971).
68. L. S. Katz and E. O. Price, *Dev. Psychobiol.* **19,** 197 (1986).
69. G. D. Gray, H. N. Davis, and D. A. Dewsbury, *Horm. Behav.* **7,** 317 (1976).
70. D. F. Hardy and J. F. DeBold, *Physiol. Behav.* **7,** 643 (1971).
71. D. Q. Estep, E. O. Price, S. J. R. Wallach, and M. R. Dally, *Appl. Anim. Behav. Sci.* **24,** 287 (1989).
72. M. R. Murphy, *Behav. Neural Biol.* **30,** 323 (1980).
73. E. R. Stockman, R. S. Callaghan, and M. J. Baum, *Physiol. Behav.* **34,** 409 (1985).
74. J. P. Signoret, *in* "Pig Production" (D. J. A. Cole, ed.), p. 295. Butterworth, London, 1971.
75. A. J. Tilbrook and D. R. Lindsay, *Appl. Anim. Behav. Sci.* **17,** 129 (1987).
76. E. O. Price, L. S. Katz, S. J. R. Wallach, and J. J. Zenchak, *Appl. Anim. Behav. Sci.* **21,** 347 (1988).

# [3] Component Analysis of Male Sexual Behavior

## John T. Clark

> In medical writings, sexual function is all too often dealt with as a unidimensional continuum extending from hypersexuality to impotence/frigidity. This tendency, common among both medical practitioners and clinical researchers, is deleterious to patient care and medical progress. A first and vital step in the assessment process is to attempt to distinguish between libido and potency factors.
>
> *Julian M. Davidson (1)*

## Introduction

Why does one measure sexuality? The importance of copulation for the propagation of the species is readily accepted. Also important, however, are the quality of life aspects of sexual function. Our understanding of the complex interplay between neurochemical, hormonal, muscular, and psychosocial factors involved in the control of sexual function, in any species, remains incomplete. To some people the idea that neuroendocrine mechanisms controlling human sexuality are essentially similar to those of "lower" animals is distasteful. Nonetheless, the use and utility of animal models to aid in our quest for biological determinants of sexuality is dictated by a common biology as well as ethical barriers to human experimentation. More is known about the biology of rat sexual behavior, and this chapter presents methods for assessing male rat sexual behavior. Of course, studies of animal sexual behavior can only assess biological (neurochemical, hormonal, etc.) aspects, but this does not preclude the importance of psychosocial factors in human sexuality.

## The Duality of Sexual Behavior

How does one measure sexuality? The simple presence or absence of copulatory patterns is not a sufficient index, although it is an important first step. In 1956 Beach (2) formalized a hypothesis regarding the duality of sexual arousal and performance. Even at that time, however, this was not a revolu-

*Methods in Neurosciences, Volume 14*

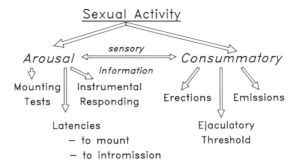

FIG. 1   Relationships between the categories of male sexuality required for sexual activity, arousal and consummatory behavior (analogous to libido and potency), and methods useful in assessing these categories. "Sensory information" denotes effects of physiological sexual responses that participate in the postulated positive feedback relationship between arousal and consummation.

tionary concept; it had been suggested for human sexuality by a number of earlier investigators and popular writers (e.g., Shakespeare). Accumulated evidence over the past decades has clearly validated this concept. The separability of sexual behavior into arousal and consummatory (or intromission–ejaculation) mechanisms, each with its own threshold for excitation, stands on a solid experimental foundation. Because this chapter deals with the measurement of male sexual behavior, we focus on this. However, the contribution of the female is also important. The female must, minimally, cooperate with the male in order for copulation to occur (see next chapter). Of particular value in attempting to identify mechanisms underlying discrete aspects of sexual behavior is the relative ease with which rat sexual behavior can be experimentally dissected into its component parts (a situation not paralleled in the human). In addition to the various parameters of sexual behavior monitored during mating tests, we make use of a number of quantitative tests that allow the assessment of the separate components of male sexual behavior in isolation.

In assessing sexuality, the libido–potency/arousal–copulatory dualism is not the only dichotomy observed. For male rats, sexual bahavior is readily divisible into two parts: the active pursuit of a female and subsequent copulation, and reflexive activity which can occur in the absence of any overt sexual stimulation (reflex–nonreflex duality). A final (for our discussion) duality comes to mind when we consider neuroanatomical substrates influencing sexuality, namely, central versus peripheral control. The use of these dualistic concepts facilitates our interpretation of sexual behavioral data (Fig. 1) (e.g., Refs. 1–5).

## Description of Copulatory Behavior

The copulatory pattern of the male rat has been well characterized (see, e.g., Refs. 5–9). A complete sequence of male sexual behavior would include precopulatory behavior, copulation culminating in ejaculation (or multiple ejaculations), and sequelae to copulation. Precopulatory behavior in male rats includes locating, approaching, and sniffing or licking the perineal region of the female, as well as ultrasonic vocalizations. In inexperienced males precopulatory behavior or mounting itself may depend, to some extent, on the soliciting (proceptive, see following chapter) behavior of the female.

Copulatory behavior in the male rat is characterized by the repeated occurrence of three readily distinguishable behaviors: mounts, intromissions, and ejaculation. Mounting activity can be in the proper orientation or not. Typically, only mounts dorsally from the rear are considered and counted in tests of male copulatory behavior. The forelimbs of the male grasp and palpate the flanks of the female prior to and during pelvic thrusting. This stimulation either induces or intensifies the receptive posture of the female. Without the female rat elevating her perineal region, the male does not have access to her vagina. In some studies, mounts that are not accompanied by palpation and pelvic thrusting are not counted, or are identified as incomplete mounts.

Intromissions are mounts in which penile insertion into the vagina is achieved. Most studies do not verify this directly but rely on characteristic behavioral patterns associated with insertion, which are readily detected by the experienced observer. These are well-defined muscle movements, including a deeper pelvic thrust than that seen in mounting. Following an intromission there is a rapid dismount and autogrooming of the genitalia. This grooming may or may not occur after mounts without insertion. After an intromission there is a brief (10–90 sec or so) interval during which no copulatory activity ensues. The male may engage in a variety of activities but tends to maintain pxoximity to the female. The intercopulatory interval (or interintromission interval) may conclude with behavior similar to that seen prior to the initiation of copulation.

Following a number of intromissions (typically 8–12 in young male rats), the male displays an ejaculatory pattern. This is characterized by an even deeper, more forceful pelvic thrust, a prolonged duration of insertion, and a slower dismount. The ejaculation is followed by autogrooming and a prolonged period of sexual inactivity (typically 4–7 min) which commonly is associated with general inactivity and recumbency. The interval from ejaculation to the resumption of sexual activity is referred to as the postejaculatory interval.

Ejaculation, or ejaculatory behavior, warrants further discussion. Technically, two processes are occurring: seminal emission, that is, movement of the ejaculate into the urethra, and ejaculation, that is, forceful contraction of the muscles of the pelvic floor normally resulting in expulsion of the ejaculate from the penis. However, seminal emission may also be used to describe occasional seepage of seminal fluid outside of the copulatory situation *(ex copula),* which may be associated with pelvic thrusting and autogrooming (sometimes referred to as *ex copula* ejaculation or spontaneous ejaculation, see below). It is, perhaps, germane to point out that the postejaculatory interval is not due simply to the expulsion of seminal fluid, since drug-induced seminal emission (10, 11), prevention of seminal emission (12), or electroejaculation (13, 14) is not associated with the "normal" postejaculatory interval or activity.

## Measures of Male Rat Copulatory Behavior

Analysis of copulatory behavior must consist of reliable and valid measures constituting a quantitative description (15). The assessment of sexual behavior relies on the astuteness of the observer. As Benjamin Sachs has written, "Our eyes and ears must usually read the behavior directly. . . . The absence of technical metering systems does not preclude reliable or precise measurement, but it can . . . sometimes give rise to two illusions: that behavior is easy to measure or, conversely, that behavior is impossible to measure" (5, p. 1396). The measurement of copulatory behavior in male rats is precise when performed by experienced observers, and most observers can make the necessary distinctions between mounts (without insertion), intromissions (sometimes called successful mounts, or successful copulations), and ejaculations.

Typically, the measurement of copulatory behavior in male rats involves counting the number of mounts and intromissions preceding an ejaculation. These are historically referred to as mount and intromission frequencies (since it is the number per ejaculation), but these terms are sometimes used differently (number per unit time). Some investigators utilize the terms number of mounts and number of intromissions to ejaculation. Less useful measures are the total number of mounts (without distinguishing between mounts, intromissions, and ejaculations), or intromissions, during a specified test period. In addition to numbers of copulatory acts, temporal aspects are also recorded, namely, the latency from introduction of the female to the initial mount (mount latency) and/or intromission (intromission latency), the time from the first intromission to the ejaculation (ejaculation latency), and the interval from the ejaculation to the next intromission (postejaculatory interval). A typical temporal relationship between these measures is diagrammatically represented in Fig. 2, which also gives "normative" values.

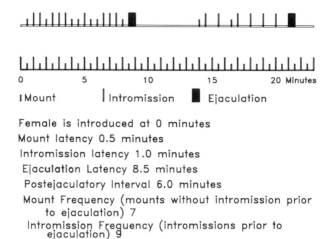

Female is introduced at 0 minutes
Mount latency 0.5 minutes
Intromission latency 1.0 minutes
Ejaculation Latency 8.5 minutes
Postejaculatory Interval 6.0 minutes
Mount Frequency (mounts without intromission prior
    to ejaculation) 7
Intromission Frequency (intromissions prior to
    ejaculation) 9

FIG. 2   Diagrammatic representation of testing copulatory behavior in male rats. The female is introduced into the testing chamber after the male has adapted to the chamber for 3 to 5 min. Values for latencies and intervals are given in minutes. In this example, the intercopulatory interval would be 8.5 min divided by 9 intromissions preceding ejaculation, and copulatory efficiency would be 9 intromissions divided by 16 and multiplied by 100. Alternatively, values may be in seconds. Normative values from 653 male rats on the third copulatory test for the initial ejaculatory series are as follows: mount latency, $32.8 \pm 3.0$ sec; intromission latency, $62.7 \pm 4.9$ sec; ejaculation latency, $511.1 \pm 12.6$ sec; mount frequency, $7.1 \pm 0.3$ prior to ejaculation; intromission frequency, $9.4 \pm 0.2$ prior to ejaculation; intercopulatory interval (average time between intromissions), $64.2 \pm 1.9$ sec; copulatory efficiency, $64.3 \pm 1.9\%$; and postejaculatory interval, $352.2 \pm 2.7$ sec.

In addition to these direct measures, several useful measures can be derived. From the direct measures of mount and intromission frequencies, one can calculate the copulatory efficiency (also referred to as the hit rate; see, e.g., Ref. 5). This measure indicates how often the male achieves intromission (success) relative to the number of times he attempts. It is calculated by dividing the number of intromissions preceding ejaculation by the number of mounts plus the number of intromissions preceding ejaculation. A final useful measure is the average interval between successive intromissions (the intercopulatory or interintromission interval), which is calculated as the ejaculation latency divided by the number of intromissions preceding the ejaculation. Alternatively, the reciprocal is referred to as the copulatory rate.

Logistically, one can perform behavioral observations with a timer/stopwatch and a keyboard. Historically, analog event recorders have been utilized to allow the observer(s) to devote all their attention to the behavioral observation. The availability of low-cost personal computers has prompted a number of investigators to produce programs to replace the analog event recorder. A number of programs designed specifically to record sexual behavior in rodents are available from individual investigators (see, e.g., Refs. 16–19). In addition, there are generic programs to assist in the quantitative analysis of animals (e.g., Ref. 20; several event recorder programs are available commercially from Life Sciences Associates, Bayport, NY). A useful adjunct to the computer keyboard is a separate numerical keypad. Several models are available from Genovation, Inc. (Irvine, CA). Videotaping of behavioral sessions allows for post hoc verification, but it is only occasionally used.

## Test Conditions

The measures of copulatory behavior may vary, sometimes considerably, depending on the time of day of testing, the size of the test chamber, the time of adaptation to the test chamber, the "quality" of the female partner, temperature, and other variables. Male rats are nocturnally active, and they copulate more readily in the lights-off period. Typical tests of male copulatory behavior are conducted from 1 to 6 hr after lights off in dimly lit chambers (e.g., a 7-W light bulb or red illumination). On introduction of a male rat to a new environment, he will typically investigate for a variable period of time. To eliminate the influence of environmental conditions on measures of copulatory behavior, males are allowed to adapt to the test chamber. This adaptation period is typically 2 to 5 min.

Suitable test chambers for rats (and other rodents) include glass aquaria (10 or 20 gallon sizes work well), boxes with plexiglass or glass fronts, and glass cylinders, usually with sawdust/wood shavings covering the floor. Many investigators use semicircular chambers to minimize female utilization of corners. Alternatively, initial observations could be made from above, with the rats in any type of enclosure. Changing test chamber size can increase the effect of the female on the pacing of copulation.

Measures of copulatory behavior vary depending on the experience of the rat. Some rats, for unknown reasons, do not mate in standard copulatory tests. With repeated testing, measures become stable for extended periods of time. Thus, it is suggested that, except where experimental design precludes experience, animals be screened for the display of copulatory behavior (at least three tests with ejaculation), and that experimental groups be balanced for measures of copulatory behavior.

The duration of copulatory tests can be extremely variable. One standard set of conditions is for the test to be terminated immediately after the postejaculatory interval. Alternatively, the male is considered negative if he fails to intromit within 15 min of introduction to the female, if more than 15 min elapses between successive intromissions, or if more than 15 min elapses after ejaculation without intromission. There are circumstances where longer test periods may be warranted, for example, when the males are motor-impaired or after some drug treatments. Similarly, under certain conditions one may wish to assess more than one ejaculatory series (e.g., ending the test after two or more ejaculations). Tests may also be conducted until sexual satiety is reached. Classic criteria for reaching sexual satiety in male rats are 30 to 60 min without mounting activity (5).

Alternatively, one may utilize time-limited tests, as are commonly used in the assessment of other types of behaviors. However, these tests often yield data that may be difficult to interpret. Allowing all animals to ejaculate the same number of times allows for all animals to be assessed relative to comparable behavioral measures. It has been repeatedly observed that the latency to initiate copulation is not correlated with other measures. Taking an extreme example, in a test limited to 20 min a rat intromitting within 1 min and one intromitting at 19.5 min would not be comparable on anything but latency. If, however, the test were terminated after the postejaculatory intromission one could compare all measures.

## Interpretation of Copulatory Behavioral Data

It is useful to interpret changes in measures of copulatory behavior in terms of the arousal–performance duality. The question arises as to which measures are associated with which component. Factor analyses of ten measures of copulatory behavior indicate that four relatively independent factors, possibly representing independent processes, underlie sexual performance in male rats: an initiation factor, a copulatory rate factor, a copulatory efficiency (or hit rate) factor, and an ejaculatory threshold (intromission count) factor (21, 22). The initial factor analysis studies were done using limited numbers of animals. We have replicated the essence of these findings for eight standard measures using archival data on 653 experienced male rats (23). The measures associated with these factors are as follows: initiation factor: latencies to first mount and/or intromission (accounting for 22% of the variance); copulatory rate factor: ejaculation latency, intercopulatory interval, and postejaculatory interval (accounting for 38% of the variance); ejaculatory threshold factor: number of intromissions to ejaculation, intercopulatory interval, and number of mounts (without intromission) preceding ejaculation (accounting for 15% of the variance); and copulatory efficiency factor: copulatory efficiency and the number of mounts preceding ejaculation (23). The proposed relationship between components, measures, and factors is depicted in Fig. 3.

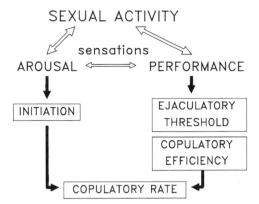

FIG. 3 Hypothetical relationship between the categories of male sexuality required for sexual activity, namely, arousal and performance (analogous to libido and potency), and the factors associated with them. We view the link between arousal and performance as an additional category (i.e., the maintenance of behavior until ejaculation has occurred, which is dependent on positive feedback effects of "sensations" evoked by physiological manifestations of sexual activity) represented by the copulatory rate factor (copulatory rate as a measure can be derived as the reciprocal of the intercopulatory interval). Experimental data indicate that if the intercopulatory interval is elongated beyond a point, ejaculation is not achieved in copulatory tests. The measures associated with the factors are detailed in the text.

The interpretation of data obtained in copulatory tests is not always straightforward, however. Questions that arise include the following: Does a failure of mating indicate that performance would be (or is) impaired? Does a reduction in the ejaculatory threshold indicate a general stimulation of sexual behavior (as a number of investigators have suggested) or the induction of prematue ejaculation? Is a failure to mount, intromit, or ejaculate due to motivational deficits, to sensory deficits, to the inability to achieve erection, or to the disconnection of motivational aspects of behavior from reflexive/ performance aspects? To address these questions specialized tests are utilized.

## Evaluation of Genital Function

### Penile Reflex Tests

The erectile reflex test was originally introduced by Hart for use with spinally transected animals (24), and it was subsequently applied to intact rats (25, 26). The recommended procedure for evoking penile reflexes and the criteria

FIG. 4   Penile reflex testing in male rats. Reflexes are evoked by retraction of the preputial sheath to expose the glans penis. The glans is maintained in position by light pressure on a wooden applicator. The use of a clear plexiglass tube allows for the concomitant evaluation of grooming responses (restricted to the forepaws and upper body). The penis is in a flaccid (quiescent/nonerect) state.

for scoring are modified from those of Davidson *et al*. (25). Briefly, the animal is coaxed into a clear plastic cylinder (typical internal diameter for sexually mature 3- to 6-month-old rats is 80 to 85 mm); the rat is subsequently rolled over onto his back and gently restrained. The penile (preputial) sheath is retracted using a wooden applicator (Fig. 4). The size of the cylinder allows the rat to groom his head and forepaws, and to flex his neck enough to lick his anterior ventrum. The use of a clear cylinder allows for the additional observation of upper body grooming during the test. This grooming may serve as an indicator of stress levels and of the possible occurrence of reflexes during the test period (27). Modifications of this technique include using opaque cylinders or restraining the rat using toweling mounted on a board.

"Hands off" penile reflexes can be observed following retraction of the sheath in rats that have had the suspensory ligament excised (which reduces the chance of the glans being retracted into its sheath during the test), with restraint maintained by strips of paper tape over the abdomen, the base of the hind legs, and the tail (see, e.g., Refs. 5 and 27). Tests are considered positive if the rat exhibits a penile erection within 15 min of sheath retraction. Latencies to all genital responses and the number of such responses, including seminal emission, which occur within 10 to 30 min of the first erection are scored via a keypad connected to a microcomputer as follows: erection: increased tumescence of the glans penis, partial or complete, with subsequent detumescence (a flaccid penis is illustrated in Fig. 5a and an erect penis in Fig. 5b); cup: an intense erection with flaring of the engorged glans penis (Fig. 5c); quick flip: a rapid dorsiflexion of the penis, representing erections of the penile body; and long flip: a more gradual and sustained dorsiflexion of the penis, representing a pronounced erection of the penile body (Fig. 5d). Note that flips (penile body erections) can occur with the glans engorged or nonengorged. These penile reflexes typically recur rhythmically in clusters, with an approximate frequency of 1 to 4 min. Recently, Sachs and co-workers have suggested that a different nomenclature be applied, with E1 corresponding to "slight–moderate" tumescence, E2 corresponding to "moderate" erections, and E3 corresponding to cups. Seminal emissions (plugs) are collected, allowed to air dry for at least 3 days, and weighed. Additionally, the latency to and duration of facial grooming during the test can be recorded. If appropriate, the latency to and the number of defecations (fecal boli) and urinary frequency may be quantified.

In these *ex copula* tests, most rats display erectile reflexes within 15 min. However, some rats do not. An interesting observation regards strain differences in penile reflexes, with reflexes readily seen in intact Long-Evans (Simonsen Labs, Charles River Labs, Blue Spruce Labs), Sprague-Dawley (Charles River Labs, Harlan Sprague-Dawley), Wistar (Charles River Labs), and Wistar-Furth (Harlan Industries) but not in intact Fisher 344 rats. Many rats may be negative for reflexes on the first exposure to the test situation but exhibit reflexes in subsequent tests (see Ref. 27 for a discussion). Unlike copulatory behavior, there is no apparent circadian periodicity for penile reflexes (28).

## Erection in Freely Moving Male Rats

A proposed alternative method for monitoring penile erection has been utilized (e.g., 29–31). Studies are carried out in the lights-on period, and the number of "penile erections" are counted over 60 min. A "penile erection" is considered to have occurred when the following characteristic behaviors are exhibited: repeated pelvic thrusts followed by assumption of an upright

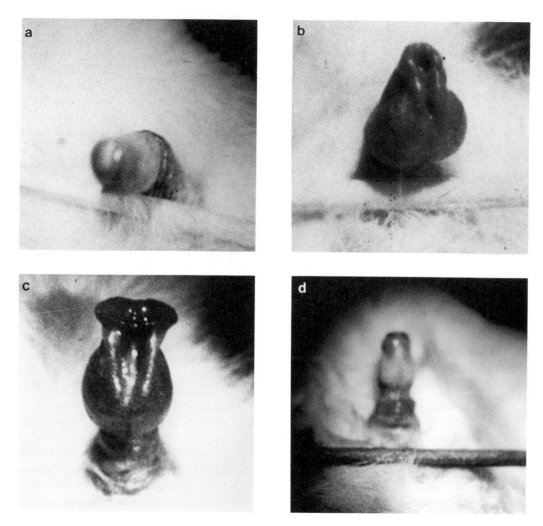

FIG. 5   (a) Close-up photograph of the glans penis in its quiescent state. (b) Erection in which the glans is slightly flared. Note that this is readily distinguished from a flaccid penis as well as from a cup. (c) A cup, or intense erection of the glans penis (compare to a less intense erection of the glans penis in b). (d) A long flip, or erection of the penile body. Note that flips can occur in conjunction with tumescence (erection of the glans penis). (Photographs were generously provided by Dr. Benjamin D. Sachs of the University of Connecticut at Storrs.)

posture and visualization of the emerging, engorged penis by the observer, that is immediately followed by autogrooming (31). There are several problems with this test; (1) normal, untreated rats display very few of these penile episodes in a 1-hr observation period, and (2) most erections would be missed. It is our opinion that this is, probably, a measure of *ex copula* ejaculation, or seminal emission with penile extrusion, and should not be equated or confused with erections measured in reflex tests. This measure is, perhaps, more comparable to the spontaneous seminal emission test (see below). Despite these reservations, this method has been successfully utilized to demonstrate convincingly the effects of drugs on this behavior. To our knowledge direct comparisons of the various methodologies has not been performed.

## Sexual Function in Anesthetized Rats

At first glance, assessment of sexual function in anesthetized rats seems inappropriate. However, the difficulties associated with interpretation of data obtained in conscious, behaving rats are formidable. There are questions that are best addressed in reduced preparations. These concern the effects of peripheral mechanisms, local release of chemical agents, local changes in blood flow, etc.

A model has been proposed that suggests the utility of such a reduced preparation (32, 33). Anesthetized animals (urethane, 1.2–1.5 g/kg, s.c.) are used. The trachea is cannulated for artificial ventilation. The carotid artery is cannulated for measurement of blood pressure and the jugular vein for intravenous administration of drugs. The pelvic organs are exposed and the urethra cannulated via an incision in the bladder (PE-50 tubing). The tubing is connected to a pressure transducer for assessment of intraurethral pressure and to a syringe pump for infusion of fluid into the urethra. Intrapenile pressure is recorded from a needle placed in the corpus spongiosum, at the base of the penis, or into the corpus cavernosum. Electromyographic recordings are obtained from the ischiocavernosus and bulbospongiousus muscles by placing two thin (0.003 inch) bared silver wire electrodes 1–2 mm apart in the muscles. Neural recordings are made using bipolar silver hook electrodes spaced about 1 mm apart from the pelvic, hypogastic, cavernous, pudendal sensory, and/or pudendal motor nerves. The exposed area is filled with warm mineral oil, and the rats are maintained at 38°C.

A reflex, termed the urethrogenital reflex, is elicited by urethral stimulation or urethral distention. The reflex is characterized by erection, emission, and rhythmic contractions of the ischiocavernosus and bulbospongiosus muscles. Urethral stimulation is accomplished by insertion of a small catheter (PE-10 tubing) into the urethra via the meatus, only as far as the urethral bulb, and physically moving the catheter. Urethral distention is accomplished by occluding the meatus and perfusing the urethra with saline (0.25–1.0 ml/min). Though the physiological significance of this reflex is hypothetical, it would allow the testing of hypotheses regarding peripheral mechanisms involved in sexual response/climax. Interestingly, this reflex appears unaltered by castration (33).

## Evaluation of Emission and *ex Copula* Ejaculation

### *Spontaneous Seminal Emission*

The spontaneous seminal emission test is based on the observation (14, 34) that male rats, housed without females, normally show seminal emission at least once during each night (the period of lights on for the rat). Since rat semen/seminal emissions readily coagulate, this reflex can be measured by simply counting and weighing seminal plugs. However, the plugs are normally consumed by the rat or his cage mate(s), necessitating some modifications. For this test, we use rats caged in isolation in wire-bottomed cages and fitted with special "corsets" to prevent genital grooming. The corsets are designed from cut cylindrical thin plastic (a good method for recycling round shampoo bottles), which is wrapped in gauze. The corset must be fit tightly enough to prevent escape but not tightly enough to interfere with breathing.

The circadian periodicity of this reflex has been worked out (28), and it is usually sufficient to check the rat, and papers under the cage, twice daily. Around the clock observations indicate that stereotyped ejaculatory behavior does not always accompany seminal emission *ex copula* (28). The plugs are collected, stored at room temperature exposed to air for 2–3 days, and weighed. Do not attempt to store in closed containers. To allow for some variation, we have adapted a 3-day corset test for spontaneous seminal emission. The parameters assessed are number and weight of plugs per day and per 3-day test period. Note that animals that have copulated to ejaculation (or multiple ejaculations) will not normally show nocturnal emissions for the next several days.

### *Decapitation-Induced Ejaculation/Seminal Emission*

In the course of studies on the neuroendocrine substrates underlying sexual, ingestive, and hormonal functions, we have accumulated a large body of data on ejaculation/seminal emission in response to decapitation. The effect of massive sympathetic discharge, induced by guillotine or hanging, has been anecdotally recorded since (at least) the French Revolution. We simply check the penis of males after decapitation, allow seminal material to air dry for several days, and weigh it. This response does not occur in anesthetized animals or in animals that are castrated. The weight of the collected plugs is directly proportional to testosterone levels in castrated, testosterone-treated rats. Plug weight is reduced in food-deprived rats and is also decreased as a function of aging. Plug weight is greater in sexually experienced than in sexually naive males. We suggest that this may be an easy method to assess ejaculatory function if animals are being sacrificed by decapitation without anesthesia for other purposes (e.g., hormonal or neurochemical studies).

## Evaluation of Sexual Motivation

The above-mentioned tests assess reflex like behavioral components of the sexual patterns of male rats. As genital responses arising in the absence of sexual stimulation, they can be viewed as analogous to a pure performance (potency) response in the human uncontaminated by sexual motivation (libido). Although there are reports where increased incidences of penile erection/emission/penile grooming or decreases in the number of intromissions to ejaculation have been interpreted as a global stimulation of male sexual behavior, and implications of increased sexual motivation, these are difficult to justify. Decrements in sexual motivation may be indicated by elongated intervals to the initiation of copulation, but this may also indicate motor or sensory impairment. Thus, sexual motivation (arousal/libido) is more properly assessed by additional tests: mounting after procedures which prevent intromission; schedule-induced performance with a receptive female as the reward; partner and place preference tests; the strength of aversive stimuli necessary to prevent the occurrence of mounting; or the speed with which a male will approach a female through a maze or down and alley. We now discuss the first three of these, which are currently in use.

### Mounting Tests

As an index of sexual motivation we use a test in which genital responses and sensory input from the genitalia are eliminated. This is accomplished by the local application of an anesthetic agent. We have found that tetracaine (Pontocaine, Breon Laboratories; 2% tetracaine hydrochloride solution) works best. With an assistant holding the rat, the preputial sheath is withdrawn, and Pontocaine is liberally applied to the penis using a cotton swab for 1 min. This procedure is repeated in 5 to 7 min. This procedure precludes intromissive and ejaculatory behavior by eliminating sensory input from the penis (35, 36). Following the second application, the male is placed in the mating arena, and a receptive female is introduced 2 to 3 min later. The frequency of mounting is then scored (manually, with the aid of a counter, or with an event recorder/personal computer) during a 15-min exposure to the female. This measure is taken as a good indicator of the propensity of the male rat to initiate the copulatory sequence. Other local anesthetics (e.g., lidocaine) have been utilized, but they have generally not prevented intromissive behavior for the entire 15-min test. One difficulty with this testing paradigm is that it cannot be repeated at short intervals.

## Schedule-Induced Instrumental Responses

Everitt and colleagues at the University of Cambridge have developed a method for assessing sexual, and other, motivation independent of genital stimulation using intrumental behavior (37–39). This approach was developed in order to obtain a reliable measure of the appetitive behavior of a male rat in a sexual context that is independent of performance or of genital reflexes. This method involves a relatively complex second-order conditioning schedule where the instrumental behavior is maintained by sexual reinforcement (the acquisition of a receptive female and the opportunity to copulate). Males are allowed to copulate with receptive females in the presence of a discrete and arbitrary light stimulus (the conditioned stimulus). Thereafter, they are tested in a modified operant chamber in which they have previously copulated. Most rats learn quickly, pressing a lever at high rates to earn the presentation of the conditioned stimulus, which signals that a receptive female will be presented at the end of the session.

This method generates a high baseline level of behavior, changes in which can be monitored as a consequence of experimental manipulations. This procedure has convincingly demonstrated that sexual motivation is not abolished by castration, nor by lesions of the diencephalic medial preoptic area. Although male rats fail to copulate, they continue to exhibit high levels of instrumental behavior to gain access to a receptive female.

## Place and Partner Preference

Place preference is another measure which can be utilized to assess some aspects of the motivation of the male to seek out a sexual partner (40–44). Place preference is conditioned by the repeated exposure of males to a receptive female, a nonreceptive female, a castrated male, or an aggressive male in distinct environments. Following this conditioning procedure, the approach to the environment previously associated with prior sexual behavior (reinforcement) is quantified. Behavioral measures are considered as related to those obtained with schedule-induced instrumental responses (40).

For partner preference tests, the choice of a male rat for a receptive versus an anestrus female is assessed. Here, the information gained is relevant to the incentive properties of the female and may be useful in discriminating the strength of the rewarding or unrewarding/aversive effects of social interactions.

## Other Considerations

### Choice of Female

An important decision to be made when conducting studies of male rat sexual behavior is the choice of the female to be used as the incentive to copulate. From our empirical observations, it does not appear to be of paramount importance whether males and females are of the same strain or size. However, strain differences do exist in male behavior. It has been our experience that Long-Evans, Sprague-Dawley, and Wistar rats mate consistently well with Long-Evans, Sprague-Dawley, Wistar, or Fisher females. However, the behavior of these strains is qualitatively different. Fisher females are less active, and lordosis is induced quite readily. The high level of soliciting behavior seen in Long-Evans females is a deterrent for behavior of Fisher males.

A second question relates to how one induces sexual behavior in females (45)? For most studies of male sexual behavior, one wants to induce maximal behavior in the female. This can be accomplished (in either intact or ovariectomized females) by subcutaneous injection of 20–100 μg estradiol benzoate (in 0.1 ml of sesame oil, peanut oil, corn oil, etc.) followed 48 hr later by a subcutaneous injection of 250–1000 μg progesterone. Maximal receptivity and proceptivity (with minimal rejection) is observed 4 to 12 hr after progesterone. Lower doses of estradiol are somewhat less consistent. Alternatively, one may wish to use naturally receptive females. Female rats typically exhibit estrous cycles of 4 to 5 days' duration. Stages of the cycle can be monitored by taking daily vaginal lavages. Most females become behaviorally receptive in late vaginal proestrus, when vaginal smears are characterized by the preponderance of nucleated epithelial cells (46).

## Age of the Male

Prior to initiating studies of male sexual behavior it is important to have some knowledge of the age of the animal and the characteristics of normal sexual behavior for that age. According to Sachs and Meisel (47, 48), mounting behavior and *ex copula* erections (E1 or E2) are first evident at a mean age of 40–41 days, intromissive behavior and *ex copula* flips are first evident at 44 days of age, and ejaculatory patterns and *ex copula* cups (intense erections with flaring of the glans penis) are first evident at 47–48 days of age. For reference, note that preputial separation normally occurs at 44–45 days of age, and peak testosterone levels are observed at 50–51 days of age (47).

Following the initial appearance of the behavioral patterns, only minor changes are evident until early middle age (approximately 11 months of age). In a longitudinal study in male Long-Evans rats (49), we observed that significant changes in mating behavior first appeared at 11 months of age: increments in mount latency, intromission latency, ejaculation latency, intercopulatory interval, and postejaculatory interval. Conversely, the ejaculatory threshold (number of intromissions preceding the ejaculation) was reduced. At this age, one also observes detectable decreases in circulating levels of testosterone, whereas decrements in gonadotropin were not evident until 19 months of age. Repeatedly, in our laboratories, we have seen that the percentage of males ejaculating decreased by 25 to 50% by 16 months of age.

## Gonadal Status

One drawback of using intact male rats for studies of sexual behavior is that changes in parameters in intact males may be difficult to distinguish (e.g., if a male achieves intromission within 10 sec of introducing the female a facilitative effect may be difficult to document) or interpret (e.g., what does a decreased number of intromissions prior to ejaculation really mean?). Sexual behavior in the male rat, namely, the occurrence of mounting, intromissive behavior, and ejaculation *in copula,* as well as erectile reflexes *ex copula,* is dependent on gonadal steroids. In certain situations, it may be advantageous to work with castrated animals, either supplied with exogenous gonadal hormones or not. In this way, experimental effects on pituitary–gonadal hormone production/secretion are avoided, and a submaximal level of behavior can be maintained, which can be either facilitated or inhibited.

Sexually experienced male rats that are castrated display a steady decline in copulatory behavior (50, 51). Typically, ejaculatory patterns are the first to disappear. Within 14 days of castration 75–100% of males typically mount, whereas 50–60% retain ejaculatory behavior. By 28–35 days after castration, 90–100% of males typically fail to ejaculate in mating tests.

There are two general paradigms for hormone replacement: maintenance and restoration. In the maintenance paradigm hormone is replaced at the time of castration, whereas in the restoration paradigm hormone is replaced after the behaviors have waned. There are several methods of systemic hormone replacement: (1) subcutaneous injection; (2) Silastic capsule implantation; (3) pellet implantation; and (4) implantation of osmotic minipumps containing hormone. Sexual behavior can be maintained, or reinstated, in castrated male rats by peripheral administration of testosterone. An injection of 12.5 $\mu$g/day/100 g body weight of testosterone propionate is sufficient to maintain sexual behavior at precastration levels, whereas 100 $\mu$g/day/100 g body weight is required for restoration of full copulatory behavior when 2 months have elapsed since castration (52).

A method not associated with the stress of daily injections, which has been extensively utilized in studies of male rat sexual behavior, is the implantation of Silastic capsules containing crystalline testosterone. These capsules are made by cutting predetermined lengths of Silastic tubing (0.62 inch internal diameter, 0.125 inch external diameter; Dow Corning Corporation, Medical Products Division, Midland, MI). One end of the cut tubing is sealed with Silastic medical grade adhesive (type A), which is allowed to dry overnight. The unsealed end of the tubing is then filled with testosterone by tapping into a pile; 25–30 taps are usually more than sufficient. After packing, the open end of tubing is sealed with Silastic adhesive. After the adhesive has dried, the capsules may be stored indefinitely. Prior to use the capsules are placed in phosphate-buffered saline for 3 days, changing the solution two to three times daily. This eliminates an initial pulse in hormone release from the capsule. These capsules maintain constant circulating levels for at least 8 weeks. The length of the packed crystalline testosterone column predicts circulating levels (Fig. 6).

Capsule implantation is a simple procedure. The animals are anesthetized, the skin on the back is shaved and swabbed with antiseptic, a small incision is made in the skin, and a subcutaneous pocket is blunt dissected. The capsule is placed within the subcutaneous pocket, and the incision is closed using wound clips. Another rule of thumb is that capsules longer than 10–20 mm should not be implanted as a single capsule. We implant two 10-mm capsules to achieve a 20-mm implant. Reasonable behavior can be maintained by small implants (i.e., 2–5 mm). Recent data from our laboratory indicate that 5-mm testosterone-containing Silastic capsules effectively restore copulatory behavior in the majority of male rats that were castrated 9 months earlier (at 6 months of age; J. T. Clark, unpublished observations, 1992).

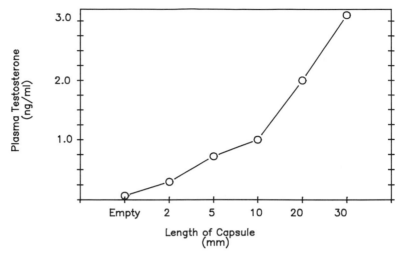

FIG. 6   Relationship between length of the testosterone column in Silastic capsules and circulating levels of testosterone. Data are means of 30 to 50 rats for each data point and were retrieved from archival data files. Testosterone was measured by radioimmunoassay on blood samples obtained by jugular venisection under ether anesthesis, 21 days after capsule implantation. The rats were castrated at 3 to 4.5 months of age, and testosterone-containing Silastic capsules were implanted subcutaneously 60 days after castration.

Finally, a large body of data has accumulated on the proposed role of testosterone metabolites, notably estradiol and dihydrotestosterone, in the regulation of male sexual behavior. Suffice it to say that treatment of castrated male rats with estradiol restores mounting behavior but not erectile reflexes, whereas treatment with dihydrotestosterone restores erectile reflexes but not mounting behavior (53) (but see Refs. 3 and 5 for discussion). These observations underscore and support the division of sexual behavior into component parts.

## Concluding Remarks

The male rat serves as a useful model for assessing some of the biological substrates underlying sexual function and behavior. The first step in assessing sexuality is to identify component parts and attack experimentally the mechanisms associated with them. We have presented our thoughts on the hypothetical framework within which sexual behavior data can be interpreted, the basic methodologies necessary for the measurement of male copulatory behavior in the male rat, and paradigms to assess independently reflexlike and motivational aspects of sexual function.

## Acknowledgments

This work is dedicated with great affection to Julian M. Davidson. My sincere gratitude is extended to Benjamin D. Sachs for freely giving of his thoughts and photographs (photographs in Fig. 5 were provided by Dr. Sachs). The technical assistance of Howard Wynder and Alphonso Keaton is gratefully acknowledged. The author was supported by awards from the National Institutes of Health (HL-02482, GM-08037, and RR-03032).

## References

1. J. M. Davidson, *in* "Principles of Geriatric Medicine" (R. Andres, E. L. Bierman, and W. R. Hazzard, eds.), p. 154. McGraw-Hill, New York, 1985.
2. F. A. Beach, *in* "Nebraska Symposium on Motivation" (M. R. Jones, ed.), p. 1. Univ. Nebraska Press, Lincoln, Nebraska, 1956.
3. J. M. Davidson, *in* "The Psychobiology of Consciousness" (R. J. Davidson and J. M. Davidson, eds.), p. 271. Plenum, New York, 1980.
4. B. D. Sachs and R. J. Barfield, *Adv. Study Behav.* **7,** 91 (1976).
5. B. D. Sachs and R. L. Meisel, *in* "The Physiology of Reproduction" (E. Knobil and J. D. Neill, eds.), p. 1393. Raven, New York, 1988.
6. C. P. Stone, *J. Comp. Psychol* **2,** 95 (1922).
7. E. E. Anderson, *J. Comp. Psychol.* **21,** 447 (1936).
8. C. Diakow, *J. Comp. Physiol. Psychol.* **88,** 704 (1975).
9. D. A. Dewsbury, *in* "Endocrine Control of Sexual Behavior" (C. Beyer, ed.), p. 3. Raven, New York, 1979.
10. J. T. Clark, M. L. Stefanick, E. R. Smith, and J. M. Davidson, *Pharmacol. Biochem. Behav.* **19,** 781 (1983).
11. M. L. Stefanick, E. R. Smith, J. T. Clark, and J. M. Davidson, *Physiol. Behav.* **29,** 973 (1982).
12. M. L. Stefanick, E. R. Smith, D. A. Szumoski, and J. M. Davidson, *Pharmacol. Biochem. Behav.* **23,** 55 (1985).
13. T. Arvidsson and K. Larsson, *Physiol. Behav.* **2,** 341 (1967).
14. F. A. Beach, W. H. Westbrook, and L. G. Clemens, *Psychosom. Med.* **28,** 749 (1966).
15. F. A. Beach, *Am. Psychol.* **15,** 1 (1960).
16. G. M. Holmes, D. G. Holmes, and B. D. Sachs, *Physiol. Behav.* **44,** 825 (1988).
17. F. Claro, A. D. Abril, S. Segovia, and A. Guillamon, *Physiol. Behav.* **48,** 489 (1990).

18. J. R. Leu and S. E. Hendricks, *Behav. Res. Methods Instrum.* **12,** 388 (1980).
19. B. Rackerd, D. A. Brigham, and L. G. Clemens, *Physiol. Behav.* **35,** 999 (1985).
20. C. A. Hendrie and S. Bennett, *Physiol. Behav.* **32,** 865 (1984).
21. B. D. Sachs, *in* "Sex and Behavior" (T. E. McGill, D. A. Dewsbury, and B. D. Sachs, eds.), p. 267. Plenum, Press, New York, 1978.
22. D. A. Dewsbury, *J. Comp. Physiol. Psychol.* **5,** 868 (1979).
23. M. Miernicki, J. B. Powers, and J. T. Clark, *Soc. Neurosci. Abstr.* **14,** 98 (1988).
24. B. L. Hart, *Science* **155,** 1283 (1967).
25. J. M. Davidson, M. L. Stefanick, B. D. Sachs, and E. R. Smith, *Physiol. Behav.* **21,** 141 (1978).
26. J. T. Clark and P. S. Kalra, *Horm. Behav.* **19,** 304 (1985).
27. B. D. Sachs, J. T. Clark, A. G. Molloy, D. Bitran, and G. M. Holmes, *Physiol. Behav.* **43,** 637 (1988).
28. M. L. Stefanick, *Physiol. Behav.* **31,** 737 (1983).
29. M. Baraldi, A. Benassi-Benelli, M. T. Barnabei, R. Cameroni, F. Ferrari, and P. Ferrari, *Neuropharmacology* **18,** 165 (1979).
30. A. Benassi-Benelli, F. Ferrari, and B. Pellegrini-Quarantotti, *Arch. Int. Pharmacodyn.* **242,** 241 (1979).
31. H. H. G. Berendsen and A. J. Gower, *Neuroendocrinology* **42,** 185 (1986).
32. S. K. Chung, K. T. McVary, and K. E. McKenna, *Neurosci. Lett.* **94,** 343 (1988).
33. K. E. McKenna, S. K. Chung, and K. T. McVary, *Am. J. Physiol.* **261,** R1276 (1991).
34. J. Orbach, *Science* **134,** 1072 (1961).
35. G. D. Gray, H. N. Davis, and D. A. Dewsbury, *Horm. Behav.* **7,** 317 (1976).
36. J. T. Clark, E. R. Smith, and J. M. Davidson, *Science* **225,** 847 (1984).
37. B. J. Everitt, P. J. Fray, E. Kostarczyk, and P. Stacy, *J. Comp. Physiol. Psychol.* **101,** 395 (1987).
38. B. J. Everitt and P. Stacy, *J. Comp. Physiol. Psychol.* **101,** 407 (1987).
39. A. M. Hughes, B. J. Everitt, and A. M. Herbert, *Psychopharmacology* **102,** 243 (1990).
40. B. J. Everitt, *Neurosci. Biobehav. Rev.* **14,** 217 (1990).
41. L. T. Dunn and B. J. Everitt, *Behav. Neurosci.* **102,** 3 (1988).
42. A. G. Phillips and H. C. Fibiger, *in* "Methods of Assessing the Reinforcing Properties of Abused Drugs" (M. A. Bozarth, ed.), p. 275. Springer-Verlag, New York, 1985.
43. D. Van Der Kooy, *in* "Methods of Assessing the Reinforcing Properties of Abused Drugs" (M. A. Bozarth, ed.), p. 299. Springer-Verlag, New York, 1985.
44. F. H. DeJonge and B. J. Meyerson, *Horm. Behav.* **16,** 1 (1982).
45. D. A. Dewsbury, *Neurosci. Biobehav. Rev.* **14,** 147 (1990).
46. J. A. Long and H. M. Evans, *Mem. Univ. Calif.* **6,** 1 (1922).
47. B. D. Sachs and R. L. Meisel, *Psychoneuroendocrinology* **4,** 287 (1979).
48. B. D. Sachs, *in* "Hormones and Behavior in Higher Vertebrates" (J. Balthazart, E. Prove, and R. Gilles, eds.), p. 86. Springer-Verlag, Berlin, 1983.
49. E. R. Smith, M. L. Stefanick, J. T. Clark, and J. M. Davidson, *Horm. Behav.* **26,** 110 (1992).

50. J. M. Davidson, *Anim. Behav.* **14,** 266 (1966).
51. C.-O. Malmnas, *Acta Physiol. Scand.* **395** (Suppl.), 1 (1973).
52. J. M. Davidson, E. R. Smith, C. H. Rodgers, and G. J. Bloch, *Physiol. Behav.* **3,** 227 (1968).
53. R. L. Meisel, J. O'Hanlon, and B. D. Sachs, *Horm. Behav.* **18,** 56 (1984).

# [4] Analysis of Female Sexual Behavior: Proceptivity, Receptivity, and Rejection

John T. Clark

## Introduction

In the previous chapter we discussed the duality (arousal–performance) of categories necessary for male sexual activity using the male rat as a model. This model has proved useful for the elucidation of biological factors involved in the regulation of sexual function. Although one could argue that the female rat is a much less appropriate model for assessing human sexual function, sexual behavior in the female rat is a hormone-dependent motivated behavior that has been utilized successfully to elucidate some aspects of neuroendocrine function (see e.g., Refs. 1–3).

## The Duality of Sexual Behavior

The separability of sexual behavior into arousal and performance (or libido–potency) categories, each with its own threshold for excitation, is not limited to the male of a given species. Rather, one may be surprised at the degree of similarity in sexual responses between the sexes. In humans, libido and potency are correlated with the majority of situations involving sexual activity in both sexes (4). Although human females do not require ovarian hormones to engage in sexual intercourse, evidenced by the retention of sexual behavior after menopause or ovariectomy, animal models such as the female rat are useful in elucidating sexually relevant biological substrates. However, we must be cautious—human females may be continuously able but are not always willing to copulate (as is also true for males) (4–6).

Regarding the arousal–performance duality as two distinct elements may be too simplistic. Davidson has suggested that viewing them as two sides of a coin might be more appropriate (4). We should always keep in mind that, in the "normal" situation, they interact, extensively modifying the other category as a function of physiological manifestations elicited by the first. The concepts, as discussed for the male, have heuristic value and distinct (but overlapping) physiological bases (1–8). Please note that, although our discussion is limited to the duality of appetitive and consummatory aspects, others [e.g., Beach (7)] have advocated the convenience of dividing feminine

*Methods in Neurosciences, Volume 14*

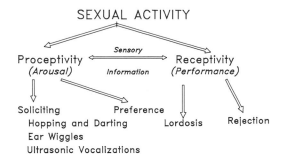

FIG. 1 Relationships between the categories of female rat sexual behavior, namely, proceptivity and receptivity, necessary for copulation to occur, and measures associated with these categories. "Sensory information" denotes the sensory effects of sexual responses that participate in the postulated relationship between sexual arousal and performance.

sexual behavior into three categories, namely, attractivity, proceptivity, and receptivity. Attractivity is difficult to explain and quantify, but, in functional terms, it refers to those qualities of the female (behavioral and otherwise) that assist the male in identifying an appropriate sexual partner (7, 8). It is our view that the behavioral aspects of attractivity can be included in proceptivity. Although we do not discuss it further, this is not a trivial point, since without some degree of attractiveness copulation will not occur.

Because the rat is the most commonly used laboratory animal, in this chapter we focus on methods of measuring female sexual responses in female rats, with a discussion of methods of interpretation of the data obtained. The utility of the rat in identifying mechanisms underlying discrete aspects of sexual behavior is enhanced by the relative ease with which rat sexual behavior can be experimentally dissected into its component parts (a situation not paralleled in the human). In addition to the various parameters of sexual behavior monitored during mating tests, we refer to several quantitative tests which allow the assessment of the separate components of female sexual behavior.

In assessing sexuality, the libido–potency/desire–ability/arousal–performance dualism is not the only dichotomy observed. For female rats, sexual behavior is readily divisible into two parts: (1) the active soliciting behavior designed to attract a male partner, and subsequent copulation, and (2) reflexive activity which can occur in the absence of any overt sexual stimulation (reflex–nonreflex duality). A final (for our discussion) duality comes to mind when we consider neuronatomical substrates influencing sexuality, that is, central versus peripheral control. The use of these theoretical concepts facilitates our interpretation of sexual behavioral data (Fig. 1) (1–8).

FIG. 2    Photograph of a male rat mounting a female in which lordosis does not occur.

## Description of Female Rat Copulatory Behavior

The copulatory pattern of the female rat has been well characterized. A complete sequence of feminine sexual behavior would include precopulatory behavior, copulation, and sequelae to copulation. Precopulatory behavior in female rats includes attracting the male and inducing him to copulate (proceptive or soliciting behaviors). The specific locomotor behavior patterns that are commonly observed are hopping and darting, after which the estrous female comes to an abrupt halt, typically assuming a crouching posture. If the male fails to investigate and mount, the female will continue soliciting. While in the crouched position, the ears of the female may vibrate quite rapidly. This is referred to as ear wiggling and is difficult to observe. Inexperienced observers may simply note that the ears are blurred. Close observation usually pays off with the observer noticing that the ears have stopped moving. Females also exhibit ultrasonic vocalizations which may serve to attract the male. In situations where the male is uncooperative, or when no male is available, the estrous female may exhibit mounting behavior.

The male rat, mounting dorsally from the rear, contacts the female flanks with his forepaws and palpates. A receptive female will rock forward, elevating her perineal region. As the male starts pelvic thrusting the female elevates her head, extends all four legs, and deviates her tail to one side. By the time the deeper thrust associated with intromission occurs, the female is in the full lordosis posture [swayed back, with the head, rump, and base of the tail elevated and the thorax lowered (7–11)]. Intromission by a male is dependent on the occurrence of lordosis. A male mounting a female in which lordosis does not occur is illustrated in Fig. 2. Lordosis and intromission are illustrated

FIG. 3  Photograph of a male rat mounting a female in which lordosis and intromission occur.

in Fig. 3. Lordosis in female rats normally terminates rapidly after dismount, but can be prolonged (Fig. 4). Sequelae to copulation are somewhat dependent on the type and extent of stimulation. After most intromissions, females are quiescent for a brief period (15–45 sec), after which they engage in proceptive behaviors. The quiescent period may be elongated after a sufficient number of intromissions have elicited ejaculation.

There can be extreme differences in different rodent species. Female hamsters and guinea pigs are stimulated to induce lordosis, but lordosis is then maintained for extended periods. There are also differences between strains of rats. Long-Evans and Sprague-Dawley rats exhibit high levels of proceptive behavior and typically terminate lordosis rapidly on male dismount. In contrast, Fisher 344 females exhibit low levels of proceptive behavior, but lordosis is easier to elicit and is prolonged.

## Standard Measures of Feminine Copulatory Behavior

The most frequently utilized measure of female rat sexual behavior is the lordosis quotient (see, e.g., Refs. 5 and 11–13). This is usually derived from observations where the female is introduced into a test chamber containing a male that is known to be sexually vigorous (rats will display sexual behavior in almost any type of testing chamber). The behavior of the female is observed until the male has mounted the female 10–20 (or more) times. Less useful are time-limited tests. The behavior is often scored manually, with L indicating

FIG. 4   Photograph of a male rat autogrooming after an intromission. Note that in this case the lordosis posture is maintained after dismount, which is atypical in Long-Evans females.

lordosis (e.g., Figs. 3 and 4) and M indicating mounts with pelvic thrusting that do not elicit lordosis (e.g., Fig. 2). The lordosis quotient (LQ) is calculated as the number of mounts with lordosis divided by the total number of mounts [i.e., $L/(L + M)$]. A number of authors report this as a percentage. Another useful index is the percentage of females displaying lordosis [or, alternatively, displaying arbitrary levels, i.e., low levels (LQ < 30) and high levels (LQ > 70)].

In tests with females that are marginally sexually receptive, it is convenient to have several cages containing sexually vigorous males, moving the female from one male to the next whenever there occurs an extended time between mounts/intromissions (or after an ejaculation). A high level of sexual interest can usually be maintained in the male by occasionally introducing a highly receptive female and allowing the male to achieve intromission.

Less frequently, investigators may report on the intensity of lordosis. These measurements usually involve rating the lordosis by assigning a value based on estimates of the observer. For example, 0 would indicate no lordosis, 1 would indicate only slight dorsiflexion of the spinal cord, 2 would indicate intermediate lordosis, and 3 would indicate maximum lordosis. Further measures associated with lordosis could include quantification of lordosis duration. However, in female rats, lordosis is typically terminated immediately after dismount. This measure is of extreme importance in other rodent species (e.g., hamsters and guinea pigs), where lordosis quotients may be relatively useless.

Although most lordosis testing utilizes males to induce the behavior, there is an alternate method. This involves scratching the flanks of the female near the hindlimbs with two fingers followed by palpation of the flanks and perineal region (3, 12, 13). Irrespective of the stimulus utilized to elicit lordosis, standardization of the stimulus is important. The lordosis response is typically unambiguous.

In addition to lordosis, the pattern of copulatory behavior in the female rat includes other components that are more difficult to measure, and thus reported less often. These include solicitation (proceptive) behaviors (hopping and darting, ear wiggling, and motionless crouching) and rejection (ranging from evading the male to facing and attacking him). One method we have used in quantifying these behaviors (14–16) follows. Proceptive behavior (soliciting) is scored for the period preceding each mount on a scale of 0–2 as follows: 0, no proceptive behavior; 1, a low level of proceptive behaviors (e.g., hop–darting, freezing, and ear wiggling); and 2, a moderate to high level of proceptive behaviors. The arithmetic mean of proceptivity scores for each individual rat is then derived. Similarly, rejection behavior (directed toward the male) is recorded for the period preceding each mount as follows: 0, no rejection or evading the male; 1, a low level of rejection or avoidance; and 2, a moderate to high level of rejection and avoidance. The arithmetic mean is calculated for each rat. As for lordosis, the incidence of proceptive or rejection behavior may be calculated. The advantage of this method is that behaviors associated with only a single mount will not contribute undue weight to the behavioral score. A similar method, utilizing a rejection quotient defined as the ratio of the total number of rejections of the male to the total number of mounts or mount attempts by the male, has also been utilized (17).

Alternatively, proceptivity has been quantified by using a standard measure of solicitations per minute. Three elements of proceptive behavior are recorded: hopping, defined as a short leap with the animal landing on all four paws, followed by assumption of a crouching posture with the female braced to receive the male's weight; darting, a run consisting of several steps, abruptly terminated by adoption of the crouching position; and ear wiggling, when the ears vibrate rapidly (usually accompanied by an upward toss of the head) (18–20). The occurrence of these behaviors is recorded on an analog event recorder or a personal computer with event recorder capabilities.

We have intentionally simplified our discussion of proceptivity. According to Beach (7), proceptive behavior refers to ". . . appetitive activities shown by females in response to stimuli received from males . . . . In the actual mating sequence, appetitive and consummatory reactions are often alternate." Behavioral measures associated with proceptivity include (1) those

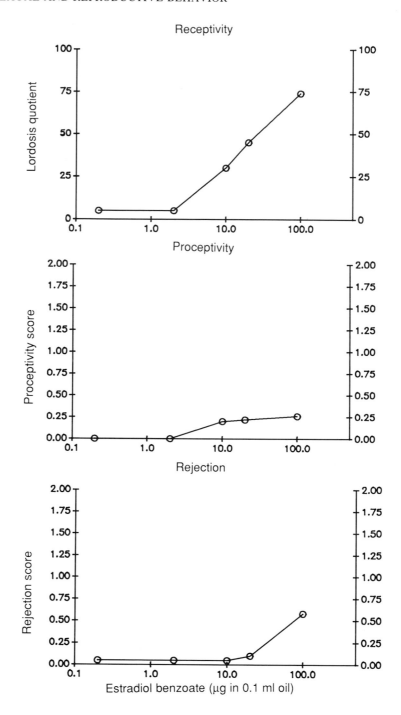

leading to establishment and maintenance of proximity to the male; (2) active solicitation; (3) alternating approach and withdrawal; (4) physical contact, to include anogenital sniffing and licking by the female; and (5) mounting by the female.

The lordosis testing, either using males or a human set of fingers, is not greatly affected by prior testing, providing that some time elapses (e.g., 30 min or more). This is advantageous in some experimental situations. For example, if one wishes to ascertain if a treatment (e.g., drug injection or electrical stimulation) has an immediate effect on lordosis, one can obtain pretreatment data, perform the treatment procedure, and obtain posttreatment data at various time intervals.

## Specialized Tests

Although most laboratories use the standard conditions discussed above, with small and restrictive testing chambers, these are far from ideal for quantifying many aspects of feminine sexual behavior. The standard conditions, using sexually experienced and vigorous males, often preclude the accurate measurement of some aspects of feminine sexual behavior. For example, a vigorous male may not allow adequate time between intromissions for the female to express proceptive behaviors. In this case, an proceptivity score of 0 would be generated, which may be inappropriate.

### Place and Partner Preference

Place preference is another measure which can be utilized to assess some aspects of the motivation of the female to seek out a sexual partner (21–25). Place preference is conditioned by the repeated exposure of females to a sexually vigorous male, to a nonreceptive female, to a receptive female, to a castrated male, or to an aggressive male in distinct environments. Following

---

FIG. 5   Dose-dependent relationship between estradiol benzoate and female rat sexual behavior (as measured by lordosis quotient, proceptivity score, and rejection score). Subjects were Long-Evans females that had been ovariectomized 12–14 days prior to hormone treatment. A single dose of EB was administered subcutaneously (in 0.1 ml sesame oil) 52 to 56 hr prior to mating tests with sexually vigorous males. Data are presented as the median of 9 to 15 rats per dose of EB. The $x$ axis is logarithmic, with 0.2, 2, 10, 20, and 100 $\mu$g EB administered to separate groups of rats.

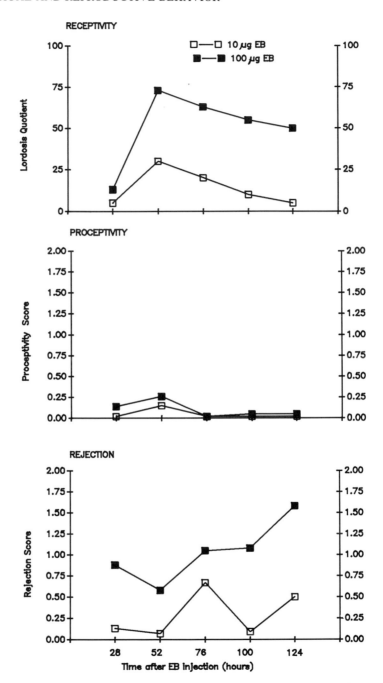

this conditioning procedure, the approach to the environment previously associated with prior sexual behavior (reinforcement) is quantified.

For partner preference tests, the choice of a female rat for a sexually active versus a castrated male is assessed. In most instances, the males are tethered or otherwise prevented from having free access to the experimental female (e.g., placed behind a wire screen). The number of approaches and the exhibition of proceptive behaviors toward the males are recorded. When these tests were conducted in conjunction with standard tests, Chambers and Phoenix (21) observed that young females showed greater preference for active males than did middle-aged females, but there was no difference in lordosis responses. The information gained in these tests is relevant to the incentive properties of the male, and it may be useful in discriminating the strength of the rewarding or unrewarding/aversive effects of social interactions.

## Special Testing Chambers

There is evidence that females in the natural environment contribute to the pacing of copulatory behavior. Recent studies suggest that specialized testing chambers, although not duplicating the natural environment, allow for additional measures of sexually relevant interactions to be quantified. Providing the female with a means of evading the male allows for the female to influence the temporal patterning of copulatory behavior (18, 26–30). One specialized chamber which appears promising was designed by Mendelson and Gorzalka (31). The chamber consists of a narrow (15 × 60.5 cm), tall (51 cm) plexiglass chamber with two levels. The two-level design allows the female an avenue of escape from the male. An additional advantage of this chamber is the maintenance of a "side-view" orientation, which facilitates accurate classification of the behaviors by inexperienced observers. The use of specialized test chambers is not widespread, but it is an area deserving more attention. Data obtained in tests that allow the female to determine the pacing of

---

FIG. 6   Effects of a single injection of estradiol benzoate on female rat sexual behavior as a function of time elapsed since hormone administration. Subjects were Long-Evans females that had been ovariectomized 12–14 days prior to hormone treatment. A single dose of EB was administered subcutaneously (in 0.1 ml sesame oil) 28 hr prior to the first mating test with sexually vigorous males. Subsequently, mating tests were administered at 24-hr intervals. Data are presented as the median of 9 to 15 rats per dose of EB.

copulation may provide important information on the level of sexual motivation, and on the effects of previous male behaviors.

## Sexual Function in Anesthetized Rats

At first glance, assessment of sexual function in anesthetized rats seems inappropriate. However, the difficulties associated with interpretation of data obtained in conscious, behaving rats are formidable. There are questions that are best addressed in reduced preparations. These concern the effects of peripheral mechanisms, local release of chemical agents, local changes in blood flow, etc.

Recently, a model has been proposed which suggests the utility of such a reduced preparation (32, 33). Anesthetized animals (urethane, 1.2–1.5 g/kg, s.c.) are used. The trachea is cannulated for artificial ventilation. The carotid artery is cannulated for measurement of blood pressure and the jugular vein for intravenous administration of drugs. The pelvic organs are exposed and the urethra cannulated via an incision in the bladder (PE-50 tubing). The tubing is connected to a pressure transducer for assessment of intraurethral pressure and to a syringe pump for infusion of fluid into the urethra. Intravaginal pressure is recorded from a pressure transducer. Electromyographic recordings are obtained from the smooth muscle of the vagina by placing two thin (0.003 inch) bared silver wire electrodes 1–2 mm apart in the muscle. Neural recordings are made using bipolar silver hook electrodes spaced about 1 mm apart from the pelvic, hypogastric, cavernous, pudendal sensory, and/or pudendal motor nerves. The exposed area is filled with warm mineral oil, and the rats are maintained at 38°C. A reflex, termed the urethrogenital reflex, is elicited by urethral stimulation or urethral distention. Vaginal and uterine contractions as well as contractions of the perineal muscles are elicited. Although the physiological significance of this reflex is hypothetical, it would allow the testing of hypotheses regarding peripheral mechanisms involved in sexual response/climax. Interestingly, this reflex is similar in males and females, and it appears to be unaltered by castration (33).

## Experimental Models

In female rats the display of sexual behavior is dependent on the action of gonadal steroids. Estrogen treatment is sufficient to induce receptive behavior (i.e., lordosis). However, in estrogen-treated females proceptive behavior is observed only infrequently. Progesterone administration to females previously treated with estradiol results in augmented receptive behavior and

TABLE I   Experimental Models Used in the Study of Female Rat Sexual Behavior[a]

| Model | LQ | %L | PS | %P | RS | %R | NM | %M |
|---|---|---|---|---|---|---|---|---|
| Intact females | | | | | | | | |
|   Diestrus I | Low | Low | Low | Low | Mod | Low | Low | Low |
|   Diestrus II | Low | Low | Low | Low | Low | Low | Low | Low |
|   Proestrus | High | Max | High | High | Low | Low | Mod | Mod |
|   Estrus | Low | Low | Low | Low | Mod | Mod | Low | Low |
|     +10 μg EB | Mod | Max | Low | Low | Low | Low | Low | Low |
|     +10 μg EB + P4 | Mod | Max | Mod | Mod | Low | Low | Mod | Mod |
|     +100 μg EB | High | Max | Low | Low | Low | Low | Low | Low |
|     +100 μg EB + P4 | Max | Max | High | Max | Low | Low | High | High |
| Ovariectomized females | | | | | | | | |
|   +Vehicle | Low | Low | Low | Low | Low | Low | Low | Low |
|   +2 μg EB | Low | Good | Low | Low | Low | Mod | Low | Low |
|   +2 μg EB × 3 | Low | High | Low | Low | Low | Low | Low | Low |
|   +2 μg EB + P4 | High | Max | Mod | Max | Low | Low | Low | Mod |
|   +10 μg EB | Low | High | Low | Mod | Low | Low | Low | Mod |
|   +10 μg EB + P4 | Good | Max | Low | Mod | Low | Low | Mod | Mod |
|   +100 μg EB | High | Max | Low | Mod | Low | Low | Low | Low |
|   +100 μg EB + P4 | Max | Max | High | Max | Low | Low | Mod | Max |

[a] For intact females, a standard 4-day estrous cycle is assumed and monitored by daily vaginal smears, with behavioral observations initiated 4–10 hr after obtaining the vaginal smear. In intact females, maximal receptivity is seen in the afternoon (i.e., 1300–1900; lights off 1130–2130) following a proestrous vaginal smear (characterized by a preponderance of nucleated epithelial cells). EB, Estradiol benzoate; P4, progesterone; LQ, lordosis quotient; %L, percentage of females displaying at least one lordosis; PS, proceptivity score; %P, percentage of females displaying proceptive behaviors; RS, rejection score; %R, percentage of females showing rejection behavior; NM, number of mounts in a 15-min test; %M, percentage of females exhibiting mounting. In an attempt to enhance utility, absolute numbers are not given for measures; rather, relative terms are used. These terms correspond for LQ and % categories to the following: Low, 0–20; Mod, 20–50; Good, 50–70; High, 70–90; and Max, 90–100. For PS and RS: Low, 0–0.5; Mod, 0.5–1.0; High, 1.0–1.5; and Max, 1.5–2.0 (see text for explanation of these scores). For MN: Low, 0–5; Mod, 5–10; High, 10–15; Max, >15. Hormones (EB, P4) were administered in oil (0.1 ml s.c.); P4 was always administered at a dose of 500 μg. Data for hormone-induced sexual behavior were derived from tests initiated 52–62 hr after EB (4–10 hr after P4) or, where multiple EB treatments are indicated, 4–10 hr after the last injection. The data were derived only from the first treatment with hormones. Data are unpublished archival data from the author's laboratory and are in agreement with previously published reports.

the display of proceptive behaviors (19, 20). When evaluating feminine sexual behavior in rats, one needs to choose a method (or multiple methods) of ensuring that adequate levels of behavior will be displayed (see, e.g., Ref. 11). Table I summarizes some commonly used paradigms for inducing sexual behavior in female rats. Females can be either intact or castrated. Intact females can be either naturally cycling, and intact or castrated females can be hormone treated. In general, female copulatory behavior has been reported to be similar across the different modes of estrus (11).

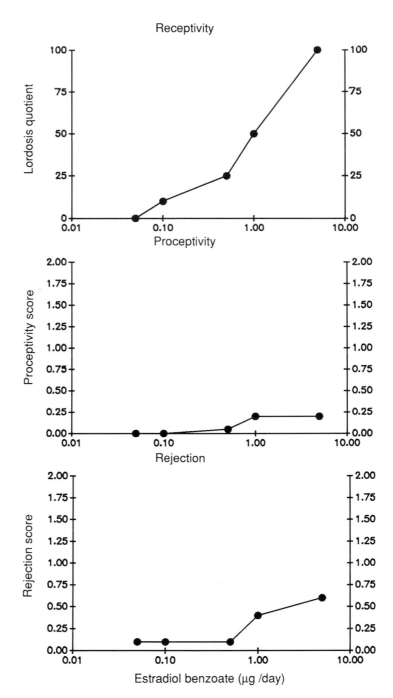

## Naturally Cycling Females

The estrous cycle of the laboratory rat can be monitored by taking vaginal smears (34). One method entails inserting a blunt dropper containing several drops of saline into the outer vagina, expelling the saline, and aspirating a sample. Alternatively, a probe (such as a thin wire loop with a water drop in the loop) is inserted into the vagina and withdrawn. Samples of sloughed cells are smeared onto a microscope slide, dried and fixed, stained (e.g., toluidine blue), and viewed under a microscope. Changing ratios (numbers) of leukocytes, nucleated epithelial cells, and cornified epithelial cells are readily apparent.

Female rats typically exhibit 4- to 5-day estrous cycles, remaining behaviorally receptive for less than 1 day. Behavioral receptivity normally occurs in late vaginal proestrus, characterized by a preponderance of nucleated epithelial cells. In mating tests with sexually vigorous males initiated 4 to 6 hr after the female exhibits a proestrous vaginal smear, lordosis quotients typically exceed 90, proceptivity scores are above 1.5, and rejection scores are below 0.25. On all other days of the estrous cycle, rejection scores are near maximal (1.5–2.0), whereas proceptivity scores and lordosis quotients are minimal. A problem with this technique is the phenomenon of constant vaginal estrus, which is commonly observed in middle-aged and old females and is not associated with high levels of proceptive or receptive behaviors (35). The advantages of naturally cycling females are obvious. The disadvantages include the possibility of pregnancy, which may be overcome by using vasectomized or castrated, testosterone-treated males (but then the females may exhibit pseudopregnancy), and the necessity of monitoring vaginal smears daily, with the possible need for last-minute schedule changes or maintenance of large numbers of rats.

## Hormone-Induced Sexual Behavior

In studies on sexual behavior hormones are typically administered subcutaneously in oil. This method results in a relatively slow, but not necessarily

---

FIG. 7   Dose-dependent relationship between daily estradiol benzoate (EB) and female rat sexual behavior (as measured by lordosis quotient, proceptivity score, and rejection score). Subjects were Long-Evans females that had been ovariectomized 12–14 days prior to the first hormone treatment. EB was administered subcutaneously (in 0.1 ml sesame oil) for 7 days. The last EB dose was injected 28 hr prior to mating tests with sexually vigorous males. Data are presented as the median of 9 to 15 rats per dose of EB. The $x$ axis is logarithmic, with 0.05, 0.1, 0.5, 1.0, or 5 $\mu$g EB/day administered to separate groups of rats.

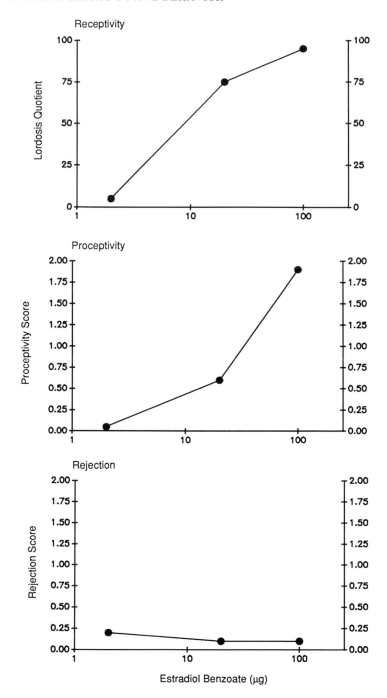

stable, release of hormone into the bloodstream. Three basic protocols are utilized to induce sexual behavior in female rats: (1) a single large (relatively) dose of estrogen; (2) multiple small doses of estrogen; and (3) estrogen treatment followed by progesterone. The most commonly used estrogen for behavioral studies is estradiol benzoate [EB; 3,17$\beta$-dihydroxy-1,3,5(10)-estratriene 3-benzoate]. Differential effects on receptivity (lordosis) and proceptivity can be assessed by choosing the appropriate hormonal treatment protocol. Thus, if an experimental manipulation is expected to produce inhibitory effects on female sexual behavior, one might choose a hormonal protocol associated with maximal levels of receptive and proceptive behaviors. Conversely, if facilitatory effects are expected, one might choose a protocol associated with low to moderate levels of behavior. It may be appropriate to use more than one hormonal treatment protocol in selected studies.

How long does it take for steroids (estrogen) to activate feminine sexual behavior? Irrespective of the hormonal treatment protocol utilized, the time from the initial estrogen administration to the onset of receptivity cannot be brought to much less than 1 to 2 days. Maximal effects are typically seen from 48 to 60 hr after a single administration.

*Single Estrogen Dose*

A single injection of EB (in 0.1 ml oil) dose-dependently induces lordosis when mating tests are conducted 48–58 hr after treatment (Fig. 5). The effects of EB on female rat sexual behavior may vary as a function of time elapsed since injection (Fig. 6), with maximal effects observed 48–58 hr after treatment. Note that, in these rats, receptivity wanes only gradually over the course of a number of days.

*Multiple Estrogen Doses*

Multiple injections of EB allow for the administration of lower doses of hormone. Repeated low doses of EB (approximating the "physiological" range) can induce levels of lordosis that are similar to those observed in intact

---

FIG. 8  Dose-dependent relationship between estradiol benzoate and female rat sexual behavior (as measured by lordosis quotient, proceptivity score, and rejection score) when a constant amount (500 $\mu$g) of progesterone was administered 48 hr after a single injection of EB. Subjects were Long-Evans females that had been ovariectomized 12–14 days prior to the first hormone treatment. Hormones were administered subcutaneously (in 0.1 ml sesame oil). Mating tests with sexually vigorous males were initiated 4–8 hr after progesterone administration. Data are presented as the median of 9 to 15 rats per dose of EB. The $x$ axis is logarithmic, with 2, 20, or 100 $\mu$g EB administered to separate groups of rats. To compare the behavioral effects of EB plus P4 to EB alone, compare these data with those depicted in Fig. 5.

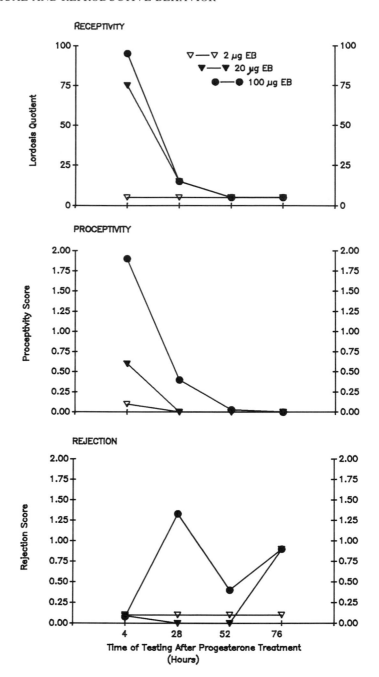

rats exhibiting a preponderance of nucleated epithelial cells in vaginal smears taken 4–8 hr prior to mating tests (36; J. T. Clark, unpublished observations). The dose-dependent relationship of multiple injections of EB is depicted in Fig. 7. A commonly used method of inducing low to moderate levels of lordosis in ovariectomized rats is to administer 2 to 5 $\mu$g EB for 3 days, with testing being initiated 28–32 hr after the third injection (37) [data from our laboratory indicate that rats ovariectomized 2 weeks previously respond with lordosis quotients of 24 $\pm$ 12, with 40% having LQ $\geq$ 10 (J. T. Clark, unpublished data)].

### Estrogen Treatment Followed by Progesterone

Administration of EB followed by progesterone (P4) dose-dependently induces sexual behavior in female rats. A commonly used protocol is to inject EB followed in 48 hr by P4, with mating tests being initiated 4 to 10 hr after P4. An alternate protocol is to administer daily injections of EB with a P4 injection 4 to 10 hr prior to testing and 24 hr following the last EB injection. Note that although female rats copulate with males (and respond to manual stimulation with lordosis) following EB alone, the display of solicitation is heavily dependent on the addition of P4 (19, 20, 27, 37–40). When ovariectomized rats are given a constant dose of EB and different doses of P4, the incidence and level of proceptive behavior increase with the increasing doses of progesterone. Typically, 50 $\mu$g P4 is the minimal dose and induces low levels of solicitation; maximal levels are observed after 200–500 $\mu$g P4 (19, 20, and J. T. Clark, unpublished observations, 1991). The effects of varying doses of EB administered 48 hr prior to 500 $\mu$g P4 on feminine sexual behavior are depicted in Fig. 8.

An intriguing observation is that, in addition to its facilitatory effects, P4 also exerts inhibitory effects on female rat sexual behavior. Data obtained in tests with ovariectomized females treated with a high dose of EB followed by 500 $\mu$g P4 or 0.1 ml oil vehicle demonstrate that P4 administration is associated with a facilitation of receptivity and proceptivity in tests conducted 4 to 10 hr after P4 but with an attenuation of receptivity in tests conducted more than 20 hr after P4 (Fig. 9).

---

FIG. 9   Female rat sexual behavior induced by sequential treatment with estradiol benzoate (EB) followed 48 hr later by progesterone (P4; 500 $\mu$g). Subjects were Long-Evans females that had been ovariectomized 12–14 days prior to EB treatment. Hormones were administered subcutaneously (in 0.1 ml sesame oil). Mating tests with sexually vigorous males were initiated 4, 28, 52, and 76 hr after progesterone administration. Data are presented as the median of 9 to 15 rats per dose of EB.

## Other Considerations

### Hormonal Treatment History

The above discussions have been limited to the first utilization of a given hormonal protocol in the female rat. An important, but often neglected, question deals with alterations in hormone-induced sexual behavior as a function of prior hormonal treatments. Hormone-induced sexual behavior can be facilitated or attenuated, depending on prior steroid exposure. Thus, administration of 5 or 10 $\mu$g EB induces a low level of lordosis in tests after the first treatment and significantly higher levels of lordosis in tests following the second through fifth EB treatments, with maximal levels being observed in all animals after the fifth EB treatment. Similar effects were seen in EB-treated rats that also received 50 $\mu$g P4 (41). Data from our laboratory demonstrate a facilitation of lordosis, but a biphasic effect on proceptivity, of prior steroid treatments (Fig. 10). Also altered is the receptivity-terminating effect of P4 (Fig. 10). These observations underscore the necessity of considering prior steroid exposure when conducting studies of female sexual behavior. Along these lines it is germane to note that variations in responsiveness to ovarian hormones have been associated with hormonal deprivation as well as hormonal treatment (42, 43).

### Age

In gonadally intact female rats, the hormonal cycle becomes disrupted and sexual behavior declines with advancing age (44–46). However, as alluded to above, categories of female sexual behavior (i.e., receptivity and proceptivity) may be under separable neural and endocrine controls. Thus, at least theoretically, these categories could be differentially affected by aging. Enhanced lordosis in response to daily EB administration has been reported in old female rats (47–50), but not in old females treated with EB plus P4 (48). Old, intact female rats that are ovariectomized and treated with EB are as sexually receptive as young females (49, 51–53). However, age-related differences in proceptivity and place preference have been observed (21). These data underscore the importance of considering the age and gonadal status (including length of hormonal deprivation and prior hormonal treatments) when conducting tests of female sexual behavior.

## Concluding Remarks

The female rat is a useful model for assessing some of the biological substrates underlying hormone-induced motivated behaviors, specifically those related

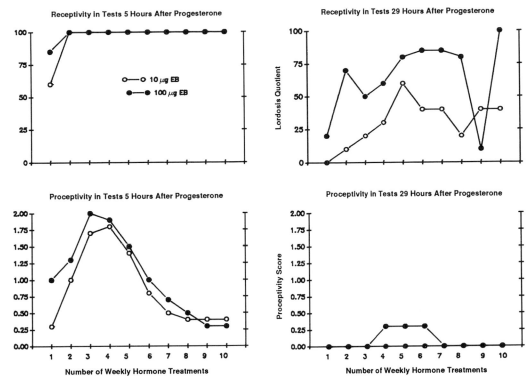

FIG. 10   Female rat sexual behavior induced by weekly administration of estradiol benzoate plus progesterone. Subjects were Long-Evans females that had been ovariectomized 12–14 days prior to EB treatment. Hormones were administered subcutaneously (in 0.1 ml sesame oil); EB was administered 48 hr prior to P4 and repeated on a weekly basis. Mating tests with sexually vigorous males were initiated 4–6 or 28–30 hr after progesterone administration. Data are presented as the median of 9 to 15 rats per dose of EB.

to reproduction. The first step in assessing sexual behavior, in either sex and in all species, is to identify component parts. Secondarily, one can experimentally examine the putative mechanisms associated with them. The utility of heuristic concepts separating female rat sexuality into proceptivity and receptivity lies in the simplification of experimental design and interpretation. The observations that measures associated with the components respond differentially to hormonal treatments, to aging, and to manipulations of the nervous system justify their continued use. In the past, increases in lordosis have been considered as indicating increased sexual motivation. Proper assessment of sexual motivation requires consideration of the categories of behavior and of the evidence, anecdotal and otherwise, that the ability to mate does not always imply a willingness.

## Acknowledgments

During the writing of this chapter the author's research was supported by awards from the National Institutes of Health (HL-02482, GM-08037, and RR-03032). Thanks go to Dr. Paul E. Micevych for theoretical discussions which refined some of my thoughts. The assistance of Alphonso K. Keaton, Artenzia Young, and Eric Floyd in collecting some of the unpublished data presented here is gratefully acknowledged.

## References

1. F. A. Beach, "Hormones and Behavior." Harper, New York, 1948.
2. N. T. Adler, "Neuroendocrinology of Reproduction: Physiology and Behavior." Plenum, New York, 1981.
3. D. W. Pfaff, "Estrogens and Brain Function: Neural Analysis of a Hormone-Controlled Mammalian Reproductive Behavior." Springer-Verlag, New York, 1980.
4. J. M. Davidson, in "The Psychobiology of Consciousness" (R. J. Davidson and J. M. Davidson, eds.), p. 271. Plenum, New York, 1980.
5. F. A. Beach, in "Reproductive Behavior" (W. Montagna and W. A. Sadler, eds.), p. 333. Plenum, New York, 1974.
6. K. Wallen, *Neurosci. Biobehav. Rev.* **14,** 233 (1990).
7. F. A. Beach, *Horm. Behav.* **7,** 105 (1976).
8. L. G. Clemens and D. R. Weaver, in "Handbook of Behavioral Neurobiology, Volume 7: Reproduction" (N. T. Adler, D. W. Pfaff, and R. W. Goy, eds.), p. 183. Plenum, New York, 1985.
9. D. W. Pfaff, C. Diakow, M. Montgomery, and F. A. Jenkins, *J. Comp. Physiol. Psychol.* **92,** 937 (1978).
10. D. W. Pfaff and C. Lewis, *Horm. Behav.* **5,** 317 (1974).
11. D. A. Dewsbury, *Neurosci. Biobehav. Rev.* **14,** 147 (1990).
12. F. A. Beach, *Physiol. Rev.* **27,** 240 (1947).
13. K. Larsson and P. Sodersten, *Psychopharmacology (Berlin)* **21,** 271 (1971).
14. J. T. Clark, J. W. Simpkins, and S. P. Kalra, *Neuroendocrinology* **44,** 488 (1986).
15. J. T. Clark, P. S. Kalra, and S. P. Kalra, *Endocrinology (Baltimore)* **117,** 2435 (1985).
16. J. T. Clark, *Physiol. Behav.* **52,** 965 (1992).
17. D. F. Hardy, *Behaviour* **41,** 288 (1972).
18. J. Madlafousek and Z. Hlinak, *Behaviour* **63,** 129 (1978).
19. B. H. Fadem, R. J. Barfield, and R. E. Whalen, *Horm. Behav.* **13,** 40 (1979).
20. B. J. Tennent, E. R. Smith, and J. M. Davidson, *Horm. Behav.* **14,** 65 (1980).

21. K. C. Chambers and C. H. Phoenix, *Neurobiol. Aging* **7,** 165 (1986).
22. L. T. Dunn and B. J. Everitt, *Behav. Neurosci.* **102,** 3 (1988).
23. A. G. Phillips and H. C. Fibiger, *in* "Methods of Assessing the Reinforcing Properties of Abused Drugs" (M. A. Bozarth, ed.), p. 275. Springer-Verlag, New York, 1987.
24. D. Van Der Kooy, *in* "Methods of Assessing the Reinforcing Properties of Abused Drugs" (M. A. Bozarth, ed.), p. 229. Springer-Verlag, New York, 1987.
25. F. H. DeJonge and B. J. Meyerson, *Horm. Behav.* **16,** 1 (1982).
26. M. K. McClintock, *Adv. Study Behav.* **14,** 1 (1984).
27. D. E. Emery and R. L. Moss, *Pharmacol. Biochem. Behav.* **20,** 337 (1984).
28. D. P. Gilman, L. F. Mercer, and J. C. Hitt, *Physiol. Behav.* **22,** 675 (1979).
29. M. S. Krieger, D. Orr, and T. Perper, *Behav. Biol.* **18,** 379 (1976).
30. J. T. Pierce and R. L. Nuttall, *J. Comp. Physiol. Psychol.* **54,** 310 (1961).
31. S. D. Mendelson and B. B. Gorzalka, *Physiol. Behav.* **39,** 67 (1987).
32. S. K. Chung, K. T. McVary, and K. E. McKenna, *Neurosci. Lett.* **94,** 343 (1988).
33. K. E. McKenna, S. K. Chung, and K. T. McVary, *Am. J. Physiol.* **261,** R1276 (1991).
34. J. A. Long and H. M. Evans, *Mem. Univ. Calif.* **6,** 1 (1922).
35. N. T. Adler and D. Bell, *Physiol. Behav.* **4,** 151 (1969).
36. J. M. Davidson, E. R. Smith, C. H. Rodgers, and G. J. Bloch, *Physiol. Behav.* **3,** 227 (1968).
37. F. A. Beach, *Proc. Soc. Exp. Biol. Med.* **51,** 369 (1942).
38. D. A. Edwards and J. K. Pfeifle, *Physiol. Behav.* **30,** 437 (1968).
39. D. F. Hardy and J. F. DeBold, *Horm. Behav.* **2,** 287 (1971).
40. R. E. Whalen, *Horm. Behav.* **5,** 157 (1974).
41. G. J. Bloch, A. M. Babcock, R. A. Gorski, and P. E. Micevych, *Physiol. Behav.* **39,** 217 (1987).
42. F. A. Beach and R. K. Orndoff, *Horm. Behav.* **5,** 201 (1974).
43. F. A. Beach, *Physiol. Behav.* **16,** 807 (1976).
44. A. A. Gerall, J. L. Dunlap, and W. E. Sonntag, *J. Comp. Physiol. Psychol.* **94,** 556 (1980).
45. H. H. Huang, R. W. Steger, J. R. Bruni, and J. Meites, *Endocrinology (Baltimore)* **103,** 1855 (1978).
46. K. H. Lu, B. R. Hopper, T. M. Vargo, and S. C. C. Yen, *Biol. Reprod.* **21,** 193 (1979).
47. R. L. Cooper, *Horm. Behav.* **9,** 321 (1977).
48. M. T. Peng, C. F. Chuong, and Y. M. Peng, *Neuroendocrinology* **24,** 317 (1977).
49. M. T. Peng, C. T. Yao, and W. C.-M. Wan, *Physiol. Behav.* **25,** 633 (1980).
50. L. R. Lee, J. Y. L. Yu, and M. T. Peng, *Neurobiol. Aging* **6,** 241 (1985).
51. C. M. Borchardt, J. R. Lehman, and S. E. Hendricks, *Age* **3,** 59 (1980).
52. R. L. Cooper and M. Linnoila, *Physiol. Behav.* **18,** 573 (1977).
53. G. D. Gray, B. Tennent, E. R. Smith, and J. M. Davidson, *Endocrinology (Baltimore)* **107,** 187 (1980).

# [5] Measurement of Hormonal and Neural Correlates of Reproductive Behavior

Dean Dluzen and Victor D. Ramirez

## Introduction

A basic goal and ultimate challenge of the neurosciences is to understand the dynamics of the cause–effect relationship between brain and behavior. Since a fundamental behavior for species survival is reproduction, this behavior is of particular importance when trying to understand brain–behavior correlations and has, therefore, received major scientific attention (1, 2).

There exist many approaches one can take in attempting to correlate an external behavioral response with internal hormonal/neural changes. Historically, the study of mating behavior and its neural/physiological correlates was pioneered largely through the efforts of Frank Beach (3, 4). His careful scrutiny of reproduction in both male and female mammals offered both temporal and discrete postural descriptions of the responses needed to compartmentalize and quantify the sequence of mating behavior. Once the behavior in question was adequately defined, various manipulations could be performed on the experimental subject to evaluate their effects on the behavioral response.

Manipulations of the animal can be performed at peripheral sites (e.g., castration, adrenalectomy) or centrally in the brain (e.g., lesions, ablations, stimulations, infusions). Resultant changes in the behavioral responses provide insights into the target site within the organism that may be controlling or at least involved with specific components of the behavioral response. An extensive number of experiments have been performed in which the effects of central nervous system lesions, ablations, stimulations, and/or infusions on reproductive behavior were examined (see Refs. 1 and 5–7). As a result of these studies, a basic understanding of the neural basis of mating behavior was achieved.

To better assess the correlations between physiological changes and reproductive behavioral responses, devices that permit repeated sampling in the freely behaving individual were sought. Uninterrupted sampling in the freely moving animal permitting temporally dependent measurements within individuals as associated with specific behavioral events represents the appeal of such procedures. Initially, this approach was applied to peripheral measurements, as in venous cannulation for sampling of blood hormonal changes.

*Methods in Neurosciences, Volume 14*

Such a dynamic approach not only reveals some of the hormonal correlates of reproductive behavior, but also provides some insights into neural components since changes in peripheral hormone levels observed during copulation, for example, are indicative of underlying modifications in central nervous system (CNS) activity and control. More recently, a number of dynamic techniques have become available which permit repeated sampling from the CNS of the freely behaving animal. The latter procedures have provided a means to evaluate changes in CNS activity associated with copulation. In this way, more direct correlations in brain–behavior relationships can be revealed.

Because these dynamic approaches provide the most effective means of correlating hormonal and neural changes with reproduction, this chapter focuses on those experimental paradigms leading to dynamic measurements of peripheral and/or central changes which occur in the freely behaving animal prior to, during, and after mating. In what follows we describe a series of experimental manipulations as examples of procedures to illustrate hormonal/neural correlates of reproductive behavior.

## Dynamic Determinations—Peripheral Measurements

### Responses in Males

An example of dynamic peripheral measurements in which repeated serum sampling was performed in the male rat during mating was reported by Kamel and Frankel (8). In this experiment, male rats were implanted with indwelling atrial cannulae, thereby allowing repeated bleeding in the freely moving animal. On the day following cannula implantation, a basal blood sample was collected followed 15 min later by the introduction of an estrous female into the cage and subsequent blood sampling at 5, 15, 30, and 60 min postfemale introduction. In sexually experienced male rats, serum luteinizing hormone (LH) and prolactin (PRL) showed significant increases at 5 min postfemale exposure, with a return to basal levels by 30 and 60 min, respectively. Serum testosterone levels showed a more gradual rise, reaching maximal levels at the 60-min sampling interval. During this period, normal mating behavior was observed, indicating that cannulae implantation did not disrupt copulation. The authors were unable, however, to indicate any specific components of the copulatory sequence that were temporally associated with precise changes in hormonal concentrations, since the overall mean latencies for display of mounting, intromission, and ejaculation were less than that of the first postfemale exposure blood sampling.

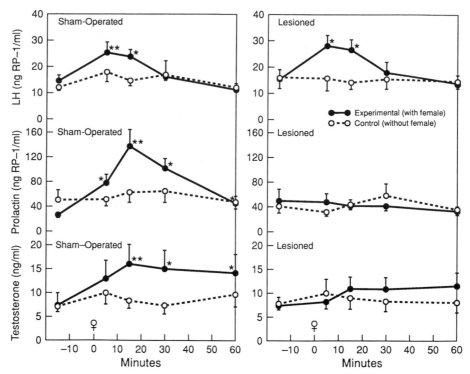

FIG. 1   Effect of medial preoptic area lesions on sexually stimulated hormone release in male rats. In sham-operated animals introduction of a receptive female led to a significant increase in blood luteinizing hormone (LH), prolactin, and testosterone, whereas in medial preoptic area lesioned rats, who failed to mount, only LH was increased. For details, see text. (Modified presentation of data from Refs. 8 and 9.)

In an attempt to achieve some additional insights into the CNS control of female-induced serum hormone concentration changes, Kamel and Frankel (9) replicated the above experiment under conditions in which male rats had received electrolytic lesions in the medial preoptic area (MPOA). In this way, these authors have, in effect, combined two approaches (i.e., lesioning with repeated *in vivo* blood sampling in the copulating animal) to understand some of the neural/hormonal components of reproduction. Interestingly, their results show that serum LH changes in response to estrous female exposure were virtually identical among intact, sham-lesioned and MPOA-lesioned animals. However, increases in PRL and testosterone observed in intact and sham-operated animals were totally abolished in MPOA-lesioned rats (Fig. 1). Behaviorally, none of the lesioned animals mounted, but they displayed normal precopulatory behaviors (e.g., anogenital sniffing and pursuit). Sham-lesioned and intact animals mated normally. These results provide an elegant partition of some of the neural/hormonal and behavioral relationships which occur in the male rat, suggesting that discriminatory neural sites and hormonal responses may be related to specific components of the copulatory sequence. Whereas the initial, precopulatory (appetitive, anticipatory, or arousal) behaviors may involve an MPOA-independent LH response, the actual copulatory (consummatory) behaviors would appear to require an MPOA-dependent PRL and testosterone response.

In addition to serum LH, blood levels of epinephrine are increased in male rats exposed to a sexually active stimulus female (10). In this experiment blood samples from cannulated male rats were collected at either 10- or 1-min intervals during encounters with either gonadectomized males or intact estrous females. Although no major changes in blood epinephrine levels were obtained in response to exposure, contact, and/or interactions with a gonadectomized male, simple exposure to the estrous female produced a marked increase in epinephrine levels. Further increases in epinephrine levels were obtained when the male was engaged in sexual activity. When samples were collected at 1-min intervals to achieve a better resolution of the epinephrine response, it appeared that epinephrine levels were enhanced during copulatory bouts and returned to lower, stable levels during intercopulatory intervals.

Repeated blood sampling via cannulation of the atrium has also been performed in male mice following exposure to either estrous or ovariectomized females or urine pooled from female mice (11). We refer to this work for two reasons: first, to recognize the technical accomplishment of both cannulation of the mouse and the ability to perform repeated sampling at 5-min intervals over a 90-min period and, second, to acknowledge an additional insight regarding hormone/neural/behavioral relationships that resulted from this work. Similar to that reported for the rat, a very rapid and robust increase in male mouse serum LH accompanied estrous female exposure. Essentially identical responses were achieved with ovariectomized females or urine from females. In addition, the data of this experiment suggested that following a socially, chemically, or spontaneously induced LH surge, there exists a refractory period during which a second LH surge is evidently precluded. Subsequent work with this model has revealed that the socially stimulated refractory period is present for 25–45 min, is of neural and not pituitary origin, and does not appear to be attributable to inhibitory actions of gonadal steroids (12). Although no definitive explanation for the refractory LH surge effect is available, it is intriguing that, behaviorally, the maximal period during which ejaculatory behavior is observed in these mice is approximately 45 min (11). It is possible that activation of neuroendocrine events with social stimuli related to reproduction may activate a hard-wired series of neural events which must be permitted to proceed to completion prior to reactivation with subsequent stimuli.

## Responses in Females

Correlations of female lordosis behavior in the rat, a spontaneous ovulator, with hormonal release has also been performed. It is well known that gonadal steroids are required for both LH surges and lordosis, and there is some degree of temporal contiguity between these two events. Intact cycling female rats show a clear surge of LH which precedes sexual receptivity. However, by altering the gonadal steroid milieu and performing serial bleeding of cannulated female rats, it was possible to demonstrate a complete abolition of an LH surge in females displaying lordosis quotients of 100% (13). This dissociation between the LH surge and lordosis has received support from data using the male rat as a model (14). Since lordosis can be induced in the gonadectomized, estrogen/progesterone-treated male rat, the issue raised in this experiment was whether an accompanying LH surge could be achieved with this treatment. Cannulated gonadectomized male rats treated with different administrations of estrogen and progesterone and sequentially bled a 1- to 2-hr intervals around the time of the LH surge failed to show any significant change in LH secretion in spite of their ability to display lordosis. Collectively, these reports show that an LH surge is not required for the display of behavioral receptivity. It would appear that the CNS sites for the activation of these two events (LH surge and lordosis) are different in the female and, apparently, nonexistent in the male.

Unlike LH, prolactin (PRL) may be required for the complete display of sexual receptivity in the female rat (15). Proestrous female rats who were treated with the dopamine agonist bromocryptine (CB-154) and bled at 2-hr intervals showed a clear abolition of a PRL surge with no alteration in serum LH. This lack of a PRL surge was associated with a significant decrease in lordosis quotients (~25% versus 80% for controls). Subsequent work with this model has indicated that PRL is required to activate adrenal steroids, in particular progestins, which are necessary for the complete display of lordosis. In this regard, the increased levels of epinephrine in female rats exposed to and mated with males should be noted (10). Such increases in epinephrine levels may be reflective of adrenal activation by PRL during sociosexual encounters. In contrast to the rat, the female rabbit, a reflex ovulator, shows marked increases in blood concentrations of LH, follicle-stimulating hormone (FSH), and PRL following mating, with each hormone showing a specific pattern of response (see Ref. 16).

## Pulse Analysis of Hormonal Response

Recent developments in the analyses of hormone release have indicated that gross changes in hormone concentrations may not provide a complete or adequate picture of brain/endocrine function. With sufficiently frequent blood sampling, additional information regarding subtle changes in the pulsatile nature of hormonal release can be appreciated. Although such an approach would be quite prohibitive in small rodents (rats, mice), it is amenable when larger animals are used.

Application of frequent blood sampling for identification of changes in hormonal pulses has been utilized in the ram. As in the male rat, the ram also shows a surge of LH with exposure to the female. Refined analysis of this response indicates that the presence of the female produces an increase in the LH peak frequency (17). This response is observed in both sexually experienced and inexperienced rams; however, maximal increases in LH peak frequency are obtained in sexually experienced rams exposed to a receptive female. In spite of the differences in LH release profiles and testosterone levels, no differences in sexual behavior were observed between experienced and inexperienced rams. The LH release profile of the ewe is also altered when exposed to the ram (18). There is an approximate 2-fold increase in the frequency of LH pulses in ovariectomized, estrogen-treated ewes when 15-min interval blood samples were collected after compared to before ram exposure.

In addition to sheep, serial blood sampling during mating has been performed in male and female ferrets (19). The ferret, like the rabbit, offers the opportunity to study mating-induced hormonal release patterns in a reflex ovulator. In the ferret, blood samples were collected at 5-min intervals over a 5-hr period and then at 15-min intervals over the subsequent 7-hr interval. In the female, a significant increase in LH pulse frequency was obtained following intromission. In contrast, male ferrets showed a significant reduction in the number of LH pulses obtained during the 1- to 5-hr period following intromission. Somewhat surprisingly, these males had increased androgen levels at 5–12 hr postintromission. Whether this androgen secretion was also pulsatile could not be determined in this study, nor could it be demonstrated in a subsequent experiment whose primary purpose was to examine whether these mating-induced androgen increases were pulsatile in nature (20).

The ultimate extension of repeated blood sampling would involve continuous sampling, a procedure which has been successfully performed in the sheep. In ewes with indwelling jugular cannulae, blood was withdrawn continuously (6–8 ml/min) for 35–45 min. Samples were collected at 1-min intervals and assayed for oxytocin during exposure to and mating with a ram (21). Although the ewes showed low overall plasma oxytocin levels, the frequent sampling intervals allowed the identification of significant pulses of oxytocin release superimposed on basal levels when the receptive ewes were exposed to the ram. This increased oxytocin pulse frequency in the presence of the ram did not appear to be associated with actual mating. Therefore, whereas salient changes in hormonal concentrations may not be apparent when attempting to identify a clear relationship between hormonal and behavioral responses, the possibility of subtle changes in various release parameters of the hormone may convey important information of physiological significance.

## Dynamic Determinations—Central Measurements

Unequivocally, a complete understanding of the neural correlates of reproductive behavior will require direct, continuous, noninvasive measurements of CNS functions in the freely moving animal. While this approach currently remains a pipe dream, application of invasive techniques for dynamic measurement of CNS activity has been utilized in animals during copulation.

Direct central measurements of neuroregulator correlates of reproductive behavior have used six different approaches: (1) electroencephalography, (2) recording of single and multiple unit activity, (3) voltammetry, (4) sampling of cerebrospinal fluid, (5) microdialysis, and (6) push–pull perfusion. Each of these techniques has its own limitations and advantages. In general, one could consider that electrical recordings provide the investigator with the fastest time resolution currently available; however, these techniques are the least specific, an exception being voltammetry, which indirectly relates an electrical signal with a defined neurochemical. The remaining invasive procedures have chemical resolution, since they are limited only by the sensitivity of the analytical procedure, but they lack adequate time resolutions. Even in cases of minute-to-minute sampling, the temporal resolutions are inadequate for precise correlations with discrete aspects of reproductive behavior. The vast majority of the work in this area has been performed in the male, but data are also available from the female hamster, rabbit, and monkey. The male has the advantage of being in a perennial reproductive basal state which can be activated as required by the receptive female or her stimuli. Such a stimulus will then produce discrete behavioral responses which can be recorded and correlated with CNS neuroregulator changes.

## Electroencephalography

Electroencephalography (EEG), pioneered by C. H. Sawyer and co-workers (22) during the late 1950s, represents the first attempt to correlate brain electrical activity with reproductive behavior. In their classic studies in the rabbit, Sawyer and Kawakami showed well-defined electroencephalographic after-reactions and behavioral phenomena in female rabbits, bearing chronic electrodes implanted in the cortical and deep regions of the brain, following copulation or after vaginal stimulation (23).

The following citation (23) clearly summarizes these events in conscious unrestrained animals: "A phase of sleep spindles lasting from several seconds to half an hour or more, during which the rabbit appears drowsy if not actually sleeping. A phase of 'hippocampal hyperactivity' characterized by an 8/sec high amplitude synchronous EEG record from the hippocampus and its projections. During this period the rabbit is further depressed behaviorally: its head on the floor, ears retracted, eyes almost closed, pupils contracted, heart and respiratory rates slowed, and its only other motion consisting of twitches of the eyelids and jaw movements. Following this phase, which lasts only a few minutes, the rabbit usually stands and starts eating either food or feces which may be extracted directly from her anus." Since these changes occurred too late to be casually related to brain-activated release of pituitary hormones, the authors concluded that changes in EEG activity may be induced "by the feedback to the nervous system of pituitary hormone release in response to the coitus or vaginal stimulus."

In another paper the authors clearly demonstrated that progesterone in estrogen-primed female rabbits or does in natural estrus induces sequential changes consisting of first lowering and later elevating the EEG arousal and the EEG after-reaction thresholds (24). These data are one of the first solid demonstrations that steroid hormones mediate part of their central effects on the CNS by altering thresholds of cerebral activity.

More recently, using a similar approach, hippocampal EEG activity was recorded from male rats who were required to perform an instrumental response (bar press) to gain access to and copulate with a receptive female (25). Changes in the frequency of rhythmical slow activity in the rat hippocampus are believed to represent different motivational states of the animal. A clear discriminatory EEG response profile was associated with noncopulatory (pursuit of the female) versus copulatory (mounting, intromission, ejaculation) behaviors. Accompanying the former behavioral responses were hippocampal rhythmical slow activities of 7–12 Hz, whereas the latter was associated with an immediate reduction of frequency to below 7 Hz. It was concluded that these two frequency patterns of EEG activity correspond to appetitive and consummatory motivational states, respectively, which can be differentiated in this paradigm.

## Single and Multiple Unit Activity

Measurements of single and multiple unit activity offer a potentially powerful approach to correlating direct central nervous system activity with reproductive behavior. For example, in the monkey, single neuron activity was recorded from the medial preoptic area and dorsomedial hypothalamic nucleus in partially restrained males who were able to copulate with a receptive female (26). The data from this report indicate that activity from these two areas show differential responses as a function of the particular component of the copulatory behavior sequence in which the male is engaged. Specifically, medial preoptic area firing rates were elevated during periods antecedent to copulation and decreased with acquisition and throughout actual mating with the female. In contrast, increased activity of dorsomedial hypothalamic neurons was closely associated with mating (mounting, intromission, and thrusting), while little activity was obtained at all other phases of copulation.

The dichotomy of behavioral/neural responsiveness observed between these two CNS sites was confirmed in a subsequent report (27). In addition, the responsiveness of the female was also examined. Neuronal activity changes in the medial preoptic area of the female were also associated with precopulatory periods. During copulation, increased neuronal activity within the female ventromedial hypothalamus was obtained.

Similar to results obtained in the monkey, multiple unit activities recorded from the medial preoptic area of the male rat are increased with the introduction of the female and during pursuit, that is, at initial, noncopulatory phases (28). This activity drops substantially with the onset of copulation. It seems clear from the data in both the monkey and the rat that electrical activity of the medial preoptic area is enhanced by the presence of the female alone and/or stimuli associated with the female or mating. Confirmation of this hypothesis has been provided from experiments in which medial preoptic area neuronal activity was measured when males were in the presence of estrous females, but not during actual copulation (29). Under such conditions, changes in medial preoptic area single unit activity were observed.

Single unit activity from the lateral mesencephalic tegmentum as a function of copulatory behavior has also been examined in the male rat (30). In this study, there did not seem to be any one specific component of the reproductive sequence that was associated with a discrete change in single unit activity. Two of the more consistent effects in this experiment were that 59% of the units showed increased firing rates during pursuit of the female, whereas 37% showed suppressed activity during the postejaculatory period. In this regard, the changes in activity of the lateral mesencephalic tegmentum during precopulatory and copulatory behavior are similar to those observed in the medial preoptic area.

An interesting series of experiments have been reported which attempt to understand the relationship among gonadal steroid hormones, brain electrical activity, and lordosis in the hamster (31, 32). In these experiments, compilations of electrical recordings from various hypothalamic and midbrain neurons were performed following gonadal steroid hormone administration and somatosensory lordosis-inducing stimuli. It appears that progesterone produces distinct changes in neuronal activity levels such that a reorganization in specific areas and activities of the brain can occur. The result of this reconfiguration is a focus of activity directed to the elicitation of lordosis, at the appropriate time and in response to the appropriate hormonal milieu.

## Voltammetry

Differential normal pulse voltammetry has been used to evaluate changes in nucleus accumbens dopaminergic activity in copulating male rats. Differential normal pulse voltammetry has the distinct advantage of enhanced time resolution. With this technique, continuous voltammetric measurements of dopamine activity and metabolism can be recorded every minute throughout behavioral tests. Initial work with voltammetry involved measurements of the dopamine metabolite 3,4-dihydroxyphenylacetic acid (DOPAC) from the nucleus accumbens (33). Very different patterns of DOPAC responsiveness were obtained as a function of social stimulus presentation. In response to a 15-min exposure to an aggressive or nonaggressive male intruder, gradual increases in DOPAC were obtained, with maximal levels reaching 40 and 15% over baseline, respectively. These increases persisted for an extended period after removal of the stimulus animal. In contrast, exposure to an intact female (irrespective of estrous cycle day) produced a much more acute response, with a rapid increase to 30% over baseline and a rapid return to baseline after removal of the stimulus female. Behaviorally, only noncopulatory activity (licking, sniffing, and chasing) was directed to the female.

When actual copulatory behavior was quantified in male rats subjected to voltammetry, it was shown that maximal dopamine activity (increase over baseline) was obtained from sexually experienced males in response to receptive females (70%), with lesser activity to nonreceptive females (50%)

and even less to intact males (30%) (34). The DOPAC activity profiles showed a similar stimulus-dependent pattern as that observed for dopamine, although the percent increases over baseline were lower for each condition. When temporal relationships between voltammetric data were analyzed in reference to latencies for onset of copulatory behavior (mounting, intromission, and ejaculation), it was concluded that the height of the voltammetric signal coincided with the beginning of mating. In the sexually naive male rat, however, mere exposure to the receptive female, in the apparent absence of copulatory behavior, produced a rapid and significant increase in nucleus accumbens dopamine activity (35). Significant increases were obtained only in response to receptive females and not to nonreceptive females or intact males. These data suggest that olfactory cues from the receptive female produce a very rapid, unlearned, reflexlike increase in dopaminergic activity within the nucleus accumbens.

## Sampling of Cerebrospinal Fluid

One of the initial attempts to correlate neuroactive substances with reproduction was to implant cannulae into the cisterna magna and sample cerebrospinal fluid (CSF) for amino acids at specific episodes of the mating sequence (36). CSF samples collected from sexually experienced male rats before sexual activity, immediately after ejaculation, and at the end of the postejaculatory refractory period were subjected to reversed-phase high-performance liquid chromatography to evaluate changes in amino acid levels (37). The data from these reports indicate that $\gamma$-aminobutyric acid (GABA), Asp, and Glu show postejaculatory increases, whereas Ser, Arg, Ala, and Leu decrease. Subsequent work utilizing this procedure revealed no changes in CSF tryptophan, 5-hydroxytryptophan, 5-hydroxytryptamine, or 5-hydroxyindoleacetic acid when male rats were sampled during the three time periods (38).

It can be appreciated that this approach has the capacity to yield a great deal of information, since 23 amino acids and/or the serotonergic metabolic system can be assayed with each sample. Moreover, these results do demonstrate a degree of specificity in the response, with concentrations of some amino acids increasing, some decreasing, and no changes in serotonergic activity being observed. Imprecision of temporal resolution between amino acid or indoleamine concentrations and behavioral responses represents a salient disadvantage of this approach. In addition, with the exception of one animal, CSF samples were taken over a series of days. In this way, although samples were obtained from individual animals, the sampling technique was not performed successively over a single mating sequence. The question of the exact site or source of these neuroregulators and the meaning or interpretation of these concentration changes represent additional issues which require consideration when evaluating these data.

## Microdialysis

At least two different laboratories have applied *in vivo* microdialysis as a means to measure neurotransmitter release during male copulatory behavior (39, 40). In both examples the microdialysis probes were implanted into the nucleus accumbens and/or striatum of sexually experienced male rats. With the introduction of the receptive female there is an approximately 60% increase in dopamine levels over baseline from the nucleus accumbens. Pleim *et al.* (39) collected microdialysis samples at 30-min intervals, with dopamine levels being significantly increased during the second and third 30-min sampling intervals. Although the augmented dopamine levels in these two intervals were essentially equivalent, behavioral responses were markedly different, with the number of mounts, intromissions, and, in particular, ejaculations being reduced from the second to the third 30-min sampling interval.

Pfaus *et al.* (40) used a slightly different behavioral paradigm and sampled dopamine at 10-min intervals within the striatum as well as the nucleus accumbens. The results show some of the more discrete components of the reproductive sequence, notably the anticipatory phase, which are associated with increased dopamine levels. Specifically, stimuli associated with the mating test produce increases in dopamine levels, with transfer to the mating test cage and exposure to the estrous female sequestered behind a screen producing 30 and 90% increases in dopamine, respectively. Actual copulation saw a further increase to only about 100% of baseline (i.e., a 10% increase over that obtained following exposure to the receptive female). By far the vast majority of the increase in nucleus accumbens dopamine levels in copulating male rats appears to be attributable to initial or anticipatory components of mating. Interestingly, changes in striatal dopamine levels were much less pronounced and increased significantly only after the later phases of copulation. Subsequent work has confirmed these results and has shown that these increases in dopamine release induced by receptive females cannot be attributed to novelty or increased locomotor activity (41).

When comparing the results of the two laboratories, two critical technical aspects regarding correlations of behavior with neuroregulator output are illustrated: (1) the importance of the behavioral paradigm and (2) the time resolution of neuroregulator determinations. Although both laboratories show a clear increase in nucleus accumbens dopamine output during mating, it would appear that these increases may have little to do with actual copulation (mounting, intromission, or ejaculation) per se, but rather with the "anticipatory" stimuli associated with mating.

## Push–Pull Perfusion

Measurement of neuropeptides from the copulating animal does not easily lend itself to application of microdialysis or voltammetry since membranous barriers and/or voltammetric measurements currently preclude adequate collection and assay of these substances. In the "open" push–pull perfusion technique there exists no barrier as perfusate samples exchange unrestricted with neuroregulator output. Such an approach has been effectively utilized to measure the activity of the neuropeptide luteinizing hormone-releasing hormone (LHRH) from the hypothalamus of the mated female rabbit. A rapid increase in the amplitude, but not the frequency, of LHRH in the doe accompanied mounting by the buck (42). Demonstration of this effect is critically dependent on the site of perfusion in the doe, since cannulae placements identified to be in the dorsal and anterior areas of the hypothalamus failed to indicate any increase in mating-induced LHRH increases but those in the tuberal region of the hypothalamus showed significant increases in LHRH levels and amplitudes. It seems likely that it is primarily the tactile vaginal stimulation by the buck that is responsible for producing the increase in LHRH activity, since similar results were obtained with artificial mechanical vaginal stimulation of the receptive doe (43).

A more detailed examination of the mating-induced changes in neuroregulators and pituitary hormone output in the doe has revealed that the copulation-dependent increase in LHRH is accompanied by increases in hypothalamic norepinephrine levels (44). In this report, the authors took advantage of the capacity to remove an aliquot from the perfusate sample and simultaneously assay an individual sample for both a neurotransmitter, norepinephrine (radioenzymatic assay), and a neuropeptide, LHRH (radioimmunoassay). In addition, insertion of an indwelling catheter into the jugular vein for sampling of peripheral adenohypophyseal hormones permitted confirmation that the mating-induced LHRH surge resulted in a corresponding increase of LH (and prolactin). In this way, a comprehensive profile of internal physiological changes which occur both centrally and peripherally during mating in the doe was achieved (Fig. 2).

Based on the push–pull perfusion experiments, the following interactive sequence of behavioral, neurochemical, neuropeptide, and hormonal events can be envisioned. Vaginal stimulation of the receptive doe resulting from mounts and intromissions of the buck produces a neurogenic signal initiating norepinephrine release. This catecholamine has the capacity to activate

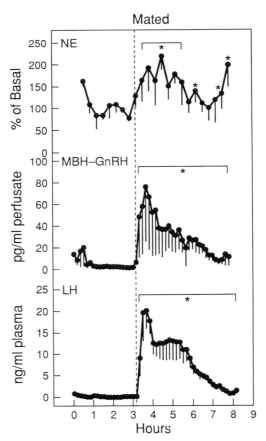

FIG. 2 Mating-induced changes in output of neuroregulators and pituitary hormones in female rabbits. Significant increases in norepinephrine (NE), gonadotropin-releasing hormone (GnRH), and luteinizing hormone (LH) are observed in female rabbits following mating. For details, see text. (Modified presentation of data from Ref. 44.)

LHRH release, resulting in stimulation of adenohypophyseal LH to generate the putative mating-induced LH surge. Further work using this approach will be required to elaborate the details of these events and to explain additional observations such as the prolactin surge (decrease in dopamine release?), the more delayed and gradual rise of FSH, and the possible role of neuropeptide Y in this phenomenon (45).

Because the data obtained from the male rat indicate that stimuli from the receptive female, most likely olfactory, can produce rapid and discriminable neuroregulator changes which can be measured *in vivo* in the freely behaving rat, we have applied the push–pull perfusion technique to measure neurotransmitter output from the olfactory bulbs (46). Presumably, this structure represents a critical first relay station for receipt and processing of such information. In the olfactory bulb we observed that norepinephrine output would increase in response to a receptive female. Unlike the dopamine response obtained from the nucleus accumbens, the increase in olfactory bulb norepinephrine levels occurred only to the second and/or third exposures to the receptive female. No changes in norepinephrine output were noted following similar exposures to an intact male rat (Fig. 3). We have speculated that these changes in norepinephrine are involved with memory/recognition processes of the receptive female, as has been indicated from other work examining memory/recognition processes of the olfactory system as related to reproduction (47). Such responses may be indicative of some of the more complex cognitive responses in the CNS associated with mating.

## Overview

Correlations of hormonal/neural responses with reproduction provides us with one approach to understand brain–behavior relationships. The two salient disadvantages of this approach are that the relationships are not necessarily causal and the time resolution between the hormonal/neural and the behavioral responses is inadequate. In spite of these limitations, this approach has procured invaluable information on brain–behavior relationships associated with reproduction and has illustrated questions and directions for future work in this area.

The analyses of reproductive behavior can be performed at a wide range of levels, from measurements of individual muscular responses (e.g., bulbospongiosus muscle), to discrete components of the reproductive sequence (e.g., anogenital investigation, a mount), to more global combinations of these discrete components (e.g., appetite or precopulatory versus consummatory or copulatory behaviors). For the most part, attempts to correlate hormonal/neural responses with reproduction have been performed using the latter level of behavioral analyses. This is due either to the limited time resolution of the hormonal/neural measurement techniques or, conversely, to the fact that when minute-to-minute hormonal/neural measurements were taken, relatively gross behavioral measures were conducted. Given that the current state of investigation of hormonal/neural correlates of reproductive behavior primarily involves attempts to correlate hormonal/neural responses to either the appetitive or consummatory components of mating, the question is what information can be derived from the analyses of these correlations.

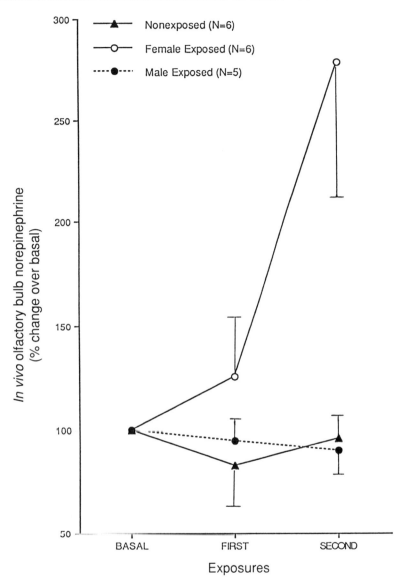

FIG. 3 Neurotransmitter output from the olfactory bulbs of male rats exposed to receptive females. Olfactory bulb norepinephrine shows substantial increases in males following a second, but not first, exposure to an estrous female. No such spontaneous increases and no increases following male exposure were obtained using this procedure. For details, see text. (Modified presentation of data from Ref. 46.)

One general issue which emerges from this analysis is that dynamic peripheral determinations, that is, hormonal responses, tend to be unary in nature with regard to correlations with appetitive versus consummatory aspects of reproduction. By unary we mean that the hormonal correlate is associated exclusively with either the appetitive or the consummatory behavioral component. For example, LH responses in the male appear to be the domain of the appetitive component of reproductive behavior. In male rats, mice, and sheep, simple exposure to the female, or stimuli from the female, results in a very rapid and acute increase of LH. A variety of evidence would seem to support the exclusive relationship of LH with the appetitive aspect of reproduction. Abolition of the consummatory component in the male, as achieved with MPOA lesions, does not eliminate the LH response. Moreover, MPOA-lesioned rats continue to display appetitive behaviors (anogenital sniffing and pursuit). In the mouse, the initial LH increase observed in response to the female or her stimuli is incapable of increasing further during the consummatory component owing to a refractory period of this hormone. Finally, although there exists some differences in LH frequency changes in sexually experienced versus inexperienced rams in response to a female, no differences in actual sexual (consummatory) behavior were observed.

In contrast to the case of LH, PRL appears to be a hormone associated with the consummatory component of reproductive behavior. From a temporal perspective, PRL, like LH, increases very rapidly on exposure to the female, which would suggest a relationship to the appetitive aspect of reproduction. However, if this PRL rise is prevented, appetitive behavioral responses continue while consummatory responses are abolished. Other dynamic peripheral hormone determinations do not appear to be as readily or exclusively categorized into this reproductive behavior dichotomy. On the basis of temporal criteria, testosterone would seem to be primarily a consummatory reproductive behavior hormone. With regard to epinephrine, owing to the possibility of a more direct neural control and/or because of the relatively short time resolution of determinations, these levels show more of a ternary responsiveness, with slight increases during appetitive periods, greater increases during consummatory phases, and return to stable basal levels during intercopulatory intervals.

For the most part, the above description can be applied to the female, with LH being appetitive, PRL consummatory, and epinephrine responses observed in both components. In addition, data from the ewe indicate that changes in oxytocin pulses may be associated with appetitive activities. Interestingly, the above description would seem to be applicable to spontaneously ovulating species (both males and females), since LH responses, for example, are clearly a part of the consummatory portion of reproduction in reflex ovulators.

In marked contrast to peripheral determinations, central dynamic determinations can be considered binary, ternary, or greater with regard to correlations with reproductive behavior. In other words, the same CNS area or function is capable of responding, albeit differentially, to both appetitive and consummatory components of reproductive behavior. Perhaps the best example of this would be the single and multiple unit activity measurements from the MPOA. In both male rats and monkeys, unit activities increase during appetitive and decrease during consummatory components of reproduction. Similarly, hippocampal EEG recordings in the male rat shift from rhythmical slow wave activity of 7–12 Hz during appetitive behavioral responses to an abrupt reduction of less than 7 Hz with consummatory behaviors.

Other areas of the CNS appear to be somewhat more selectively responsive to either appetitive or consummatory behavioral responses. Dorsomedial hypothalamic neurons of the male monkey show increased activity specifically during mating (consummatory), with little activity during other phases of copulation. Lateral mesencephalic neurons, like those of the MPOA, are active during appetitive but suppressed during other, postejaculatory periods. Data from analyses of dopaminergic systems of the nucleus accumbens and striatum, as obtained from collecting information obtained from voltammetry and microdialysis, indicate that the nucleus accumbens is clearly associated with appetitive whereas the striatum appears to be more related to consummatory aspects of the response. Still other dynamic central determinations (e.g., CSF amino acid levels and olfactory bulb norepinephrine) do not seem to show any apparent relationship to either appetitive or consummatory aspects of reproduction. In the case of the male, norepinephrine levels in the olfactory bulb seem to correlate with a specific sex recognition pattern, since the response (an increase in norephinephrine levels) occurs after the males are exposed for a second or third time to females, but not males. With the exception of the female rabbit, relatively little work in this area has been done in the female. In this reflex ovulator, CNS changes (EEG, LHRH, norepinephrine) have all been associated with consummatory aspects of reproduction.

Taken together, an overall evaluation of the hormonal/neural correlates of reproductive behavior appear to indicate a reduction in the complexity of the internal physiological response as correlated with reproductive behavior when measurements proceed from central to peripheral sites. Using the end point of serum hormone levels, distinct increases in specific hormones occur with specific aspects of reproduction. When sampling at the central level, there is an enhancement in the complexities of sites as well as differential responsiveness within a specific site which occurs in association with a specific behavior. As schematized in Fig. 4, sensory input that initiates

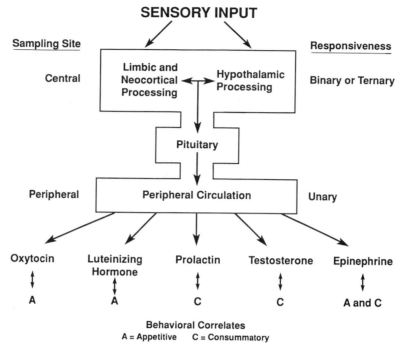

FIG. 4   Summary diagram of hormonal/neural correlates of reproductive behavior. Sensory inputs can activate simultaneously neural pathways leading to unary responses (i.e., a specific hormone in the peripheral circulation associated with a particular component of reproductive behavior) and/or binary or ternary responses (i.e., a specific component of reproductive behavior associated with more than one central measurement). For details, see text.

reproduction can simultaneously activate relatively simple unary responses through a reflexlike pathway involving the hypothalamic–pituitary–gonadal axis and relatively complex binary or ternary responses through processing at hypothalamic, limbic, and neocortical sites. The binary or ternary nature of the latter neural systems represents the multifarious integration of complex information (sensory, motor, cognitive) required for the complete display of the appetitive and consummatory components of reproductive behavior.

The fact that appetitive and consummatory components of reproduction can be partitioned on the basis of hormone responses suggests important discriminatory mechanisms which can be revealed using simultaneous central and peripheral measurement techniques. It remains to be determined whether sophisticated measurement techniques of peripheral determinations, as is being performed with pulse analysis of hormone release, will reveal more complex relationships when evaluated at the hormonal level. In addition, recent reports demonstrating the use of molecular biology techniques for measurement of gene expression correlated with both male (48) and female (49) reproductive behavior should prove to be an important approach to acquire new information regarding hormonal/neural correlates of reproductive behavior. The outstanding issues regarding these hormonal/neural correlates of reproductive behavior which need to be addressed include the exact meaning of the correlations (e.g., what does it mean that LH increases are obtained during appetitive aspects of reproductive behavior and that activity of the MPOA increases during appetitive and decreases during consummatory components of reproduction?) and the degree to which these correlations represent causal events, or most likely the simultaneous activation and/or inhibition of neural networks involved in reproductive behavior.

## Acknowledgments

We thank Dr. E. M. Hull for providing some pertinent references. We would also thank the Word Processing Department of the Northeastern Ohio Universities College of Medicine for help in typing the manuscript. This work was supported in part by grants from the National Science Foundation (NSF-BNS 91-10401 to D.D. and NSF-DCB 90-17490 to V.D.R).

## References

1. B. D. Sachs and R. L. Meisel, in "The Physiology of Reproduction" (E. Knobil and J. D. Neill, eds.), Vol. 2, p. 1393. Raven, New York, 1988.
2. D. W. Pfaff and S. Schwartz-Giblin, in "The Physiology of Reproduction" (E. Knobil and J. D. Neill, eds.), Vol. 2, p. 1483. Raven, New York, 1988.
3. F. A. Beach, in "Handbook of Experimental Psychology" (S.S. Stevens, ed.), p. 387. Wiley, New York, 1951.
4. F. A. Beach, *Physiol. Rev.* **47,** 289 (1967).
5. B. J. Everitt, *Neurosci. Biobehav. Rev.* **14,** 217 (1990).
6. D. Bitran and E. M. Hall, *Neurosci. Biobehav. Rev.* **11,** 365 (1987).
7. E. M. Kostarczyk, *Neurosci. Biobehav. Rev.* **10,** 67 (1986).
8. F. Kamel and A. I. Frankel, *Endocrinology* (*Baltimore*) **103,** 2172 (1978).
9. F. Kamel and A. I. Frankel, *Horm. Behav.* **10,** 10 (1978).
10. F. H. Bronson and E. F. Rissman, *Physiol. Behav.* **45,** 185 (1989).
11. A. Coquelin and F. H. Bronson, *Endocrinology* (*Baltimore*) **106,** 1224 (1980).
12. A. Coquelin and F. H. Bronson, *Endocrinology* (*Baltimore*) **109,** 1605 (1981).
13. P. Sodersten and P. Eneroth, *J. Endocrinol.* **112,** 133 (1987).

14. D. H. Olster and J. D. Blaustein, *Physiol. Behav.* **50,** 237 (1991).
15. J. A. Witcher and M. E. Freeman, *Biol. Reprod.* **32,** 834 (1985).
16. V. D. Ramirez and C. Beyer, *in* "The Physiology of Reproduction" (E. Knobil and J. D. Neill, eds.), Vol. 2, p. 1873. Raven, New York, 1988.
17. R. Gonzalez, P. Orgeur, P. Poindron, and J. P. Signoret, *Reprod. Nutr. Dev.* **31,** 97 (1991).
18. G. B. Martin, R. J. Scaramuzzi, and D. R. Lindsay, *J. Reprod. Fertil.* **67,** 47 (1983).
19. R. S. Carroll, M. S. Erskine, and M. J. Baum, *Endocrinology (Baltimore)* **121,** 1349 (1987).
20. G. M. Lambert and M. J. Baum, *Horm. Behav.* **25,** 382 (1991).
21. C. L. Gilbert, K. Jenkins, and D. C. Wathes, *J. Reprod. Fertil.* **91,** 337 (1991).
22. R. W. Porter, E. B. Cavanaugh, B. V. Critchlow, and C. H. Sawyer, *Am. J. Physiol.* **189,** 145 (1957).
23. C. H. Sawyer and M. Kawakami, *Endocrinology (Baltimore)* **65,** 622 (1959).
24. M. Kawakami and C. H. Sawyer, *Endorcinology (Baltimore)* **65,** 652 (1959).
25. R. Korczynski, J. Beck, and M. Biay, *Acta Neurobiol. Exp.* **49,** 255 (1989).
26. Y. Oomura, H. Yoshimatsu, and S. Aou, *Brain Res.* **266,** 340 (1983).
27. Y. Oomura, S. Aou, Y. Koyama, I. Fujita, and H. Yoshimatsu, *Brain Res. Bull.* **20,** 863 (1988).
28. T. Horio, T. Shimura, M. Hanada, and M. Shimokochi, *Neurosci. Res.* **3,** 311 (1986).
29. J. W. Mink, H. M. Sinnamon, and D. B. Adams, *Behav. Brain Res.* **8,** 85 (1983).
30. T. Shimura and M. Shimokochi, *Neurosci. Res.* **9,** 173 (1990).
31. J. D. Rose, *Physiol. Behav.* **37,** 633 (1986).
32. J. D. Rose, *Physiol. Behav.* **47,** 1201 (1990).
33. A. Louilot, M. LeMoal, and H. Simon, *Brain Res.* **397,** 395 (1986).
34. M. Mas, J. L. Gonzalez-Mora, A. Louilot, C. Sole, and T. Guadalupe, *Neurosci. Lett.* **110,** 303 (1990).
35. A. Louilot, J. L. Gonzalez-Mora, T. Guadalupe, and M. Mas, *Brain Res.* **533,** 313 (1991).
36. G. A. Qureshi and P. Sodersten, *Neurosci. Lett.* **70,** 374 (1986).
37. G. A. Qureshi and P. Sodersten, *J. Chromatogr.* **400,** 247 (1987).
38. G. A. Qureshi, G. Forsberg, I. Bendar, and P. Sodersten, *Neurosci. Lett.* **97,** 227 (1989).
39. E. T. Pleim, J. A. Matochik, R. J. Barfield, and S. B. Auerbach, *Brain Res.* **524,** 160 (1990).
40. J. G. Pfaus, G. Damsma, G. G. Nomikos, D. G. Wenkstern, C. D. Blaha, A. G. Phillips, and H. C. Fibiger, *Brain Res.* **530,** 345 (1990).
41. G. Damsma, J. G. Pfaus, D. Wenkstern, A. G. Phillips, and H. C. Fibiger, *Behav. Neurosci.* **106,** 181 (1992).
42. W. W. Lin and V. D. Ramirez, *Neuroendocrinology* **53,** 229 (1991).
43. V. D. Ramirez, J. C. Chen, E. Nduka, W. Lin, and A. D. Ramirez, *Ann. N.Y. Acad. Sci.* **473,** 434 (1986).
44. A. H. Kaynard, K.-Y. F. Pau, D. L. Hess, and H. G. Spies, *Endocrinology (Baltimore)* **127,** 1176 (1990).

45. K.-Y. F. Pau, O. Khoram, A. H. Kaynard, and H. G. Spies, *Neuroendocrinology*
    **49,** 197 (1989).
46. D. E. Dluzen and V. D. Ramirez, *Neuroendocrinology* **49,** 28 (1989).
47. P. Brennan, H. Kaba, and E. B. Keverne, *Science* **250,** 1223 (1990).
48. G. S. Robertson, J. G. Pfaus, L. J. Atkinson, H. Matsumura, A. G. Phillips, and
    H. C. Fibiger, *Brain Res.* **564,** 352 (1991).
49. B. S. McEwen, K. J. Jones, and D. W. Pfaff, *Biol. Reprod.* **36,** 37 (1987).

# Section II

## Social and Aggressive Behavior

# [6]    Measurement of Social Interactions

Margaret G. Cutler

## Introduction

Currently employed measures of social interaction are derived from early studies by ethologists, who provided an unambiguous description and classification of the naturally occurring social behavior shown by animals in the wild and in the laboratory. Social behavior, as defined in this context, comprises all activities that an animal undertakes in the presence of one of the same species. The ethological method of recording behavior lists the natural action patterns of an animal in the form of catalogs or "ethograms." These contain the series of acts (mobile) and postures (static) which form the total behavioral repertoire of the animal in question. Comprehensive ethological profiles were identified for several rodent species in the 1960s [as reviewed by Mackintosh *et al.* (1)], and it was recommended that this approach should be utilized in behavioral pharmacology (1, 2) and behavioral toxicology (3).

Sequence analysis has shown that behavioral activities are distributed in time in the form of groups or clusters of associated elements (2). This is true not only for the behavior of rodents but also for behavior of insects, fish, birds, and primates. These clusters of associated elements have been termed categories of behavior. In their early work on the behavior of rodents, Grant and Mackintosh (4) identified four major categories: nonsocial behavior, social behavior and sexual investigation, aggression, and flight. Each of the categories is considered to have causal factors in common and provides the basic structure of behavior (2, 5). Thus, measures of social interaction represent a quantitative estimate of behavior within a specific subdivision of the category of social behavior and sexual investigation.

Impairment of social interaction occurs in many human psychiatric disorders. Effectiveness of drug therapy should therefore be associated with normalization of this social impairment (2). In the preclinical testing of drugs in laboratory animals, an increase in social interaction has already been shown to provide a useful index of potential anxiolytic activity (1, 2). Quantitative assessments of social interaction which are described in this chapter include measurements of social investigation between rodents by ethological procedures (2, 6, 7) and File's paradigm, namely, the rat social interaction test (8). Testing of the effect of drugs on defined natural behavior patterns has, overall, represented a change of considerable consequence for behavioral pharmacology.

*Methods in Neurosciences, Volume 14*

TABLE I    Elements of Social and Sexual Investigation Shown by Mice, Rats, and Gerbils

| Element of behavior | Species | Definition |
|---|---|---|
| Attend | All | Direction of gaze toward other animal |
| Stretched attend | All | As in "attend," except body is elongated and head stretched forward |
| Approach | All | Walks toward other animal |
| Nose | All | Nose-to-nose contact with other animal |
| Investigate | All | Sniffing body surface of other animal, but excluding ano-genital and nasal regions [termed "sniff" by File (8)] |
| Sniff | All | Examination of anogenital region of other animal (sometimes termed "anogenital sniff" or "genital sniff") |
| Ventral sniff | Gerbil | Sniffs and licks ventral scent gland |
| Follow | All | Moves directly behind other animal |
| Push-under | All | Head and forepart of body pushed beneath other animal (sometimes termed "crawl-under") |
| Groom | All | Social grooming of other animal |
| Crawl-over | All | Climbs over back of other animal |
| Push-past | All | Two animals come into close lateral contact when moving in opposite directions |
| Huddle | Gerbil | Animals sit in contact with each other |
| Genital groom | All | Postcopulatory grooming behavior |
| Attempted mount | All | Incomplete male copulatory behavior |
| Mount | All | Complete male copulatory behavior |

## Experimental Approach

### Ethopharmacological Recording of Social and Sexual Investigation in Rodents

#### Elements of Social Behavior and Sexual Investigation

Mice and rats are the species most commonly employed in investigations of social behavior and sexual investigation, although hamsters and guinea pigs have occasionally been used (4, 5), as have Mongolian gerbils (9, 10). Table I provides an outline description of the acts and postures of social and sexual investigation shown by rats, mice, and gerbils. Figure 1 illustrates certain of these elements.

During ethopharmacological experiments, it is normal practice to record treatment-induced effects on the occurrence of all categories and elements of behavior during social interactions, so that quantitative changes to the duration and/or frequency of social investigation and its constituent elements

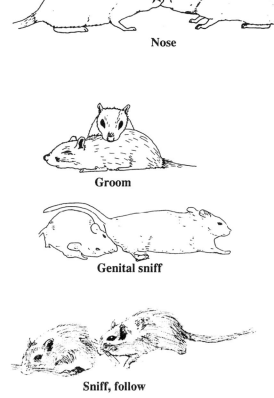

**Nose**

**Groom**

**Genital sniff**

**Sniff, follow**

FIG. 1   Some of the behavioral elements that occur during social interactions between rodents.

can be interpreted in relation to overall effects on the dynamic structure of behavior (2). The full ethological profile of behavior for mice, rats, and gerbils is summarized in Table II. This profile is based on that outlined by Mackintosh *et al.* (1) and Brain *et al.* (5). It contains additional elements that are specific for the behavior of Mongolian gerbils and thus not included in the ethogram of rats and mice (9). There are several other versions of rodent ethograms, which are derived from the original descriptive accounts (4).

*Procedures in Ethopharmacological Studies*

The housing conditions of animals prior to behavioral testing are known to have pronounced influences on subsequent behavior during the test (1, 5). Pretest housing conditions normally consist of one of three types. Where a

TABLE II  Categories of Behavior for Mice, Rats, and Gerbils and Their Constituent Elements

| Behavioral category | Species | Elements of behavior |
|---|---|---|
| Nonsocial behavior | All | Explore, scan, wash, self-groom, dig, eat, drink, scratch, shake, jump, sit, flop, stretched attend posture |
| | Gerbil | Partial upright posture, scratch walls, sawdust bathe, ventral groom, drum |
| Social investigation and sexual behavior | All | Attend, stretched attend, approach, investigate, groom, sniff, follow, nose, crawl-over, push-under, push-past, attempted mount, mount, genital groom |
| Marking behavior | Gerbil | Mark sawdust, mark walls, roll, sniff and lick surfaces |
| Aggression | All | Threat, thrust, attack, bite, chase, sideways posture, offensive sideways posture, offensive upright posture, aggressive groom, parry, kick |
| | Rat | Full aggressive |
| | Gerbil | Sidle, push-down |
| Flight | All | Flag, evade, retreat, flee, freeze, oblique upright posture, defensive upright posture, defensive sideways posture, defensive crouch, straight legs |
| | Rat, gerbil | Lie on back |
| Distance ambivalence | All | Rattle, circle, walk round, zig-zag |

reasonably high degree of social and aggressive behavior is required during the test, rodents are pair-housed for 10–14 days prior to behavioral assays. Where the tests require low levels of social interaction in untreated animals, groups of four or more animals per cage are used. Rodents are caged singly for tests, such as resident–intruder interactions, which require rodents to show a high degree of aggressiveness. Singly housed animals normally require a single pretest session for selection of those animals which will show a short latency before aggressive interactions in the test sessions.

In the period prior to drug testing, animals normally are maintained under a reversed lighting schedule for at least 10–14 days to ensure that behavior on the day of testing is recorded at a time when rodents are normally most active. A free supply of food and water remains available to the animals at all times.

Following treatment of the animals, their behavior during social encounters with a conspecific is recorded over a fixed period, normally within the range of 5 to 10 min. Examples of test situations used to detect behavioral effects of drugs include the resident–intruder paradigm in male animals (11), social encounters between unfamiliar males in a neutral cage (7, 12), and social encounters in a neutral cage between drug-treated females and untreated male partners (6). Each of these paradigms is designed to detect specific drug-induced effects on behavior (i.e., alterations to the aggressiveness of resident animals, changes to defensive behavior in intruders, changes to social investigation, aggressiveness, and exploratory activity in males that are engaged in single sex encounters in a neutral cage, and changes to social investigation and defensive behavior in the timid female paradigm). When effects of a novel compound are being examined, it is sometimes advisable to perform exploratory studies comparing drug-induced effects on behavior in different social and environmental situations (e.g., to compare behavior during encounters with a familiar animal in the home cage under low light conditions with the behavioral response of the animal in a more aversive test situation).

### Examples of Drug-Induced Effects Seen in Specific Ethological Paradigms

In the resident–intruder paradigm, the early work of Silverman (13) showed that chlorpromazine given to a resident isolated rat increased the occurrence of the social element "attend," although its major effect was to reduce aggressive and sexual behavior and to increase crouching. Spiegel and Dixon (14) demonstrated that the antidepressant drug imipramine increased social and sexual investigation in dominant pair-housed resident mice when these were confronted with an unfamiliar intruder. The resident–intruder paradigm can be seen overall to provide a sensitive measure of changes to aggressive and defensive behavior, although it is relatively insensitive to drug-induced enhancement of social investigation.

The anxiolytic compounds chlordiazepoxide, BRL 46470A [a 5-hydroxy-tryptamine$_3$ (5-HT$_3$) receptor antagonist], and buspirone (7) and the 5-HT$_1$ agonist eltoprazine (12) were found to increase the occurrence of social behavioral elements by pair-housed male mice when encountering an unfamiliar male partner in a strange neutral cage. Likewise, the 5-HT$_3$ receptor antagonist granisetron significantly increased social investigation in a neutral cage by gerbils that had been previously housed in groups of two to three (10).

The responsiveness of drug-treated female mice to untreated male partners provides an additional paradigm which appears to be highly sensitive in detecting increases of social investigation and reductions of flight by serotonergic anxiolytic agents (6). It was suggested that a conflict may be induced by this situation between "approaching toward" and "retreating away from" the unfamiliar male. A potential disadvantage of this procedure is the need to employ females at a specific phase of their estrous cycle.

*Recording and Analysis of Behavior*

Records of behavior during the encounters are made by tape recorder and videotape. Recording can be performed either by two observers, one for each animal, or by a single observer via replaying of the videotapes. The spoken commentaries of behavioral elements are then normally transcribed onto a microcomputer floppy disk by keyboard input for statistical analysis of the data.

There are several ways in which species-typical behavior can be appropriately analyzed. The most commonly employed tests record the mean frequency of occurrence and/or duration of each behavioral category and its constituent elements during social encounters within a specific test environment. For normally distributed data, the significance of the differences between mean values from treated and control groups of animals can be estimated by Student's *t*-test and the *F*-test (analysis of variance). Where the behavior is not normally distributed, results are summarized by the median values for each group, and the significance of differences between these median values from each group are calculated by the nonparametric Mann-Whitney *U* test and Kruskal-Wallis test. In all assessments of drug-induced effects on social behavior, the behavior shown by partner animals should also be recorded. This enables social investigation by the drug-treated animal to be related to the amount of investigation received, and it allows defensive behavior to be related to the amount of received aggression from the partner.

*More Specialized Methods of Behavioral Analysis*

Other more complex analyses based on the relationships between postures and the distribution of elements in time are becoming increasingly available, owing to the increased usage of videotape recording of behavior and computer-assisted techniques for data analysis (15). These more sophisticated techniques include analyses of behavioral sequences, cluster analysis to examine the interrelationships between elements, factor analysis, and principal components analysis.

Jones and Brain (16), for example, have provided a relatively simple method of sequence and cluster analysis for investigating drug-induced changes in behavioral organization independently of changes in frequency or duration of behavioral activities. The similarity of acts and postures is assessed on the basis of their position and frequency of occurrence in that position within the behavioral sequence. The behavioral organization identified is represented in the form of dendrograms. These more specialized techniques can be employed when this is warranted.

## Specific Paradigms of Social Interaction

### The Rat Social Interaction Test

The neophobia-based rat social interaction test, developed by File and Hyde (17), provides a reproducible and relatively simple preclinical screening model for detecting anxiolytic agents. It is not sensitive to the actions of antidepressants or neuroleptics (8).

The social interaction model measures the amount of time two unfamiliar rats actively interact with each other. Typically, two male rats (200–300 g) are placed in a neutral arena at a selected appropriate time after drug administration. The types of arenas that have been employed by different workers vary somewhat in size. In recent studies, for example, File and Johnston (18) used a wooden box (69 × 60 × 35 cm) with a solid floor, Costall et al. (19) employed an opaque white Perspex open-topped box (45 × 32 × 20 cm high) with 15 × 16 cm areas marked on the floor, Corbett et al. (20) used an arena measuring 50 × 50 × 30 cm, whereas Ferrari et al. (21) used a plexiglass arena measuring only 37 × 21 × 16 cm.

Several workers have used hooded Lister rats in this test (18, 22), although other strains such as Sprague-Dawley (19) and Wistar (23) rats have also been employed. Prior to the tests, animals are always allowed free access to food and water, and lights usually remain on in the animal house during daylight hours when the animals are tested.

Some workers commence the test by placing each of the animals in the center of the arena (18), while others begin the experiments with rats placed initially at opposite corners of the test box (22). The use of increased light or an unfamiliar area results in less social interaction by the rats (8). Therefore, when the test is designed to detect anxiogenic rather than anxiolytic effects, animals are placed within a dimly lit, familiar arena. Animals are made familiar to the test environment by exposure to it for a period of about 10 min on one of two successive days. Under high light conditions, illuminance on the floor of the arena is in the range of 380 lux, whereas low light conditions usually involve a dim red light of a light intensity up to 35 lux.

The amount of social interaction between control rats is influenced not only by light intensity in the arena and the degree of its familiarity to the animal, but also by the pretest housing conditions of the rats (24). Individually housed animals will show significantly higher levels of social interaction and aggression than their group-housed counterparts. A range of pretest housing conditions have been employed in previous studies, using rats that had been caged in groups of five (19, 20), in pairs (21, 23), or individually for 5 days prior to behavioral testing (18).

TABLE III    Behavioral Elements Scored in the Social Interaction Test

| Reference | Acts scored |
| --- | --- |
| Corbett et al. (20) | "Sniff" partner, "crawl over," "crawl under," "groom," "genital investigation," "follow," and "walk round" (passive social contact is not included) |
| Costall et al. (19) | "Sniff," "crawl under," "climb over," "genital investigation," and "follow" |
| File and Johnston (18) | "Sniff," "follow," "groom," "kick," "push," "wrestle," "box," "crawl under," and "crawl over" |
| Jones et al. (22) | "Follow" with contact, "sniff" (but not sniffing the hindquarters), "crawl over," "crawl under," "tumble," "box," and "groom" (sniffing of the hindquarters was excluded since it was markedly influenced by the degree of urination and defecation) |
| Kennett et al. (26) | "Groom," "self-groom," "sniff," "follow," "attempted mount," "box," and "bite" |

Effects of the drugs are generally tested by treating both members of a pair of unfamiliar rats with the same treatment at the predetermined time before testing. However, Higgins et al. (25) tested the behavior of each drug-treated rat when encountering an untreated weight-matched partner. In most tests, 6 to 10 pairs of animals were randomly allocated to each treatment group, giving a total number of 12 to 20 animals in each group. Behavior shown by each group of drug-treated animals is compared with that of an equivalent group of controls.

The duration of each social interaction test is a fixed period normally within the range of 5 to 10 min (18–20, 24). Social behavior shown by each of the rats is recorded by a video camera mounted vertically above the arena. Elements of social interaction are scored by two observers, blind to the drug treatment of the animals, who use a monitor in an adjacent room. In the experiments of File and Johnston (18), acts of social interaction were recorded onto a keyboard that fed directly into a microcomputer. During the recording of social interaction, simultaneous records are made of exploratory locomotor activity, either by counting the number of squares crossed by rats on the floor of the arena (20, 26) or by automated measurements, via, for example, a matrix of infrared cells fixed in the walls above the floor of the arena (18, 19). Records are made of the time spent by each animal in social interaction and locomotor activity.

The elements to be scored as social interaction have not yet been standardized, and this can be a considerable source of variability between findings from different workers. Table III lists the acts that have been used by different groups in recent studies. Thus, some (20, 21, 23) but not all workers have excluded elements of aggressive behavior. None of the tests of social interaction score the occurrence of "attend," "stretched attend," or "nose" which form part of the ethological category of social investigation. Data obtained in the social interaction tests can be analyzed by such statistical procedures as one-way analysis of variance (ANOVA) followed by Duncan's multiple range test, or via single factor analysis of variance followed by Dunnett's $t$-test.

Because the methods for measuring social interaction have not as yet been completely standardized, it is clearly important to select the procedure which is most appropriate for the type of pharmacological action to be examined. Buspirone, for example, shows little anxiolytic action when tested by the protocol of File (8), although anxiolytic-like effects are demonstrable via a modified procedure in which unfamiliar rats encounter each other in a familiar arena (27). Different procedures have been found to be more appropriate for the detection of mild rather than potent anxiolytic activities.

*Other tests*

Many alternative simple paradigms of social interaction have been proposed for the detection of anxiolytic drug effects. These range from a test examining drug-induced changes in the behavior of pairs of male mice when assessed via File's social interaction paradigm (28) to a test which measures the reaction of common marmosets *(Callithrix jacchus)* to confrontation by a human observer (19, 22). Preliminary studies have shown that the mouse social interaction paradigm is not a suitable model of anxiety in that species, although it is useful for assessing drug actions on social and aggressive behavior. Further studies are being carried out on the usefulness of primate models of social interaction.

## Concluding Remarks

It can be seen overall that quantitative measurements of social interactions in laboratory animals are employed in fundamental studies in psychopharmacology and in specific paradigms for the preclinical screening of compounds for anxiolytic potential. Standardization of the methodologies used in the paradigms for detecting anxiolytic compounds will considerably enhance the usefulness of these procedures.

## References

1. J. H. Mackintosh, M. R. A. Chance, and A. P. Silverman, *in* "Handbook of Psychopharmacology" (L. L. Iverson, S. D. Iversen, and S. N. Snyder, eds., Vol. 7, p. 3. Plenum, London, 1977.

2. A. K. Dixon, H. U. Fisch, and K. H. McAllister, *Adv. Study Behav.* **19,** 171 (1990).
3. A. P. Silverman, *Neurotoxicol. Teratol.* **10,** 85 (1988).
4. E. C. Grant and J. H. Mackintosh, *Behavior* **21,** 246 (1963).
5. P. F. Brain, K. H. McAllister, and S. V. Walmsley, *in* "Neuromethods, Volume 13: Psychopharmacology" (A. A. Boulton, G. B. Baker, and A. J. Greenshaw, eds.), p. 689. Humana, Clifton, New Jersey, 1989.
6. M. G. Cutler, *Neuropharmacology* **30,** 299 (1991).
7. B. Gao and M. G. Cutler, *Neuropharmacology* **31,** 207 (1992).
8. S. E. File, *Neuropsychobiology* **13,** 55 (1985).
9. J. B. Chapman and M. G. Cutler, *Neurotoxicol. Teratol.* **11,** 193 (1989).
10. M. G. Cutler and D. C. Piper, *Psychopharmacology (Berlin)* **101,** 244 (1990).
11. K. A. Miczek and M. Krsiak, *in* "Advances in Behavioral Pharmacology" (T. Thompson and P. B. Dews, eds.), Vol. 2, p. 87. Academic Press, New York, 1979.
12. B. Olivier, J. Mos, J. van der Heyden, and J. Hartog, *Psychopharmacology (Berlin)* **97,** 154 (1989).
13. A. P. Silverman, *Br. J. Pharmacol.* **24,** 579 (1965).
14. R. Spiegel and K. Dixon, *in* "Behavioural Models and the Analysis of Drug Action" (M. Y. Spiegelstein and A. Levy, eds.), p. 39. Elsevier, Amsterdam, 1982.
15. P. W. Colgan, "Quantitative Ethology." Wiley, New York, 1978.
16. S. E. Jones and P. F. Brain, *Behav. Processes* **11,** 365 (1985).
17. S. E. File and J. R. G. Hyde, *Br. J. Pharmacol.* **62,** 19 (1978).
18. S. E. File and A. L. Johnston, *Psychopharmacology (Berlin)* **99,** 248 (1989).
19. B. Costall, A. M. Domeney, P. A. Gerrard, M. E. Kelly, and R. J. Naylor, *J. Pharm. Pharmacol.* **40,** 302 (1988).
20. R. Corbett, S. Fielding, M. Cornfield, and R. W. Dunn, *Psychopharmacology (Berlin)* **104,** 312 (1991).
21. F. Ferrari, P. G. Tartoni, and V. Mangiofico, *Psychopharmacology (Berlin)* **99,** 345 (1989).
22. B. J. Jones, B. Costall, A. M. Domeney, M. E. Kelly, R. J. Naylor, N. R. Oakley, and M. B. Tyers, *Br. J. Pharmacol.* **93,** 985 (1988).
23. R. W. Dunn, R. Corbett, and S. Fielding, *Eur. J. Pharmacol.* **169,** 1 (1989).
24. R. J. M. Niesink and J. M. van Ree, *Physiol. Behav.* **29,** 819 (1982).
25. G. A. Higgins, B. J. Jones, N. R. Oakley, and M. B. Tyers, *Psychopharmacology (Berlin)* **104,** 545 (1991).
26. G. A. Kennett, P. Whitton, K. Shah, and G. Curzon, *Eur. J. Pharmacol.* **164,** 445 (1989).
27. A. P. Guy and C. R. Gardner, *Neuropsychobiology* **13,** 194 (1985).
28. R. G. Lister and L. A. Hilakivi, *Psychopharmacology (Berlin)* **96,** 181 (1988).

## [7]  Assessment of Aggressive Behavior in Rodents*

Enrico Alleva

## Introduction

After a short critical discussion of ethical principles in performing experiments on fighting animals, the general definitions of the various kinds of aggressive behavior are summarized. The common methodological problems encountered in studies of animal aggression are critically discussed and most of the current measures and indices of aggressive behavior listed. Sensory inputs regulating aggressive behavior and influencing its expression are reviewed. Seminaturalistic and laboratory assessments of mouse and rat fighting behavior and related social roles are presented, including selected ethograms for observational studies. Aggressive behavior patterns of both female (prepartum and postpartum maternal aggression) and developing rodents are briefly described. A few practical examples of scoring intermale isolation-induced and maternal aggressive behaviors are illustrated. Some warnings, statistical suggestions, and indications for selection of the genetic material and for actual assessment of dominance rank are reported. A final section deals with evaluation of drug-induced alteration of aggresive behavior.

## Ethical Issues in Studies on Animal Aggression

The increasing concern among behavioral biologists about performing experiments causing overt pain or affecting the psychological welfare of vertebrates has led to the establishment by intramural ethical committees, major scientific associations, and funding agencies (ESF, NSF, NIH, EC, APA, etc.) of a series of regulatory publications providing guidelines for ethical limits to animal studies. The Association for the Study of Animal Behaviour published a guideline paper (Huntingford, 1984) discussing some of the ethical issues raised by studies involving predator–prey and aggressive interactions in artificially staged encounters. Huntingford pointed out that, in those studies, behavior has to be recorded accurately, the possibility of collaborative experiments considered carefully, the number of subjects kept to a minimum, and

---

* To Giuseppe Montalenti, my master of natural sciences, rigor, and, hopefully, style.

*Methods in Neurosciences, Volume 14*
111

TABLE I    Factors Increasing the Possibility of
Dangerous Attacks

| |
|---|
| Small size of the testing environment |
| Difference in body size of the opponents |
| Difference in previous breeding experience |
| Lack of continuous control by the experimenter during testing |
| Difference in degree of familiarization with the test environment |
| Difference in degree of familiarization with the procedures of handling |
| Lack of hiding places to avoid being seen by the attacking individual[a] |

[a] This is a common cause of death in the case of long-duration sessions involving highly aggressive subjects.

the experiments kept as short as possible. A second report deals with pain sensitivity in animals (Bateson, 1991). Bateson (1992) also discussed in more general terms ethics in biomedical research involving animal experimentation. Accurate reading of these articles is highly recommended.

Table I indicates a series of suggestions aimed at reducing the possibility of causing bloody or even fatal attacks during the course of rodent studies. In general, male mice tend to be much more aggressive than rats. Other rodent species (e.g., *Acomys cahirinus*) are very aggressive, even against their keepers, but only when their newborn are present. Before using an unknown rodent species, the available literature on territorial behavior, parental nest defense, and major changes in intermale aggression around puberty should be checked. It is worth pointing out that attention to these issues potentially improves the science of an experiment, as well as ameliorating ethical problems.

## General Definitions

In its classic definition, aggressive behavior includes predatory behavior (intraspecific aggression and/or parasitism) and aggressive behavior directed at conspecifics (intraspecific aggression). However, Huntingford (1976) pointed out that the motivational relationships between aggressive behavior toward conspecifics, predators, and potential prey do not always depend on distinct internal factors. Moyer (1968) characterized seven classes of aggressive behavior (predatory, intermale, fear-induced, irritable, territorial, maternal, instrumental), and other authors subsequently further differentiated the taxonomy (Brain and Benton, 1981; Huntingford and Turner, 1987)

The most commonly observed patterns of intraspecific aggressive behavior regard (1) the interaction of two or more individuals, mostly males (competitive or protective aggression), (2) interactions with a lactating female (maternal aggression), or (iii) nest defense (parental aggression, including the maternal category). Mice and rats are the most commonly used animals in aggressivology (Archer, 1988; Dixon *et al.,* 1990). However, mirrors have been used with monkeys and fish to evoke aggressive responses. Adult rodents, being macrosmatic nocturnal-type mammals, rely primarily on olfactory cues for intraspecific communication. Thus, it is possible to elicit aggressive arousal by simply exposing these animals to olfactory cues (e.g., litters, feces, urine, synthetic pheromones) containing the appropriate triggering stimuli (more on this later). Any method either reducing or eliminating actual expression of unnecessarily high levels of aggressive or predatory behavior should be adapted.

## Philosophy of Recording Aggressive Behavior

Intraspecific aggressive behavior can be evaluated using either fighting pairs or social groups of different sizes. The latter methodology appears to reproduce a social setting more similar to the natural one (Blanchard and Blanchard, 1990; Mackintosh, 1981). However, in wild mice social settings are highly flexible, from demes composed of several male and female subjects to isolated individuals attempting invasion into defended territories. Single territories are rare. The size of a territory and the ability to exclude other mice are highly variable, depending on a variety of social and environmental factors such as fighting experience, invasion pressure, habitat structure, and intermale alliances (Barnard *et al.* 1991; Bronson, 1979).

Groups can be unisexual but more often include both sexes (referred to as population cages). Groups can be maintained either in laboratory cages, arenas, or enclosures (usually ranging from 0.4 to 4 m². The latter can be located indoors or outdoors (Adams and Boice, 1989; Bishop and Chevins, 1987; Blanchard and Blanchard, 1990; Hurst, 1990; Lagerspetz and Sandnabba, 1982).

Placing a conspecific intruder into an ''established'' social setting is an easy way of producing aggressive behavior of the ''territorial'' type. In fact, the odor cues emitted by an unknown or vaguely familiar conspecific are reportedly one of the major triggering factors (Adams, 1980). The intruder is often selected according to its physiological/social condition, to reduce variability in responses (Brain *et al.,* 1981). Subjects without previous sexual experiences or belonging to socially stable groups are preferred. However, Dixon and Mackintosh (1976) reported that young mice (4–6 weeks) barely induce aggressive behavior in adult conspecifics. Moreover, mice older than 10 weeks are often involved in social competition, and their status varies accordingly (Poole and Morgan, 1973). A careful examination of the social role played by the intruder in its original social setting in the period immediately prior to introduction into a new social group is essential.

It is best to use intruders or "standard" opponents whose social history is sufficiently well known since birth, particularly during the preweaning period (Poole and Morgan, 1973). Prenatal and postnatal hormonal influences, such as the relative intrauterine position, are also important determinants of the adult pattern of expression of aggressive behavior (Flannelly *et al.*, 1984; Rines and vom Saal, 1984). The prenatal hormonal influences can easily be controlled by measuring anogenital distance at birth. The factoring of the "challenging" individuals (isolated versus group-living, males with or without breeding experiences, dominant versus subdominant versus subordinate, home resident versus intruder, etc.) allows for an otherwise difficult microdissection of the actual expression of the species-specific pattern of aggressive behavior.

Other investigators have used castrated (low-aggressive) subjects (Uhrich, 1938), but they discovered that such subjects do not elicit high levels of aggression in their counterparts (Homady and Brain, 1982). Anosmic, and consequently low-aggressive, opponents were fashionable in the 1970s and 1980s (anosmia was mainly produced either by intranasal irrigation with zinc sulfate or through bilateral removal of olfactory bulbs; Alberts, 1974; Parmigiani and Brain 1983). Frischknecht and colleagues (1982) preferred the use of a genetically nonfighting mouse opponent. An ICR mouse introduced into a group of five C57BL/6 mice was attacked by subjects considered of high social status (fighters), but not by lower-ranking nonfighters.

Expression through abnormal behavior is not, however, a useful tool if the rather subtle determinants or subcomponents of aggressive behavior are to be analyzed. All these kinds of freak "standard" opponents often produce abnormal responses in their counterparts (mostly hypo- or supranormal reactions) that are of limited value for studies aimed at exhaustive evaluation of aggressive behavior.

## Gross Scores of Aggressive Behavior

The most widely used index of aggressive tendency is the "attack" category, as described in the classic ethological paper of Grant and Mackintosh (1963): ". . . a rapid approach which is carried on over the back of the other animal, the head comes in contact with the far flank of the other animal, and a Bite is usually given." Such a description is interchangeably valid for both mice

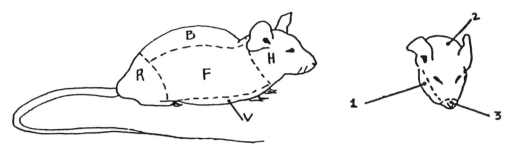

FIG. 1   A valuable indirect method of assessment is the use of wound maps. Species and sex differences are revealed by different targets and by the degree of intensity of offensive patterns. B, Back; R, rump; F, flanks; 2, dorsal, eyes, and ears; 3, snout. Lactating females bite at the nearest body part and vulnerable regions, such as head (H) and ventrum (V); females are not inhibited by the usual male inhibitory signals. In contrast, males bite only certain areas (B, R, and F). In golden hamsters areas 2 and 3 are wounded, while cheeks (area 1) are targeted only during juvenile play fighting. See Parmigiani *et al.* (1988) for mouse wound patterns, Pellis and Pellis (1988) for golden hamster wound patterns, and Pellis (1988) and Blanchard and Blanchard (1981) for wound patterns for rats and other rodent species. (Drawn by S. Bigi.)

and rats (Adams, 1980; Jones and Nowell, 1972; Blanchard and Blanchard, 1990; Grimm, 1980; Poshivalov, 1981; Winslow and Miczek, 1984). Attacks are measured in terms of frequency, duration, latency time to first appearance, or total time spent in attacking. Brain *et al.* (1981) proposed an intensity scale, ranging from a minimum of 1 (rapid biting with physical contact for a maximum of 20 sec) to a maximum of 6 (deep biting with hemorrhage, contact lasting for more than 150 sec). The Lagerspetz (1964) scale was similar, with a score of 1 indicating absence of aggressive behavior and a score of 7 (maximum) for the case of a biting attack producing hemorrhage (see also Lagerspetz and Sandnabba, 1982).

Other methodological studies considered which parts of the opponent's body were targeted. In rats, wounds are mainly located on the head, back, and flanks (Blanchard *et al.*, 1977), whereas life-threatening bites are directed at the ventral parts. In mice, the trend is similar for males, but lactating females tend to bite the head and the ventral region of the intruder. The latter behavioral pattern is considered highly defensive (Archer, 1988; Parmigiani *et al.*, 1988), and a similar pattern has been reported by Brain *et al.*(1981) in the case of isolated male mice attacking a dominant intruder. A detailed analysis of targeted biting areas ("wound maps," see Fig. 1) can provide useful indications about the offensive as well as defensive attitudes of the attacking animal. I do not like to encourage people to use the "wound maps" technique routinely, since by definition it involves injury. Field studies on wild-trapped mice can profitably use them. It is better to observe attacking animal/target attacks briefly. Anyway, the attack element remains a natural unit of offense having descriptive power, while latency to attack in an indirect measure of offense per se but gives information on the sensitivity and responsiveness of an animal to attack-provoking situations.

In close association with attacks are the "offensive postures," either upright or lateral, the latter being widely diffused among several rodent species (Grant and Mackintosh, 1963). These postures usually last for a few seconds, with the two animals pushing each other with their forepaws (Adams, 1980). Dixon *et al.* (1990) suggest that they are good indices of an ambivalent offensive tendency, caused by a strong motivation to escape mixed with a prevailing tendency to initiate an attack episode (see Table IV). Postures and acts of four laboratory species (rat, mouse, guinea pig, and golden hamster) are compared in Grant and Mackintosh (1963).

## Sensory Inputs Regulating Aggressive Behavior

Visual and tactile stimuli appear to guide the attacking performance (Katz, 1976; Thor, 1976). In particular, biting of the dorsal region seems to be induced by ventral stimulation. Animals assume the "upright defensive posture" or one of the submissive postures in order to avoid any contact, even accidental, with the ventral parts of the opponent (Adams, 1976; Blanchard *et al.*, 1977). Selective deprivation of acoustic cues (by means of ear plugs) or of visual cues (through contact lenses) affects mouse agonistic interactions (Strasser and Dixon, 1986).

Volatile compounds influencing the expression of aggressive behavior are contained in the urine of both subordinate and dominant mice (Sandnabba, 1986b; Harvey *et al.*, 1989); synthetic compounds of urinary origin active synergically in promoting intermale fighting have also been characterized (Novotny *et al.*, 1989). Furthermore, it has been found that a compound promoting aggressive behavior is produced by the preputial glands (particularly in dominants) (Jones and Nowell, 1973a,b), a production enhanced by intense fighting (Mugford and Nowell, 1971). According to Jones and Nowell, the urine of fighting male mice contains an "aversive factor," whose aversive potency depends on the combination of urine and coagulating gland secretions (Jones and Nowell, 1973a,b, 1989). Female urine elicits quite different responses in males (see, e.g., Dixon *et al.*, 1984; Dixon and Mackintosh, 1976).

## Dominance and Subordination

Dominance is currently defined as the possibility of attacking a conspecific without being counterattacked (Klopfer, 1974) or as the product of a history of agonistic encounters among individuals modifying the course of future agonistic encounters (Bernstein, 1981). Despite their widespread operational use, the concepts and the terminology surrounding the dominance/subordination phenomenology are still unresolved (Bernstein, 1981; Benton, 1982; Richards, 1974; Syme, 1974). It should be remembered that, in rodents, aggressive behavior is also a common phenomenon under natural conditions: 45% of wild male mice show wounds somehow attributable to intraspecific fighting (Southwick, 1958; Mackintosh, 1981).

Dominants are characterized by emission of olfactory cues in the urine which are different from those produced by subordinates (Harvey *et al.,* 1989). These signals have aversive effects on both dominants and subordinates. Also, only dominants release a female-attracting odor (Jones and Nowell, 1974). The preputial gland of male mice is a known source of olfactory signals that are considered a reliable indicator of social dominance (Bronson and Marsden, 1973). An electron microscopy study has characterized the preputial glands of dominant, isolated, and subordinate male mice: those of subordinates were less developed and had fewer, smaller lipid droplets and fewer cellular organelles (Brain *et al.,* 1983). More detailed analysis of olfactory communication in the course of aggression is reported in the final part of the section on Choice of Strain and Genetics.

The acquisition of a social role (rank) is also reflected by changes in neuroendocrine status, particularly evident in the enlargement of the adrenal gland in subordinates (Bigi *et al.,* 1992; Brain, 1972, 1978; Huckelbridge *et al.,* 1981; Raab *et al.,* 1986). "Dominant" DBA mice showed splenomegaly and other hematological changes following three 24-hr triadic encounters (Turney and Harmsen, 1984). A reciprocal relationship between adrenocortical activity and gonadotropin secretion was found only in the early phases of the establishment of dominance/subordination, whereas the secretory patterns for follicle-stimulating hormone (FSH), lutenizing hormone (LH), and corticosterone functioned independently thereafter (Bronson, 1973).

Subordination is also accompanied by significant changes in the immune system: subordinate male rats of the Long-Evans strain had lower T-cell proliferation and interleukin 2 (IL-2) production. However, mild agonistic encounters were less effective than intense fighting (Hardy *et al.,* 1990). Dantzer and co-workers found that these subordinates had higher plasma corticosteroid concentrations, smaller thymuses, and reduced spleen lymphocyte response to mitogenic stimulation *in vitro* (Raab *et al.,* 1986). We found that intermale fighting in CD-1 mice causes the activation of mast cells (and therefore peritoneal histamine release), resulting from salivary release of nerve growth factor (NGF). The specificity of such a mechanism is revealed by the marked inhibition produced by pretreatment with anti-NGF antibodies (Alleva and Aloe, 1989; De Simone *et al.,* 1990). For a general overview of the effect of NGF-activated coping mechanisms on rodent psychosocial stress, see Alleva and Aloe (1991) and Levi-Montalcini *et al.* (1990). It is worth pointing out that these facts have implications for how we house our subjects as well as to the quality of the data we collect in our experiments.

## Scoring Mouse Aggressive Behavior

Mice tend to arrange their social settings in hierarchies (Uhrich, 1938). Poole and Morgan (1973) showed that, in laboratory cages, the stability of the "hierarchical order" depends on the number of caged subjects and that the social situation of groups of 9 to 12 individuals is highly unstable. It is important to emphasize that hierarchical roles were not found in mice belonging to the same litter (Poole and Morgan, 1973), that is, in the case of subjects with high familiarity during critical stages of behavioral development. Mice maintained in 1.8 × 1.8 m enclosures show pronounced territorial behaviors, with only a few adult males defending the borders of their own territories (Crowcroft and Rowe, 1963; Mackintosh, 1970). The occurrence of fighting episodes is inversely proportional to the size of the enclosure; moreover, it decreases the day following introduction of the animals (Poole and Morgan, 1975) as they undergo habituation to the social and olfactory environment of their groupmates (Archer, 1968; Kimelman and Lubow, 1975; Mackintosh and Grant, 1966).

The intruder provokes a marked increase in aggressive interactions, while rendering the former social setting unstable (Poole and Morgan, 1975; Urhich, 1938). Lagerspetz and Sandnabba (1982) suggested that the decrease in fighting episodes is caused by familiarization with colony individuals (association of individual odor cues with number and outcome of suffered attacks). However, these authors used genetically aggressive or nonaggressive Turku mice, and their data need to be confirmed by studies using other inbred, outbred, or random-bred strains.

For the mouse, the information now available from a variety of naturalistic (Cox, 1984; Barnard *et al.,* 1991), seminaturalistic (Poole and Morgan, 1973; Hurst, 1989, 1990), and laboratory studies (Benton and Brain, 1979; Bigi *et al.,* 1992; Bishop and Chevins, 1987; Maestripieri *et al.,* 1990; Dixon *et al.,* 1984, 1990) makes it possible to distinguish among the different social "statuses" depicted in Table II (Barnard *et al.,* 1991; Bishop and Chevins, 1987; Burg and Slotnick, 1983; Evans and Mackintosh, 1970; Hurst, 1990). Usually, the intruder is attacked by the dominant mouse while it attempts to escape or displays species-specific submissive postures aimed at inhibiting the attacking counterpart. Several authors refer to the "home cage effect" when describing the peculiar pattern of aggressive behavior displayed by the

TABLE II   Social Roles among Wild House Mice[a]

| Group | Social roles |
|-------|--------------|
| Adult males | Exclusive territorial |
| | Dominance territorial |
| | Subdominant |
| | Subordinate |
| | Young-adult subordinate |
| Adult females | Breeding, unrestricted access to resources |
| | Breeding, restricted access to resources |
| | Breeding, partial access to resources |
| | Nonbreeding, unrestricted access to resources |
| | Nonbreeding, restricted access to resources |
| Juveniles | (Up to puberty, less than ~50 days) |
| Subadults | (Not yet reproductively active, subadult weight; ~45–90 days) |

[a] From Hurst (1989) and Barnard et al. (1991).

resident animal (Burgh and Slotnick, 1983; Poole and Morgan, 1975; Uhrich, 1938).

## Scoring Rat Aggressive Behavior

In rats, dominance hierarchies (threatening postures and biting attacks) do not appear before day 160, and they also depend on cage size. Grant (1963) provided an ethological description of male rat aggressive behavior, which includes sequence and pathway analysis, displacement and ambivalence activities, and features of sociosexual behaviors. A subsequent version was illustrated in Miczek and Krsiak (1979). An excellent and updated rat ectogram is reported in Lehrman and Adams (1977).

Play fighting and actual aggressive behavior are often difficult to distinguish in this species (Adams and Boice, 1983, 1989; Takahashi and Lore, 1983). Unlike the case for mice, in rats the intruder often does not elicit increased fighting among colony members (Blanchard et al., 1988a,b). On the other hand, the possibility of digging into the substrate of the arena makes rats more likely to perform a biting attack (Blanchard et al., 1985). In the case of colonies composed of individuals younger than 150 days (i.e., in the absence of an established social setting), all males participate with the same role in the attack directed at the intruder (Adams and Boice, 1989).

During the course of agonistic interactions, male rats emit 22–48 kHz ultrasonic vocalizations. These signals could be used for quantitative assessment of aggressive behavior. One type of ultrasound consists of short pulses (lasting about 3–65 msec), which are emitted at 50 kHz frequency. Another type consists of a series of repeated long (800–1600 msec) pulses, at 25 kHz frequency (Sales, 1972). Subordinates mainly emit vocalizations of the latter type. However, Takeuchi and Kawashima (1986) found that rat ultrasonic signals do not inhibit the initiation of aggressive behavior, and therefore they do not appear to have intraspecific communicative value.

## Social Roles within a Male Pair: A Reductionist Shortcut

To keep social variability under control, a number of authors have used pairs of rodents of known social history, instead of observing established colonies on introduction of an intruder conspecific challenging their social stability. Such a methodology appears to be a shortcut in mimicking the natural occurrence of aggressive behavior; in fact, it represents the only realistic way to date to examine neurobiological, pharmacotoxicological, or even broader physiological factors influencing the expression of intraspecific aggressive behavior. A second phase of this type of analysis, including assessment of social behavior of various interacting individuals, is advisable. However, the first phase (necessarily limited to dyadic or triadic encounters) including multidose treatment and a wide range of control groups (for handling, handling plus injection procedure, vehicle(s) treatment, anesthesia effects, etc.] cannot be eliminated.

The strategy we followed to analyze the physiological role played by NGF in adult rodents has been (1) assessing whether a single agonistic session was sufficient to produce salivary NGF release (Aloe et al., 1986; Alleva and Aloe, 1989); (2) evaluating such NGF release in repeated agonistic encounters, leading to a clear differentiation of a "dominant" and a "subordinate" role (Maestripieri et al., 1990); and (3) characterizing the effects of prolonged exposure to exogenous, highly purified murine NGF on the same agonistic behavior (Bigi et al., 1992). The latter experiment was aimed at verifying the existence of a regulative loop explaining how behavioral and neurobiological factors interact in transforming "the average male mouse" into a fully "dominant" or, alternatively, into an easily defeatable "subordinate" individual. Hormones, substance P, NGF, and behavioral experience cooperate in such a "transmutative" process (Alleva and Aloe, 1992); Bigi et al., 1991; 1992; Levi-Montalcini et al., 1990). Another investigation was aimed at characterizing central nervous system (CNS) changes (increased production of NGF and NGF mRNA in hypothalamic subareas; see Spillantini et al., 1989; Aloe et al., 1990) as a result of intermale aggressive behavior and studied the intermale fighting produced in isolated mice by the usual social isolation procedure.

TABLE III   Ethogram for Scoring Mouse Aggressive Behavior[a]

---

Fighting behavior
  Attacks (number, intensity, frequency, duration, latency to first
    episode)
  Tail rattling
  Aggressive grooming
Species-specific displays for aggression-inhibiting purposes
  Defensive upright posture
    Submissive upright posture
    Submissive crouched posture
Escape responses
  Flee
  Evade
Displacement or stereotypic elements for contextual evaluation of
    agonistic performance
  Self-grooming
  Bar holding[b]
  Digging
  Freezing
  Patrolling

---

[a] These elements are described in Bigi *et al.* (1992), Laviola *et al.* (1991), and Maestripieri *et al.* (1990). The ethogram is by Luigi De Acetis.

[b] See also Fig. 4 for the use of "displacement" activities in assessing aggressive behavior.

Repeated encounters between two previously unfamiliar mice of a widely used outbred strain (the Swiss-derived CD-1 strain, characterized by high levels of intraspecific fighting) provide evidence of a "dominant" and a "subordinate" role within the mouse pair (Bigi *et al.*, 1992; Maestripieri *et al.*, 1990). Both subjects are individually housed for 4–6 weeks. The dominant becomes responsible for 100% of the attacks as of the fifth or sixth encounter (each daily encounter lasts 20 min). In about 20% of the pairs the dominant is already evident in the first or the second encounter. However, in about 15% of the pairs a shift in the respective roles occurs during the course of the two subsequent encounters. A reliable assessment should involve repeated encounters (at least six to seven short or three long with daily agonistic sessions lasting 15–30 min). There is no need to score or videotape all the sessions; for example, recording of the first, second, third, fifth, and tenth sessions could suffice for a reliable analysis of the social relationships occurring within a dyad. Alternatively, all sessions could be videotaped, but scoring could be limited (at least, initially) to some sessions; then, in case of unclear results, the remaining sessions could be assessed.

A simple six-category ethogram used for monitoring behavioral changes over repeated encounters is presented in Table III, and its practical application is illustrated in Figs. 2 and 3. It derives from the original description given in Grant and Mackintosh (1963) and is fully explained in Maestripieri *et al.* (1990) and Bigi *et al.*, (1992). It provides a reliable picture of the two social roles while differentiating two very different statuses (a "*semper* winner," or constantly defeating subject, and a "*semper* loser," or repeatedly defeated animal). However, a direct comparison of such a "dominant" animal with the social roles of Table II is not easy, at least so far. Dixon's complete ethogram for scoring mouse social behavior (Table IV) provides a detailed list of items and allows for interpretation of "ambivalent" elements resulting from a motivationally conflicting situation during agonistic encounters. A brief description of the aggressive behavior of cohabiting pairs of adult rats is reported in Flannelly and Lore (1975).

In general, repeated encounters lead to a dyadic escalation, while the opposing counterparts learn their personal coping style with the attacking or attacked conspecific (Cairns and Scholtz, 1973). A "learned helplessness" strategy is often activated by the repeatedly defeated animal (Maier and Seligman, 1976). Enquist (1985) discussed the theoretical framework of communication during aggressive interactions, with particular reference to variation in individual strategy (Bateson and Klopfer, 1987). With the increasing interest in recent years in the issue of "intentional" signaling in animal communication (Hauser and Nelson, 1991), a reliable interpretation of aggressive behavior has to take into consideration individual coping styles, how their relative frequency is affected by such factors as drug treatment or social condition (for the latter, see overview by Lott, 1991), and the concomitant behavioral changes occurring in the opponent individual (Fig. 4).

## Female Aggressive Behavior

A paper by Haney *et al.* (1989) compared the maternal (nest-defending) behavior of female mice and rats confronting either a male or a female intruder. Such postpartum behavior, called "maternal aggression," is the most widely used method of assessment. In fact, the lactation period is associated with heightened levels of female aggressive behavior, barely observed in nonbreeding female rodents. Flannelly and Flannelly (1985) analyzed the role of the size of the opponent in eliciting maternal aggression, whereas Svare *et al.* (1981) characterized some situational and experiential determinants. Hood (1984) provided an elegant characterization of interfemale aggressive behavior in rats as a function of the estrous cycle. We found that litter size influences maternal aggression (Maestripieri and Alleva, 1990) and suggested that male mice use parental care as a buffering strategy against attacks from lactating females (Maestripieri and Alleva, 1991).

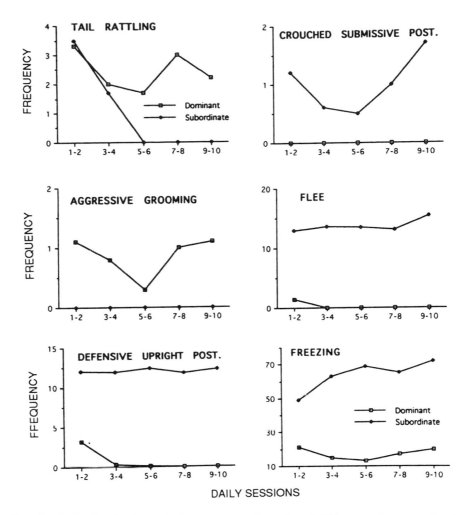

FIG. 2   A simple six-element ethogram reveals profound differences between dominant and subordinate male subjects in a mouse pair (both animals were isolated for 4–6 weeks; recording sessions lasted 20 min). The elements include a threatening item (tail rattling), aggression-inhibiting submissive postures, and conflicting or displacement activities. Freezing scores are in units of percent time. See Tables III and IV for more extended behavioral scoring. [Redrawn from data in Maestripieri *et al.* (1990).]

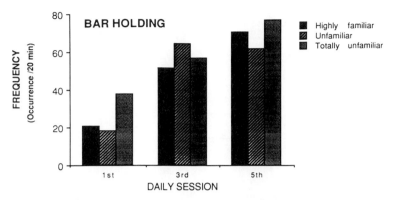

Fig. 3   Behavioral scores other than those specifically used in recording aggressive behavior can provide useful information in terms of general behavioral arousal and interpretation of an animal's reactions. For example, the number of episodes of "bar holding" (attempts at escaping from the testing cage by clutching upside-down to the bars of the metallic top, a behavioral item displayed by adult CD-1 male mice confronting an isolated conspecific of the same strain but with a different degree of olfactory familiarization) reveals that significant behavioral arousal during repeated encounters occurs only in the case of totally unfamiliar dyads. Other measurements (not shown), such as frequency and duration of "digging," showed inverse trends (decreasing on repeated encounters), but again were consistently higher in totally unfamiliar dyads. "Self-grooming" was inversely correlated to the degree of previous olfactory familiarization. In the last session, its mean level was 10-fold higher in totally unfamiliar dyads as compared to the other two groups. [Results shown ($N = 10$ mice per gorup) are from E. Alleva, unpublished.]

Aggressive behavior also increases in female rodents between the second week of pregnancy and parturition (prepartum aggression) (Mann *et al.,* 1984). Buhot-Averseng (1983) proposed a graphic representation of the effects of cohabitation on aggression in pregnant mice. Intruders of different reproductive status elicit different types of aggression in either early or late pregnant mice (Goyens and Noirot, 1977).

## Developmental Changes in Aggressive Behavior Patterns and the Effects of Isolation on Them

The differences in aggressive behavior between mice and rats become more distinct during ontogeny. It is generally agreed that developing rats exhibit a higher level and a wider spectrum of playful interactions than mice (Poole and Fish, 1976). In rat play, "rough-and-tumble" (including "pinning") or "crossover" solicitation (Table V) are good indicators of aggressive-like

TABLE IV  Classification of Behavioral Elements in Mice[a]

| Category | Behavioral elements |
| --- | --- |
| Nonsocial | Explore, scan, wash, self-groom, scratch, dig, push-dig, kick-dig, shake, jump, eat, drink, sit, turn, flop, displacement, stretched attend posture, substrate, leave, on-bars, off-bars |
| Social investigation | Attend, stretched attend, approach, investigate, nose, groom |
| Sexual activity | Follow, sniff, attempted mount, mount, genital groom, push-under, crawl-over, push-past |
| Offense | Threat, attack, bite, chase, aggressive-groom, full aggressive |
| Offensive ambivalence | Offensive-sideways, offensive-upright, sideways posture |
| Defensive ambivalence | Oblique, upright posture, defensive-upright, defensive-sideways, parry |
| Arrested flight | Crouch, freeze, straight legs, kick, on back |
| Escape | Flag, evade, retreat, flee |
| Distance ambivalence | Circle, zig-zag, walk round, rattle |

[a] Kindly provided by K. Dixon. For details and practical applications, see Dixon (1982), Dixon *et al.* (1990), and Mackintosh *et al.* (1977).

interactions. Rat play fighting is described in Meaney and Stewart (1981), Poole and Fish (1976), and Takahashi and Lore (1983). Hood (1988) provides an accurate description of the development of female aggressive behavior in rats.

Mouse locomotor–rotational and social play was initially described for wild subjects, and it is particularly evident in physically complex environments (Mendl and Paul, 1990; Wolff, 1981). However, the idea that young mice exhibit social play is gaining attention, and a more exhaustive description of play in laboratory strains has appeared (Walker and Byers, 1991). The elements of juvenile play fighting in various rodent species are described in Table V.

In mice, the effects of social isolation (onset between days 21 and 84) are age dependent (Cairns *et al.,* 1985; Goldsmith *et al.,* 1976). It has also been found that olfactory cues emitted in urine by juvenile females seem to protect them from aggressive conspecifics (Dixon and Mackintosh, 1976). The effects of social isolation are age dependent in rats also: Although early isolation (onset between days 16 and 41) was found to exert a significant and sustained impact on later agonistic behavior, isolation during a later developmental stage (days 41–68) did not result in consistent alteration (Wahlstrand *et al.,* 1982). For developmental statistics see Chiacotti *et al.,* 1987.

## Choice of Strain and Genetics

To reduce genetic variability in the expression of aggressive behavior, various investigators used highly consanguineous, inbred strains (Benus *et al.,* 1991a; Frischknecht *et al.,* 1982; Jones and Brain, 1987; Siegfried *et al.,* 1981, 1984).

TABLE V    Elements of Juvenile Play Fighting in Rodents

| Species and studies | Behavioral items | Description |
|---|---|---|
| Wild mouse<br>Wolff (1981) | Aggressive play:<br>"chase" | Two mice alternate in chasing one another, exhibiting exaggerated jerking movements |
| | Pouncing play:<br>"pounce" | One mouse crouches and pounces on another approaching individual when within striking distance, then "pouncer" usually exhibits erratic running behavior while moving away from "pouncee" |
| Rat<br>Meaney and Stewart (1981)<br>Panksepp and Beatty (1980),<br>Thor and Holloway (1983) | Rough-and-tumble play<br>(a) "Wrestle/box" | Animals cling together ventro–ventrally, struggling and rolling over one another, or stand upright facing each other and box with forepaws |
| | (b) "Pinning" | One rat is held down on its back, fully exposing its ventral surface, by partner, who stands above it in dominance stance |
| | Play solicitation:<br>"crossover" | Traversal by subject over or under median plane of partner; includes crawl-over, crawl-under, pounce, and any other gross movement that results in traversal over or under partner |
| Northern grasshopper mouse<br>Davies and Kemble (1983) | Rough-and-tumble play | See description of analogous behaviors observed in rats |
| | Play solicitation:<br>"wriggle" | One animal approaches another individual, turns over on its back, and wriggles underneath it with movements similar to dust-bathing movements but accompanied by vigorous kicking with all four paws (absent in rats) |
| Yellow-bellied marmot<br>Nowicki and Armitage<br>(1979) | Rough-and-tumble play<br>(a) "Mouth-spar" | Quadrupedal jabbing of head at head, shoulders, or chest of other animal while partner does likewise; mouths are always open, but biting motions occur infrequently |
| | (b) "Grapple" | Two marmots rear up on their hindlegs and push at one another with forepaws; mouths are open at all times, possibly with more biting motion than seen in mouth-sparring; frequently ends with one animal losing its balance and falling ("nose push," "slap," "wrestle," "chase," etc.) |

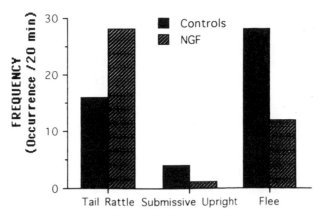

Fig. 4   Accurate inspection of the "standard opponent" animal confronting a treated mouse can reveal, or help in interpreting, behavioral changes caused by treatment. In this case, adult CD-1 males received NGF intraperitoneally for 6 consecutive days (30 $\mu$g), and alteration in their behavior (significant increase in latency to the first "attack," lowered levels of "total attacking time" and of "aggressive grooming," etc.) was mirrored by a concomitant change in the "dominant-like" profile of the standard opponents. All animals were isolated for 6 weeks before testing. [Redrawn from data in Bigi *et al.* (1992).]

Strain-dependent differences in the effects of isolation were noticed (Siegfried *et al.,* 1981), with strains such as C57BL/6 showing no increased aggressive tendency, even after 8 weeks of individual housing. However, the use of a single inbred strain can cause misinterpretation. Kimelman and Lubow (1974) found that subjects belonging to the C57/BL10 inbred mouse strain do not show increased aggressive tendency on isolation, but these data were not confirmed by subsequent analyses using other strains, for example, outbred CD-1 mice. Moreover, subordinate C57BL/6 subjects, when repeatedly defeated, enter a depressive-like state, displaying prolonged freezing and supranormal levels of nose-in-the-corner behavior (Kudryavtzeva *et al.,* 1991). The effects of social isolation are, in fact, strain dependent (for the C57 strains, see Jones and Brain, 1987; Siegfried *et al.,* 1981).

It should be remembered that most commercially available inbred strains were originally selected for physical traits, such as coat color, and that other physical or neurochemical characteristics have only been randomly included in the original selective breeding process. Thus, it may be difficult to attribute behavioral changes (including strain-dependent aggressive behavior) to specific changes in one or more particular brain structures. Behavioral geneticist Hanspeter Lipp has pointed out that the correlation between structure and function that occurs in inbred strains can demonstrate the influence of minor individual differences arising from epigenetic inferences (Lipp *et al.*, 1989). "Phylogenic trees" of a number of common inbred strains have recently been made available (Atchley and Fitch, 1991), and they could be fruitfully used for new insight into the genetic determinants of rodent aggressive behavior.

Some mouse strains have been specifically selected because of the quantifiable characteristics of their intermale aggressive behavior [short attacking latency (SAL) and long attacking latency (LAL) strains (Benus *et al.*, 1991a; van Oortmerssen and Bakker, 1981), Turku aggressive (TA) and Turku non-aggressive (TNA) strains (Sandnabba, 1986a)]. The divergence between the behavioral characteristics being selected emerges only after about 10 generations of breeding and appear to be rather specific (e.g., do not concern interfemale aggression). The two lines differed in a more general way than just their attack behavior, a fact having implications for response to test conditions. Changes in developmental rate and maternal care have been observed in strains selected for aggressive traits (Eleftheriou *et al.*, 1974; Mendl and Paul, 1990; Roubertoux and Carlier, 1988; van Oortmerssen and Bakker, 1981). Female mice have also been selected according to their levels of agonistic behavior (Ebert and Hyde, 1976). Finally, differences in intermale aggression have been reported for two karyotypic races of wild-trapped mice (Capanna *et al.*, 1984).

After work with opponents from different inbred strains, Roubertoux and co-workers suggested that, at least for some strains, the observed differences can be the result of higher testosterone sensitivity and greater production of olfactory secretions (Francois *et al.*, 1990; Robertoux and Carlier, 1988). It is conceivable that most of the genetically grounded behavioral differences actually concern sensory–motor coordination, CNS changes (Imperato *et al.*, 1991; Puglisi-Allegra and Cabib, 1988; Spillantini *et al.*, 1989), and specific aspects of the olfactory communication system. With regard to the latter, it is important to emphasize that the *t*-complex genotype determines the chemical characterization of the urinary volatile profile (Jemiolo *et al.*,1991), which is also influenced by the male social role (Coopersmith and Lenington, 1992), and that the chemosensory recognition of mouse phenotypes seems to be determined by *Tla* and *H-2K* regions of chromosome 17 (Yamaguchi *et al.*, 1981; Yamazaki *et al.*, 1982; see also critical discussion in Francois *et al.*, 1990).

## A Few Final Warnings

To promote high levels of aggressive behavior, animals are often kept in social isolation (i.e., individually housed) for some weeks, according to a procedure defined as the "isolation syndrome" and originally described by Valzelli (1973). However, the singly housed animal (particularly in the case of prolonged isolation) could react in a rather unnatural way during agonistic interactions. The actual significance of the isolation condition, and its relationships with the natural social role exerted by a rodent within its colony or deme, is still a matter of speculation (Benton and Brain, 1979; Brain, 1975; Grimm, 1980; Mainardi *et al.*, 1977). Moreover, isolation effects are strictly strain dependent (see, e.g., Siegfried *et al.*, 1981) and are reportedly influenced by adrenal activity (Goldsmith *et al.*, 1976; Brain *et al.*, 1971; Brain and Poole, 1974).

As previously stated, a longitudinal analysis of progressive familiarization between two unfamiliar subjects or the reaction to an intruder introduced into an established social setting is highly advisable. However, variables such as the following have to be carefully evaluated: (1) housing and test conditions (Burright *et al.*, 1988; Parmigiani and Brain, 1983); (2) physical exhaustion, which is inversely proportional to animal size (Schmidt-Nielsen, 1985); (3) habituation (Winslow and Miczek, 1984); and (4) fighting experience of the challenging individual (Brain *et al.*, 1981; Parmigiani and Brain, 1983; Burgh and Slotnick, 1983). Finally, it has to be remembered that prolonged and repeated defeat by a dominant conspecific produces profound physiological and behavioral alterations (Maier and Seligman, 1976; Williams and Lierle, 1988).

## Statistical Evaluation of Dominance

The book by Martin and Bateson (1986) is the best manual for behavioral biologists, and it is rich in indications for experimental design and data evaluation. Despite erratic attempts at using the theoretically more informative sequence analysis for assessing rodent aggressive behavior (Donàt *et al.*, 1991), a multivariate analysis taking into account several behavioral items is recommended as an alternative to the usual proliferation of analysis of variance (ANOVA) tests (Loggi *et al.*, 1991). A very useful index for evaluating combined repeated latency measurements (latencies recorded in intervals and inspection of trends over tests) has been proposed by Theobald and Goupillot (1990). Statistical analyses of rat ethopharmacological studies arc summarized in Silverman (1965).

General methodological indications for assessing dominance status are reported in Richards (1974). Benton *et al.* (1980) compared six current measures of dominance (display of submissive posture, tail wounds, access to water, tube-test dominance, territorial aggression, female sniffing). However, as pointed out by Huntingford (1984), a test reproducing the natural style of coping with social stress is much more advisable than a quick, but not very informative, method.

To this end, actual territory formation by laboratory mice should be scored (Mackintosh, 1970), taking into account at least the main elements of the complex network of their olfactory communication system, such as urine marking and investigation within family groups (Dixon and Mackintosh, 1976; Hurst, 1989) or intermale signaling (Bishop and Chevins, 1987; Hurst, 1990). Desjardins et al. (1983) described a way of revealing urinary marking patterns of adult male mice by ultraviolet visualization of filter paper placed on cage floors during 12-hr overnight testing. By this method each animal can be assigned a social rank.

## Drug Effects on Aggressive Behavior

Dixon and co-workers have provided exhaustive guidelines for an appropriate analysis of the effects of psychoactive drug treatment on rodent social and aggressive behavior (Dixon, 1982; Dixon et al., 1984, 1990). Figure 5 shows an example. Earlier analyses are also valuable, being indicative of the different methods used in the past by psychopharmacologists to induce either inter- or intraspecific aggressive behavior in rodents [cerebral lesions, painful stimulation, selected hormonal or pharmacological treatment, "muricidal rats," or "locusticidal mice" (Jones and Brain, 1987; Powell et al., 1973; Valzelli, 1973)]. Most of these tests, as well as automated devices recording audible vocalizations or producing aggressive reactions by repeated footshocks in 5-rat batteries (Brunaud and Siou, 1958), are presently regarded as unethical and, above all, of very little value in understanding agonistic interactions and the effect of drugs on them (Dixon et al., 1990; Huntingford, 1984; Mackintosh et al., 1977). Early attempts to find drugs inhibiting aggression regarded specificity only in terms on unitary measurements and non-ataxic doses. These drug agents failed probably because the investigators did not take into account interplays between aggression and flight elements, which were not measured.

Krsiak (1975) discussed the predictory value of isolated mice in psychopharmacological assessments. Matte (1979) proposed a method of quantifying aggressive behavior, revealing possible dissociation of motor activities and "true" aggression. However, the lack of an observational analysis renders such a study (and scores of similar investigations) highly questionable. A catalog of drug-induced modifications in rodent agonistic behaviors is reported in Miczek and Krsiak (1979) and Miczek et al. (1984).

It has to be pointed out that the increasing availability of low-cost high-performance videotape systems, allowing single-frame evaluation, makes

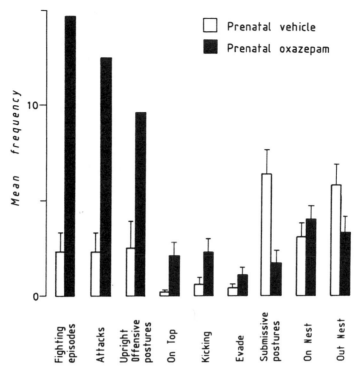

FIG. 5   Long-term effects of late-prenatal benzodiazepine exposure are revealed by a test of maternal aggression in adult females receiving oxazepam *in utero*. The treatment caused a slight and transient delay in neurobehavioral development. The specificity of the long-term effect on adult agonistic response (a 6- to 7-fold increase in the number of "fighting episodes" and "attacks" and a concomitant marked decrease in "submissive postures," while nest attending remained unaffected) was confirmed by the absence of major changes in other behavioral end points recorded at the adult stage (varimex locomotor activity, habituation to a novel environment, active or go/no go avoidance learning, water maze learning, etc.). Results are given for $N = 16$ mice per group; sessions lasted 5 min. [From Laviola *et al.* (1991), by permission.]

ethopharmacological analyses the unique choice for careful quantitative and qualitative assessment of the behavioral alterations induced by drug treatment. Moreover, these systems, supplemented by commercially available software [Observer (Noldus, 1991), Keybehaviour, etc.], eliminate most of the biases due to inter- or intraobserver reliability, while facilitating multicentric studies following standardized methodologies, which are becoming particularly important for regulatory purposes.

## Acknowledgments

Felicity Huntingford and Aubrey Manning critically read the manuscript. I acknowledge the invaluable advice provided by A. Keith Dixon (who also provided selected bibliographic materials and a complete mouse ethogram). Paola Corridi collected a number of useful papers, while Livia Terranova critically discussed developmental aspects. A special thanks is due my dear friend Andrea Puccini, who provided the gorgeous backdrop of "Villa Donna Stefania" (Maratea, Italy) for the writing of the first draft of this chapter. Work has been supported by the Subproject on Behavioral Pathophysiology (Project on Noninfectious Pathology) of the Instituto Superiore di Sanità and by the Subproject "Stress," Project for Prevention and Control of Disease Factors (FATMA–CNR).

## References

Adams, N., *Behaviour* **56,** 286 (1976).

Adams, N., *Aggressive Behav.* **6,** 295 (1980).

Adams, N., and Boice, R., *J. Comp. Psychol.* **97,** 24 (1983).

Adams, N., and Boice, R., *Behav. Processes* **19,** 127 (1989).

Alberts, J. R., *Physiol. Behav.* **12,** 657 (1974).

Alleva, E., and Aloe, L., *Int. J. Comp. Psychol.* **2,** 147 (1989).

Alleva, E., and Aloe, L., *Biol. Psychiatry* **29,** 140S (1991).

Alleva, E., and Aloe, L., *in* "Biotech Growth Factors" (C. Lenfant, R. Paoletti, and A. Albertini, eds.), p. 80. Karger, Basel (1992).

Alleva, E., Aloe, L., Bigi, S., and De Acetis, L., *Physiol. Behav.* **49,** 715 (1991).

Aloe, L., Alleva, E., Bohm, A., and Levi-Montalcini, R., *Proc. Natl. Acad. Sci. U.S.A.* **83,** 6184 (1986).

Aloe, L., Alleva, E., and De Simone, R., *Behav. Brain Res.* **39,** 53 (1990).

Archer, J., *J. Mammal.* **49,** 572 (1968).

Archer, J., "The Behavioural Biology of Aggression." Cambridge Univ. Press, Cambridge (1988).

Atchley, W. R., and Fitch, W. M., *Science* **254,** 554 (1991).

Barnard, C. J., Hurst, J. L., Aldhous, P., *Biol. Rev.* **66,** 379 (1991).

Bateson, P., *Anim. Behav.* **42,** 827 (1991).

Bateson, P., *New Sci.* (in press) (1992).

Bateson, P. P. G., and Klopfer, P. H. (eds.), "Perspectives in Ethology, Alternatives," Vol. 7. Plenum, New York (1987).

Benton, D., *Aggressive Behav.* **8,** 104 (1982).

Benton, D., and Brain, P. F., *Behav. Processes* **4,** 211 (1979).

Benton, D., Dalrympe-Alford, J. C., and Brain, P. F., *Aggressive Behav.* **8,** 106 (1980).

Benus, R. F., Bohus, B., Koolhaas, J. M., and Van Oortmerssen, G. A., *Experientia* **47,** 1008 (1991a).

Benus, R. F., Bohus, B., Koolhaas, J. M., and Van Oortmerssen, G. A., *Behav. Brain Res.* **43,** 203 (1991b).

Bernstein, I. S., *Behav. Brain Sci.* **4,** 419 (1981).

Bigi, S., Maestripieri, D., Aloe, L., and Alleva, E., *Physiol. Behav.* **51,** 337 (1992).

Bishop, M. J., and Chevins, P. F. D., *Behav. Processes* **15,** 233 (1987).

Blanchard, R. G., and Blanchard, D. C., *in* "The Biology of Aggression" (P. F. Brain and D. Benton, eds.), p. 529. Sijthoff and Noordhoff, Alphen aan den Rijn, The Netherlands (1981).

Blanchard, D. C., and Blanchard, R. J., *Neurosci. Behav. Rev.* **14,** 455 (1990).

Blanchard, R. J., Blanchard, D. J., Takahashi, T., and Kelley, M. J., *Anim. Behav.* **25,** 622 (1977).

Blanchard, R. J., Blanchard, D. C., and Flannelly, K. J., *Behav. Processes* **11,** 209 (1985).

Blanchard, R. J., Flannelly, K. J., and Blanchard, D. C., *Physiol. Behav.* **43,** 1 (1988a).

Blanchard, R. J., Hori, K., Tom, P., and Blanchard, D. C., *Aggressive Behav.* **14,** 195 (1988b).

Brain, P. F., *Psychonomic Sci.* **28,** 260 (1972).

Brain, P. F., *Life Sci.* **16,** 187 (1975).

Brain, P. F., "Hormones and Aggression," Vol. 2. Eden Press, Montreal (1978).

Brain, P. F., and Benton, D., "Multidisciplinary Approaches to Aggression Research." Elsevier/North Holland, Amsterdam (1981).

Brain, P. F., and Poole, A. E., *Aggressive Behav.* **1,** 39 (1974).

Brain, P. F., Nowell, N. W., and Wouters, A., *Physiol. Behav.* **6,** 27 (1971).

Brain, P. F., Childs, G., and Parmigiani, S., *Neurosci. Lett.* **53,** 529 (1979).

Brain, P. F., Benton, D., Childs, G., and Parmigiani, S., *Behav. Processes* **6,** 319 (1981).

Brain, P. F., Homady, M. H., and Mainardi, M., *Boll. Zool.* **50,** 173 (1983).

Bronson, F. H., *Physiol. Behav.* **10,** 947 (1973).

Bronson, F. H., *Q. Rev. Biol.* **54,** 265 (1979).

Bronson, F. H., and Mardsen, H. M., *Behav. Biol.* **9,** 625 (1973).

Brunaud, M., and Siou, G., *C. R. Seances Acad. Sci.* **210,** 282 (1958).

Buhot-Averseng, M. C., *Aggressive Behav.* **9,** 253 (1983).

Burg, R. D., and Slotnick, B. M., *Aggressive Behav.* **9,** 49 (1983).

Burright, R. G., Freeman, M. J., and Donovick, P. J., *J. Comp. Psychol.* **102,** 303 (1988).

Cairns, R. B., Hood, K. E., and Midlam, J., *Anim. Behav.* **33,** 166 (1985).

Cairns, R. C., and Sholtz, S. D., *J. Comp. Physiol. Psychol.* **85,** 540 (1973).

Capanna, E., Corti, M., Mainardi, D., Parmigiani, S., and Brain, P. F., *Behav. Genet.* **14,** 195 (1984).

Chiarotti, F., Alleva, E., and Bignami, G., *Neurotoxicol. Teratol.* **9,** 179 (1987).

Cox, T. P., *Anim. Behav.* **32,** 1068 (1984).

Coopersmith, C. B., and Lenington, S., *Ethology* **90,** 1 (1992).

Crowcroft, P., and Rowe, *Proc. Zool. Soc.* (*London*) **140,** 517 (1963).

Davies, V. A., and Kemble, E. D., *Behav. Processes* **8,** 197 (1983).

De Simone, R., Alleva, E., Tirassa, P., and Aloe, L., *Brain Behav. Immun.* **4,** 74 (1990).

Desjardins, C., *et al. Science* **182,** 939 (1983).

Dixon, A. K., *Triangle* **21,** 95 (1982).

Dixon, A. K., and Mackintosh, J. H., *Z. Tierpsychol.* **41,** 225 (1976).

Dixon, A. K., Huber, C., and Kaesermann, F., *in* "Ethopharmacological Aggression Research" (K. A. Miczek, M. R. Kruk, and B. Oliver, eds.), p. 81. Alan R. Liss, New York (1984).

Dixon, A. K., Fish, H. U., and McAllister, K. H., *Adv. Study Behav.* **19,** 171 (1990).

Donàt, P., Sulcovà, A., and Krsiak, M. (eds.), "Ethopharmacology Conference: Abstract Book." Skalsky dvur, Lisek, Czechoslovekie (1991).

Ebert, P. D., and Hyde, J. S., *Behav. Genet.* **6,** 421 (1976).

Eleftheriou, B. E., Bailey, D. W., and Denenberg, V. H., *Physiol. Behav.* **13,** 773 (1974).

Enquist, M., *Anim. Behav.* **33,** 1152 (1985).

Evans, C. M., and Mackintosh, J. H., *J. Endocrinol.* **71,** 91 (1976).

Flannelly, K. J., and Flannelly, L., *Psychol. Rep.* **57,** 883 (1985).

Flannelly, K. J., Blanchard, R. J., and Blanchard, D. C., (eds.), "Biological Perspectives on Aggression." Alan R. Liss, New York (1984).

Flannelly, K. J., and Lore, E., *Aggressive Behav.* **20,** 129 (1975).

Francois, M. H., Nosten-Bertrand, M., Roubertoux, P. L., Kottler, M.-L., and Degrelle, H., *Physiol. Behav.* **47,** 1181 (1990).

Frischknecht, H.-R., Siegfried, B., and Waser, P. G., *Behav. Processes* **7,** 235 (1982).

Goldsmith, J. F., Brain, P. F., Benton, D., *Aggressive. Behav.* **2,** 307 (1976).

Goyens, J., and Noirot, E., *Aggressive Behav.* **7,** 111 (1977).

Grant, E. C., *Behaviour* **21,** 260 (1963).

Grant, E. C., and Mackintosh, J. H., *Behaviour* **21,** 247 (1963).

Grimm, V. E., *Int. J. Neurosci.* **11,** 115 (1980).

Haney, M., DeBold, J. F., and Miczek, K. A., *Aggressive Behav.* **15,** 443 (1989).

Hardy, C.-A., Quay, J., Livnat, S., and Ader, R., *Physiol. Behav.* **47,** 1245 (1990).

Harvey, S., Jemiolo, B., and Novotnoy, M., *J. Chem. Ecol.* **15,** 2061 (1989).

Hauser, M. D., and Nelson, D. A., *Trends Ecol. Evol.* (*TREE*) **6,** 186 (1991).

Homady, M. H., and Brain, P. F., *Aggressive Behav.* **8,** 137 (1982).

Hood, K. E., *in* "Biological Perspectives on Aggression" (K. J. Flannelly, R. J. Blanchard, and D. C. Blanchard, eds.), p. 181. Alan R. Liss, New York (1984).

Hood, K. E., *Int. J. Comp. Psychol.* **2,** 27 (1988).

Huckelbridge, F. H., Gamal-el-Din, L., and Brain, P. F., *Behav. Neural Biol.* **33,** 345 (1981).

Huntingford, F. A., *Anim. Behav.* **24,** 485 (1976).

Huntingford, F. A., *Anim. Behav.* **32,** 210 (1984).

Huntingford, F., and Turner, A., "Animal Conflict." Cambridge Univ. Press, Cambridge (1987).

Hurst, J. L., *Anim. Behav.* **37,** 705 (1989).

Hurst, J. L., *Anim. Behav.* **40,** 209 (1990).

Imperato, A., Puglisi-Allegra, S. Casolini, P., and Angelucci, L., *Brain Res.* **538,** 111 (1991).

Jemiolo, B., Xie, T.-M., Andreolini, F., Baker, A. E. M., and Novotnoy, M., *J. Chem. Ecol.* **17,** 353 (1991).

Jones, R. B., and Nowell, N. W., *Physiol. Behav.* **10,** 221 (1972).

Jones, R. B., and Nowell, N. W., *Physiol. Behav.* **11,** 35 (1973a).

Jones, R. B., and Nowell, N. W., *J. Endocrinol.* **59,** 203 (1973b).

Jones, R. B., and Nowell, N. W., *Anim. Learn. Behav.* **2,** 141 (1974).

Jones, R. B., and Nowell, N. W., *Aggressive Behav.* **15,** 291 (1989).

Jones, S. E., and Brain, P. F., *Behav. Genet.* **17,** 87 (1987).

Katz, R. J., *Behav. Biol.* **17,** 399 (1976).

Kimelman, B. R., and Lubow, R. E., *Physiol. Behav.* **12,** 919 (1974).

Klopfer, P. H., "An Introduction to Animal Behavior: Ethology's First Century." Prentice-Hall, Englewood Cliffs, New Jersey (1974).

Krsiak, *Br. J. Pharmacol.* **55,** 141 (1975).

Kudryavtzeva, N. N., Bakshtanovskaya, I. V., and Koryakina, L. A., *Pharmacol. Biochem. Behav.* **38,** 315 (1991).

Lagerspetz, K. M., *Ann. Acad. Sci. Fenn.* (*Series B*) **131,** 3 (1964).

Lagerspetz, K. M., and Sandnabba, K., *Aggressive Behav.* **8,** 319 (1982).

Laviola, G., De Acetis, L., Bignami, G., and Alleva, E., *Neurotoxicol. Teratol.* **13,** 75 (1991).

Lehman, M. N., and Adams, D. B., *Behaviour* **61,** 238–275 (1977).

Levi-Montalcini, R., Aloe, L., and Alleva, E., *Prog. NeuroEndocrinImmunol.* **3,** 1 (1990).

Lipp, H. P., Schwegler, H., Crusio, W. E., Wolfer, D. P., Leisinger Trigona, M. C., Heimrich, B., and Driscoll, P., *Experentia* **45,** 845 (1989).

Loggi, G., Laviola, G., Alleva, E., and Chiarotti, F., *Pharmacol. Biochem. Behav.* **38,** 817 (1991).

Lott, D. F., "Intraspecific Variation in the Social Systems of Wild Vertebrates." Cambridge Univ. Press, Cambridge (1991).

Mackintosh, J. H., *Anim. Behav.* **18,** 177 (1970).

Mackintosh, J. H., *Symp. Zool. Soc. London* **47,** 337 (1981).

Mackintosh, J. H., and Grant, E. G., *Z. Tierpsychol.* **23,** 584 (1966).

Mackintosh, J. H., Chance, M. R. A., and Silverman, A. P., *in* "Handbook of Psychopharmacology" (L. L. Iversen, S. D. Iversen, and S. H. Snyder, eds.), Vol. 7, p. 3. Plenum, New York (1977).

Maestripieri, D., and Alleva, E., *Ethology* **84,** 27 (1990).

Maestripieri, D., and Alleva, E., *Anim. Behav.* **41,** 904 (1991).

Maestripieri, D., and De Simone, R., Aloe, L., and Alleva, E., *Physiol. Behav.* **47,** 161 (1990).

Maier, S. F., and Seligman, M. E. P., *J. Exp. Psychol. Gen.* **105,** 3 (1976).

Mainardi, D., Mainardi, M., Parmigiani, S., and Pasquali, A., *Accad. Naz. Lincei Rend.* **28,** 120 (1977).

Mann, M. A., Konen, C., and Svare, B., *Horm. Behav.* **18,** 140 (1984).

Martin, P., and Bateson, P., "Measuring Behaviour: An Introductory Guide." Cambridge Univ. Press, Cambridge (1986).

Matte, A. C., *Psychopharmacology* (*Berlin*) **60,** 247 (1979).

Meaney, M. J., and Stewart, J., *Anim. Behav.* **29,** 34 (1981).

Mendl, M., and Paul, E. S., *Behaviour* **116,** 11 (1990).

Miczek, K. A., and Krsiak, M., *in* "Advances in Behavioral Pharmacology" (T. Thompson and P. Dews, eds.), Vol. 2 p. 87. Academic Press, New York (1979).

Miczek, K. A., DeBold, J. F., and Thompson, M. L., *in* "Ethopharmacological Aggression Research" (K. A. Miczek, M. R. Kruk, and B. Oliver, eds.), p. 1. Alan R. Liss, New York (1984).

Moyer, K. E., *Commun. Behav. Biol. A* **2,** 65 (1968).

Mugford, R. A., and Nowell, N. W., *Physiol. Behav.* **6,** 247 (1971).

Noldus, L. P. J. J., *Behav. Res. Methods Instrum. Comput.* **23,** 415 (1991).

Novotny, M., Harvey, S., Jemiolo, B., and Alberts, J. R., *Proc. Natl. Acad. Sci. U.S.A.* **82,** 2059 (1989).

Nowicki, S., and Armitage, K. B., *Z. Tierpsychol.* **51,** 85 (1979).

Panksepp, J., and Beatty, W. W., *Behav. Neural Biol.* **30,** 197 (1980).

Parmigiani, S., and Brain, P. F., *Behav. Processes* **8,** 45 (1983).

Parmigiani, S., Brain, P. F., Mainardi, D., and Brunoni, V., *J. Comp. Psychol.* **102,** 287 (1988).

Pellis, S. M., *Aggressive Behav.* **14,** 85 (1988).

Pellis, S. M., and Pellis, V. C., *Aggressive Behav.* **14,** 437 (1988).

Poole, T. B., and Fish, J., *J. Zool. London* **179,** 249 (1976).

Poole, T. B., and Morgan, H. D., *Anim. Behav.* **21,** 788 (1973).

Poole, T. B., and Morgan, H. D. R., *Anim. Behav.* **23,** 470 (1975).

Poshivalov, V. P., *Aggressive Behav.* **7,** 195 (1981).

Powell, D. A., Walters, K., Duncan, S., and Holley, J. R., *Psychopharmacologia* **30,** 303 (1973).

Puglisi-Allegra, S., and Cabib, S., *Behav. Neural Biol.* **50,** 98 (1988).

Raab, A., Dantzer, R., Michaud, B., Mormede, P., Taghzouti, K., Simon, H., and Le Moal, M., *Physiol. Behav.* **36,** 223 (1986).

Richards, S. M., *Anim. Behav.* **22,** 914 (1974).

Rines, J. R., and vom Saal, F. S., *Horm. Behav.* **18,** 117 (1984).

Roubertoux, P. L., and Carlier, M., *Behav. Genet.* **18,** 175 (1988).

Sales, G. D., *Anim. Behav.* **20,** 88 (1972).

Sandnabba, N. K., *Aggressive Behav.* **12,** 103 (1986a).

Sandnabba, N. K., *Behav. Processes* **12,** 349 (1986b).

Schmidt-Nielsen, K., "Scaling: Why Is Animal Size so Important?" Cambridge Univ. Press, Cambridge (1985).

Siegfried, B., Alleva, E., Oliverio, A., and Puglisi-Allegra, S., *Behav. Brain Res.* **2,** 211 (1981).

Siegfried, B., Frischknecht, H.-R., and Waser, P. G., *Behav. Neural Biol.* **42,** 91 (1984).

Silverman, A. P., *Br. J. Pharmacol.* **24,** 579 (1965).

Spillantini, M. G., Aloe, L., Alleva, E., De Simone, R., Goedert, M., and Levi-Montalcini, R., *Proc. Natl. Acad. Sci,. U.S.A.* **86,** 8555 (1989).

Southwick, C. H., *Proc. Zool. Soc. London* **131,** 163 (1958).

Strasser, S. T., and Dixon, A. K., *Physiol. Behav.* **36,** 773 (1986).

Svare, B., Betteridge, C., Katz, D., and Samuels, O., *Physiol. Behav.* **26,** 253 (1981).

Syme, G. J., *Anim. Behav.* **22,** 931 (1974).

Takahashi, L. K., and Lore, R. K., *Aggressive Behav.* **9,** 217 (1983).

Takeuchi, H., and Kawashima, S., *Physiol. Behav.* **38**, 545 (1986).
Theobald, C. M., and Goupillot, R. P., *Anim. Behav.* **40**, 484 (1990).
Thor, D. H., *Aggressive Behav.* **2**, 39 (1976).
Thor, D. H., and Holloway, W. R., *Anim. Learn. Behav.* **11**, 173 (1983).
Turney, T. H., and Harmsen, A. G., *Physiol. Behav.* **33**, 559 (1984).
Uhrich, J., *J. Comp. Psychol.* **25**, 373 (1938).
Valzelli, L., *Adv. Pharmacol.* **5**, 79 (1967).
Valzelli, L., *Psychopharmacologie* **31**, 305 (1973).
Van Oortmerssen, G. A., and Bakker, C. M., *Behav. Genet.* **11**, 115 (1981).
Wahlstrand, K., Knutson, J. K., and Viken, R. J., *Aggressive Behav.* **9**, 29 (1982).
Walker, C., and Byers, J. A., *Anim. Behav.* **42**, 891 (1991).
Williams, J. L., and Lierle, D. M., *Anim. Learn. Behav.* **16**, 477 (1988).
Winslow, J. T., and Miczek, K. A., *Aggressive Behav.* **10**, 103 (1984).
Wolff, R. J., *J. Zool. London* **195**, 405 (1981).
Yamaguchi, M., Yamazaki, K., Beauchamp, G. K., Bard, J., Thomas, L., and Boyse, E. A., *Proc. Natl. Acad. Sci. U.S.A.* **78**, 5817 (1981).
Yamazaki, K., Beauchamp, G. K., Bard, J., Thomas, L., and Boyse, E. A., *Proc. Natl. Acad. Sci. U.S.A.* **79**, 7828 (1982).

# [8] Resident–Intruder Paradigms for the Study of Rodent Aggression

Ernest D. Kemble

## Introduction

Various forms of resident–intruder paradigms have become standard tools for the study of agonistic behavior and are being applied to an increasingly wide range of behavioral and biological phenomena (see, e.g., Refs. 1–6 for recent reviews). These confrontations between resident laboratory animals and unfamiliar intruders consistently evoke a wide range of stereotyped attack and defense behaviors that are virtually identical to those shown by wild counterparts (7–10). During such encounters the behaviors of residents and intruders typically become sharply differentiated into easily recognizable offensive and defensive patterns that clearly reveal the specificity, or lack thereof, of an experimental manipulation on aggression (11–19). Furthermore, judicious manipulation of antecedent experience and/or subject characteristics makes it possible to vary the intensity of aggression over a considerable range (15, 20–25).

This chapter describes the most commonly employed resident–intruder paradigms and suggests major variables that may determine the occurrence and/or intensity of agonistic interactions. Because the vast majority of resident–intruder encounters employ either mice or rats as subjects, discussion is restricted to those species. Information on the aggressive behavior of several other species is available and may be consulted if investigators wish to extend the range of species investigated (see, e.g., Refs. 2, 26, and 27). It should be emphasized, however, that this material provides only guidelines. For any given paradigm, procedures will probably need to be modified to meet the specific needs of an investigator. In addition, considerable caution should be exercised in drawing conclusions about motivational/emotional states which are based on changes in resident–intruder aggression (28, 29). If such questions are of interest, addi-

*Methods in Neurosciences, Volume 14*

tional testing procedures more suited to such analysis are available (30–33).

# General Methodological Considerations

## Behavioral Observations

Accurate observations of the various postures comprising offensive and defensive attack patterns are an essential prerequisite for the use of resident–intruder paradigms. Several excellent reviews of the social and agonistic behaviors of rats (7, 26, 34–38) and mice (9, 26, 37, 39) are available and should be consulted for detailed description and illustration. In addition, preliminary observations of videotaped agonistic encounters are quite helpful in recognizing these behaviors. With the exceptions noted below, the agonistic behavior of rats and mice is similar, and the behavior of both is described only briefly.

Resident males respond to the introduction of an unfamiliar male with social investigatory behaviors consisting largely of anogenital, body, and nose sniffing with occasional bouts of grooming behavior. This initial phase becomes considerably attenuated with repeated attack experience by residents. Social investigation is followed by piloerection (in rats but not mice) and may be accompanied by tooth chattering and/or tail rattling. If the back of the intruder is accessible, the resident may then deliver a bite without further preliminaries. If the back is not available, the resident may then tightly circle the intruder or crowd it with a distinctive lateral attack approach until the back is exposed and bitten.

After biting, the behavior of residents and intruders becomes sharply polarized into stereotyped offensive and defensive patterns of behavior which are virtually nonoverlapping in their occurrence. Intruders may respond to initial attack by delivering a retaliatory bite to the head or snout of the resident or, more frequently, by flight. Residents respond to flight by chasing and, if possible, further biting. Episodes of flight often end with intruders facing the attacker in a defensive upright posture and using their forepaws to fend off (''box'') the resident. Attackers typically respond to the defensive upright by either launching a lateral attack or assuming an offensive upright posture. In either case, the resident will actively seek to knock or shove the intruder off balance and to gain further access its back.

After repeated attack, rats may assume an on-the-back posture while holding the resident off with their forepaws and/or kicking with the hindpaws. The attacking animal responds to this tactic by standing over (pinning) the intruder and attempting to turn it over to gain access to the

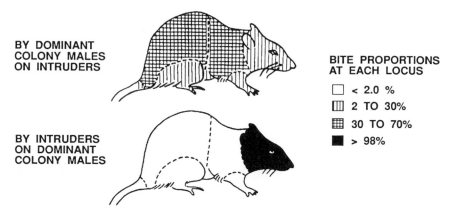

BY DOMINANT
COLONY MALES
ON INTRUDERS

BITE PROPORTIONS
AT EACH LOCUS

☐ < 2.0 %

▥ 2 TO 30%

▦ 30 TO 70%

■ > 98%

BY INTRUDERS
ON DOMINANT
COLONY MALES

FIG. 1   Distribution of bite wounds inflicted by dominant colony residents on intruders (top) and by intruders on dominant residents (bottom). [Adapted from Blanchard and Blanchard (42) and reprinted by permission of Sijthoff & Noordhoff International Publishers.]

back. This on-the-back defense and pinning offense is rarely seen among mice (40). These behaviors are virtually nonoverlapping in occurrence (e.g., defensive behaviors are extremely rare in attacking animals) and will be repeated in varying order many times during a 10- to 15-min encounter. During the intervals between fighting bouts the participants will engage in various social (e.g., sniffing, approaching) and nonsocial (e.g., autogrooming, digging) behaviors.

The distinctiveness of attack and defense postures and the high degree to which particular offense–defense behavioral dyads are synchronized between resident and intruder greatly facilitate the recognition of offensive and defensive patterns. Identification is further aided by the characteristic targeting of bites. Among both laboratory and wild rats the vast majority of offensive bites are delivered to the dorsal surface of the intruder, whereas defensive bites are preferentially directed to the head and snout of the attacker (7, 8, 41). This pattern of bite targeting is illustrated in Fig. 1, which depicts the distribution of bite wounds delivered by an aggressive resident male rat to a defensive male intruder during a resident–intruder test (42). This distribution of wounds is nearly identical to that observed in populations of wild rats (8). Similar targeting is seen in mice except that defensive bites are somewhat more randomly distributed, and the strong inhibition of biting directed toward the ventrum seen in rats (7) seems somewhat weaker among mice (40). The latter probably accounts for the absence of on-the-back and pinning behaviors seen in mice.

Although the attack behavior of nonlactating females is generally similar to that of males, it is more difficult to elicit, appears more defensive in nature, and displays some distinctive characteristics. Females show less lateral attack and piloerection and are more likely to lunge at intruders from a distance (jump attacks) and deliver bites to the snout and head. The latter behaviors are rarely seen among attacking males. Females also characteristically deliver nonwounding bites (43).

Whatever resident–intruder procedure is employed, it must be emphasized that its effectiveness is defined by the elicitation of the full offensive and defensive repertoire of the participants. If major elements of the offensive or defensive patterns described above are absent during a test, additional procedural changes are clearly needed.

## Recording

If possible, behaviors during resident–intruder encounters should be videotaped and sound recorded. Because behaviors often shift rapidly, systems having a slow motion capability are also quite useful. Such records increase interobserver reliability and may facilitate the detection of unusual behaviors induced by an experimental manipulation. A number of computer software packages for behavioral observations are also available that permit the recording of frequency and duration of behavior at any desired level of specificity (see, e.g., Ref. 44 for a recent review) and are quite useful in summarizing and analyzing data.

If albino rats or mice are used, subjects should be dye marked for easy identification. If recording is carried out under dim red light, piloerection and, in the absence of sound recording, bites may be difficult to discern on videotapes. In this case, it may be necessary to score these behaviors directly as a supplement to recordings. If this is not possible, the number of wounds and wounding sites for each participant should be recorded at the end of an encounter. In addition to aggressive acts, nonagonistic social (e.g., sniffing, grooming, approaching) and nonsocial behaviors (e.g., activity, autogrooming, digging) should also be measured. Shifts in these behaviors may be extremely valuable for detecting generalized experimental effects (e.g., hyperactivity) or sedation.

Having recorded an agonistic encounter, the investigator is faced with the thorny issue of choosing the most appropriate level of behavioral analysis. Unless an experimental manipulation is quite powerful, dividing social and agonistic behaviors into too many separate categories will increase intersubject variability and may obscure changes in levels of offense and/or defense. On the other hand, premature lumping of individual behaviors into broader categories is not only less ethologically meaningful, but it may also prevent the detection of selective behavioral changes or the occurrence of atypical behaviors that may be of great interest. It must be admitted at the outset that no universally accepted answer to this question is available. I shall describe general procedures that we have found to be useful and point out some major issues. The reader should also consult more extended discussions of this important issue (2, 6, 7, 26, 35, 45).

We record the frequency and duration of major social (anogenital, nose, and body sniffing, grooming), offensive (chase, lateral attack, pin, bite), and defensive behaviors (flee, defensive upright, on-the-back, bite) from videotape records. The behavior of only one animal is recorded depending on whether primary interest is on offensive (alpha resident) or defensive (intruder) behavior. For most applications this should provide a sufficiently detailed picture of an agonistic encounter. Effects of experimental manipulations on the individual behaviors described above should be presented and the results of statistical comparisons reported. The value of such detailed analysis is well illustrated in research by Olivier and co-workers (18), who report a highly selective reduction in attack, but not defense, by treatment with a novel phenylpiperazine compound that does not reflect generalized changes in other behaviors.

If the data form part of a larger or continuing project, the behaviors reported should remain constant from one experiment to the next, and all comparisons (even those yielding nonsignificant changes) should be reported to facilitate interexperiment comparisons. If the above analyses yield consistent but weak effects, then the individual social, offensive, and defensive behaviors may be legitimately combined for further comparison. This should not be substituted for analyses of the individual behaviors, however. We also find it very useful to score videotapes of the above behaviors at least twice. This not only increases the reliability of scoring, but also aids in the identification of atypical behaviors that may alter interpretation of results. If behaviors that are unusual during an agonistic encounter (e.g., defensive burying, excessive freezing, unusual attack or defense postures) are noted, videotapes should be reviewed further to identify the behavioral context in which they occur. This information may be quite valuable in suggesting procedural improvements and/or may suggest further research.

## Subject Variables

Wide variability in the occurrence and intensity of conspecific aggression has been reported among various strains of rats (23, 46, 47) and mice (48) as well as among experiments involving the same strain (49). Choice of subjects having appropriate characteristics and antecedent experience is therefore of critical importance. Aggression in rats has most often been studied in Long-Evans rats (7, 25, 49–51), although Lister (52), S3-Tryon Maze Dull (53), Wistar (18), and 1- to 2-year-old Holtzman strains (54) have also been used. Swiss-Webster (40), CD-1 (28), and TO (55) are commonly employed mouse strains. Several other strains might also be selected, but large differences in attack behavior may exist between (48), or within (56), these populations.

Even among highly aggressive strains, however, considerable individual variability in attack behavior may occur. It may be advisable to screen individuals prior to testing by briefly introducing a smaller male intruder to identify those which are most likely to become dominant following colony formation and readily attack intruders. Using such procedures Blanchard *et al.* (57) were able to identify highly aggressive males which tended to become dominant during colony formation and to readily attack intruders. If this procedure is used, however, investigators should be aware that imposing very stringent criteria for the selection of aggressive animals will limit the generality of conclusions that can be drawn from data and may even distort the effects of manipulations such as drug treatment.

Whatever strain is selected, a number of other variables may enhance the aggressiveness of residents. Among these, age is perhaps most important. Male aggression begins to emerge shortly after puberty but typically remains at low levels and intensity for some time thereafter. In rats, levels of aggression are high at 200 days of age (58) and continue to increase up to at least 600 days (59). Among mice, little aggression will be seen among subjects prior to 40 days of age, and 70- to 90-day-old subjects are probably optimal. In addition, cohabitation with females (16, 17) and the presence of lactating females (60) are reported to increase the intensity of male aggression toward unfamiliar male intruders. Resident attack is also increased by rearing in an environment that permits burrowing (61–64), by repeated testing (65–67), especially if unfamiliar intruders are used for each test (60), by selecting males from small (68) or mixed-sex litters (69), and by using intruders that have been reared in social isolation (23) or that are slightly smaller than residents (70). In the latter case, however, care should be exercised. Adult intruders that are markedly smaller than residents may be seriously wounded or killed during even brief (10 min) encounters.

Finally, if it is desirable to study the attack of residents uncontaminated by intruder counterattack, "standard opponents" can be prepared by intranasal zinc sulfate infusion (19, 48). Intruders treated in this way are quite docile but still elicit attack by residents. If it is important to observe attack in the total absence of defense, intruders may be anesthetized prior to their exposure to residents.

## Colony Models of Aggression

Colony models seek to approximate the social structure found in naturally occurring populations of the species. Mixed-sex groups of animals are housed together and given some time to form stable social relationships. During colony formation agonistic behaviors among males are initially intense, and clearly recognizable dominance and subordination patterns will emerge within several days. These interactions during colony formation may be of considerable experimental interest in their own right. The behavior of subordinate males, for example, is similar to that induced by exposure to predators and shows a number of interesting parallels to the behaviors seen in human clinical depression (71). If the elicitation of prompt and vigorous attack and defense is desired, colony formation is followed by the introduction of unfamiliar conspecific intruders.

Although there is some variation in the size of colonies (e.g., 54, 72), groups of three adult males, three females, and their prepubertal offspring are common (1). These colonies may be housed in a simple enclosure of plywood and hardware cloth of approximately 1.0 m$^2$, with free access to food and water provided. Offspring remain in the colony until 28–30 days of age but should be removed prior to intruder tests to avoid wounding. A closer approximation of the natural environment may be provided by surrounding an open central arena with a visible burrow system consisting of a series of tunnels and chambers that are similar to the burrow systems found among wild rats (see Ref. 73 for a description and illustration of such an environment). The latter apparatus is not only more ethologically relevant, but is also extremely valuable in revealing details of behavior during the establishment of colony social organization. If this apparatus is used, however, the longevity of subordinate males will be dramatically reduced, and a substantial number of deaths may occur within 4 months of group formation (64, 71). Although the life expectancy of subordinates in enclosures not having burrows (71) is also shorter than that of dominants, subordinates may reach 500 days of age or more. If extended colony housing is required (e.g., studies of chronic stress), an environment which does not contain a burrow system is advised.

When animals are initially placed in the colony enclosure, considerable fighting among males will occur, and one male will emerge as the dominant (alpha) member of the colony within several days. As dominance is established, fighting will most often decrease in frequency and intensity but not disappear. After social behaviors within the colony have stabilized, alpha males will begin to show attack on intruding males, but not females, while females will preferentially attack like-sex intruders (43). If female offense is to be observed, male members of the colony should be removed prior to introduction of the intruder since their presence has been reported to inhibit female aggression (43).

The period of colony housing may need to be several weeks in duration before intruder tests begin, although this process may well be speeded in burrow environments. Since prompt and vigorous attack by residents is facilitated by experience, intruders should be introduced for 10–15 min at approximately 1-week intervals until attack behavior stabilizes (24). Intruders should be slightly, but not markedly, smaller than residents to facilitate attack while avoiding serious wounding. Although prolonged (e.g., 20–24 hr) exposure of intruders to residents may yield more reliable measures of attack behavior (49), up to 42% of intruders may be killed during such an encounter (50). If longer duration tests are essential, a tunnel or nest box having a single narrow entrance may be incorporated within the colony enclosure. Such refuges are reported to reduce serious wounding during protracted tests but not interfere with the establishment of clear dominance (49).

If the colony model is to be applied to mice, a suitably scaled down version of the enclosures described above may be employed and similar procedures followed. Because male mice are highly intolerant of other males, however, formation of colonies containing multiple males may quickly result in the death of all subordinates (74). As male mice may be behaviorally isolated in natural settings (75), groups consisting of a single male and 2–3 females is probably a reasonable approximation of naturally occurring social colonies.

## Isolation-Induced Aggression

Whereas the colony model may be employed with either rats or mice, social isolation is clearly the most popular and consistently successful method of inducing intermale aggression among mice. The procedures are extremely simple yet yield prompt and intense agonistic interactions. Adult males (70–90 days old) are individually housed with free access to food and water in standard cages having a solid floor and sawdust substrate. After a period of social isolation, unfamiliar male intruders are introduced into the resident's home cage and aggressive interactions recorded for 10–15 min.

The intensity of attack by residents increases with longer periods of social isolation, but authors disagree on the duration which produces maximal attack. Mackintosh (9) finds most intense attack after 9 days of isolation, whereas Goldsmith et al. (23) report increases in aggression for 56–58 days. We have found 3 weeks of isolation to be effective, but, clearly, the optimal period of isolation will need to be established for the strain and testing conditions employed. This period may also be shortened if an artificial burrow system is included within the housing enclosure. As in rats, the vigor and intensity of attack can be increased by repeated testing with unfamiliar males (60, 66, 76) and the use of slightly smaller (1–4 g) intruders. If less vigorous attack is desired, intruders that are somewhat larger or that have previously defeated intruders in their own cages may be employed (70, 76, 77).

Although some (50, 72) have reported success in inducing attack by social isolation in male rats, others (67) report considerable difficulty in applying this technique. If social isolation is to be used with rats, a careful screening of males to be used as residents, inclusion of a burrow system, and an extended period of social isolation may be required to elicit reliable attack. Unless there are compelling considerations, however, use of the isolation-induced paradigm with rats is of questionable ethological relevance and is not recommended.

## Maternal Aggression

Unlike the previous paradigms, maternal aggression can be reliably evoked in a wide range of strains of both rats and mice (see, e.g., Refs. 9, 70, 78, and 80). The primary determinant of such aggression is the time since birth of offspring (78), with the age of the female having little effect on the intensity of attack (79). Although there is some disagreement about the exact period of maximal maternal aggression following the birth of offspring (78, 79), females will generally show high levels of attack during the first postpartum week. Females should be tested in their home cage and all offspring removed prior to testing to prevent injury to them by the intruder. It is important that this removal occur shortly before intruder introduction, however, since absence of offspring for any substantial interval may substantially reduce maternal attack (78). Intruders should be young adults that are significantly smaller than female residents (70). The sex of the intruder is also important. Long-Evans female rats are reported to attack female intruders most intensely, whereas Swiss-Webster mice preferentially attack males (79).

A related technique is used for studies of male infanticide. If a young pup (1–4 days old) is introduced into the home cage of an unfamiliar adult male, the resident will either attack the pup or will engage in parental behavior (81). Since the attacks of infanticidal males are very rapid, it is often not possible to remove unprotected pups before they are seriously wounded, and this procedure may result in substantial levels of mortality. Recently, however, a humane test for infanticide has been described (82) that protects pups from injury. When an infant mouse or rat is placed inside a small tube of wire screening with the ends pinched shut, infanticidal males will attack the tubing but will not be able to injure the pup within. Parental males will grasp the tubing and attempt to incubate it. This procedure is strongly recommended for studies of infanticidal behavior.

## Exposure to Potential Predators

If the primary interest of an investigator is in fear and defensive behavior, brief exposure of rats or mice to a potential predator (cat) is an extremely effective method for studying this behavior. To avoid injury to experimental subjects, the procedure should be carried out only with groups of animals occupying a visible burrow system (see Refs. 72 and 83 for descriptions of testing procedures). When a cat is placed briefly in the central chamber of a burrow environment, subjects will flee to the burrows and remain immobile. Following removal of the cat ultrasonic vocalizations are emitted at a high rate for 1–2 hr. The subjects will begin to tentatively explore the area previously occupied by the cat. This risk assessment behavior is accompanied by a strong inhibition of eating, drinking, and sexual behavior. Even after subjects have emerged from the burrows, eating bouts will be shortened and alternate with frequent environmental scanning (vigilance behavior), behaviors similar to those seen in natural settings (72). These behavioral changes are quite marked and may persist for 16–18 hr after predator removal; they parallel the behavioral characteristics of anxiety in potentially important ways (83).

The procedures used in the test are quite straightforward. Mixed-sex groups of animals are housed within a burrow environment for 8–14 days. A cat is placed in the central chamber of the apparatus for 15 min and then removed. The behaviors evoked by the cat are exclusively defensive but are quite different from the defenses seen during conspecific agonistic encounters. Freezing, approach to burrow entrances, tentative exploration of the chamber from the burrow entrance, and cautious emergence from the burrow accompanied by stretched attention and environmental scanning are prominent. Fuller description of behavioral scoring is available in previous publications (72, 83), which should be consulted for details. Because the behaviors induced by the predator persist for many hours, videotaping is essential. This should be combined with time sampling (e.g., recording behavior for every fourth hour) to reduce the number of observations to manageable proportions. If a briefer behavioral test of antipredator behaviors is required, several of the risk assessment behaviors described above may be evoked by brief presentation of odors from a cat (84) or dominant conspecific (54). These briefer tests may be particularly useful for pharmacological studies of anxiety (84).

## Acknowledgments

The author thanks Drs. R. J. and D. C. Blanchard for helpful comments on an early draft of this chapter and S. M. Weiss for the preparation of Fig. 1.

## References

1. D. Blanchard and R. Blanchard, *Annu. Rev. Psychol.* **39,** 43 (1988).
2. R. Blanchard, P. Brain, D. Blanchard, and S. Parmigiani (eds.), "Ethoexperimental Approaches to the Study of Behavior." Kluwer Academic Publishers, Dordrecht, The Netherlands, 1989.
3. P. Brain and D. Benton (eds.), "Multidisciplinary Approaches to Aggression Research." Elsevier/North-Holland, Amsterdam, 1981.
4. K. Flannelly, R. Blanchard, and D. Blanchard (eds.), "Biological Perspectives on Aggression." Alan R. Liss, New York, 1984.
5. K. Miczek, M. Kruk, and B. Olivier (eds.), "Ethopharmacological Aggression Research." Alan R. Liss, New York, 1984.
6. R. Rodgers and A. Waters, *Neurosci. Biobehav. Rev.* **9,** 21 (1985).
7. R. Blanchard and D. Blanchard, *Behav. Biol.* **21,** 197 (1977).
8. R. Blanchard, L. Pank, D. Fellows, and D. Blanchard, *Psychol. Rec.* **35,** 329 (1985).
9. J. Mackintosh, *Symp. Zool. Soc. London* **47,** 365 (1981).
10. L. Takahashi and R. Blanchard, *Behav. Proc.* **7,** 49 (1982).
11. D. Albert, R. Jonik, and M. Walsh, *Physiol. Behav.* **49,** 673 (1991).
12. B. Bermond, *Aggressive Behav.* **8,** 335 (1982).
13. D. Blanchard and S. Takahashi, *Physiol. Behav.* **42,** 613 (1988).
14. P. Brain and A. Poole, *Aggressive Behav.* **1,** 39 (1974).
15. V. O'Donnell, R. Blanchard, and D. Blanchard, *Behav. Neural. Biol.* **32,** 89 (1981).
16. K. Flannelly, R. Blanchard, R. Muraoka, and L. Flannelly, *Physiol. Behav.* **29,** 381 (1982).
17. K. Flannelly and R. Lore, *Anim. Behav.* **25,** 654 (1977).
18. B. Olivier, J. Mos, A. van der Poel, P. Krijzer, and M. Kruk, *in* "Biological Perspectives on Aggression" (K. Flannelly, R. Blanchard, and D. Blanchard, eds.), p. 261. Alan R. Liss, New York, 1989.
19. S. Parmigiani and P. Palanza, *Neurosci. Biobehav. Rev.* **15,** 511 (1991).
20. R. Blanchard, K. Hori, P. Tom, and D. Blanchard, *Aggressive Behav.* **14,** 195 (1988).
21. P. Brain, D. Benton, D. Howell, and J. Bolton, *Aggressive Behav.* **4,** 201 (1978).
22. K. Flannelly, A. Hiraoka, and L. Flannelly, *Psychol. Rep.* **51,** 52 (1982).
23. J. Goldsmith, P. Brain, and D. Benton, *Aggressive Behav.* **2,** 307 (1977).
24. R. Blanchard, L. Taksahashi, and D. Blanchard, *Anim. Learn. Behav.* **5,** 365 (1977).

25. D. Luciano and R. Lore, *J. Comp. Physiol. Psychol.* **88,** 917 (1975).
26. K. Miczek and M. Krsiak, *in* "Advances in Behavioral Pharmacology" (T. Thompson and O. Dews, eds.), Vol. 2, p. 87. Academic Press, New York, 1979.
27. P. Leyhausen, "Cat Behavior: The Predatory and Social Behavior of Domestic and Wild Cats." Garland STPM Press, New York, 1979.
28. E. Kemble, *in* "Ethoexperimental Approaches to the Study of Behavior" (R. Blanchard, P. Brain, D. Blanchard, and S. Parmigiani, eds.), p. 484. Kluwer Academic Publishers, Dordrecht, The Netherlands, 1989.
29. G. Griebel, M. Saffroy-Spittler, R. Misslin, E. Vogel, and J. Martin, *Psychopharmacology (Berlin)* **102,** 498 (1990).
30. L. Blumstein and J. Crawley, *Pharmacol. Biochem. Behav.* **18,** 37 (1983).
31. S. File, *Neuropsychobiology* **13,** 55 (1985).
32. R. Lister, *Life Sci.* **41,** 1481 (1987).
33. S. Pellow, P. Chopin, S. File, and M. Briley, *Neurosci. Methods* **14,** 149 (1985).
34. S. Barnett, "The Rat: A Study in Behavior" (Revised Ed.). Univ. of Chicago Press, Chicago, 1975.
35. P. Brain, *in* "Multidisciplinary Approaches to Aggression Research" (P. Brain and D. Benton, eds.), p. 53. Elsevier/North-Holland, Amsterdam, 1981.
36. E. Grant, *Behaviour* **21,** 260 (1963).
37. E. Grant and J. Mackintosh, *Behaviour* **21,** 246 (1963).
38. J. Scott, *Am. Zool.* **6,** 683 (1966).
39. P. Brain and N. Nowell, *Commun. Behav. Biol.* **5,** 7 (1970).
40. R. Blanchard, V. O'Donnell, and D. Blanchard, *Aggressive Behav.* **5,** 341 (1979).
41. R. Blanchard, C. Kleinschmidt, K. Flannelly, and D. Blanchard, *Aggressive Behav.* **10,** 309 (1984).
42. R. Blanchard and D. Blanchard, *in* "The Biology of Aggression" (P. Brain and D. Benton, eds.), p. 529. Sijthoff & Noordhoff, Alphen aan der Rijn, The Netherlands, 1981.
43. D. Blanchard, C. Fukunaga-Stinson, L. Takahashi, K. Flannelly, and R. Blanchard, *Behav. Proc.* **9,** 31 (1984).
44. P. Donat, *Neurosci. Biobehav. Rev.* **15,** 447 (1991).
45. P. Brain, K. McAllister, and S. Walmsley, *in* "Neuromethods, Volume 13: Psychopharmacology" (A. Boulton, G. Baker, and A. Greenshaw, eds.), p. 689. Humana, Clifton, New Jersey, 1989.
46. R. Blanchard, K. Flannelly, M. Layng, and D. Blanchard, *Physiol. Behav.* **33,** 857 (1984).
47. S. Barnett, *Proc. Zool. Soc. London* **134,** 611 (1960).
48. P. Brain and S. Parmigiani, *Biol. J. Linn. Soc.* **41,** 257 (1990).
49. R. Lore, M. Nikoletseas, and L. Takahashi, *Aggressive Behav.* **10,** 59 (1984).
50. D. Thor and K. Flannelly, *Behav. Biol.* **17,** 237 (1976).
51. R. Lore, K. Flannelly, and P. Farina, *Aggressive Behav.* **2,** 175 (1976).
52. P. Brain, D. Benton, A. Howell, and S. Jones, *Anim. Learn. Behav.* **8,** 331 (1980).
53. A. Van der Poel, J. Mos, M. Kruk, and B. Olivier, *in* "Ethopharmacological

Aggression Research'' (K. Miczek, M. Kruk, and B. Olivier, eds.), p. 115. Alan R. Liss, New York, 1984.

54. J. Williams, P. Worland, and M. Smith, *J. Exp. Psychol.: Anim. Behav. Processes* **16,** 345 (1990).
55. P. Brain, D. Benton, C. Cole, and B. Prowse, *Physiol. Behav.* **24,** 1003 (1980).
56. M. Krsiak, *Br. J. Pharmacol.* **55,** 141 (1975).
57. R. Blanchard, K. Hori, P. Tom, and D. Blanchard, *Aggressive Behav.* **14,** 195 (1988).
58. L. Takahashi and R. Lore, *Physiol. Behav.* **29,** 1013 (1983).
59. R. Blanchard, K. Flannelly, and D. Blanchard, *Physiol. Behav.* **43,** 1 (1988).
60. D. Luciano and R. Lore, *J. Comp. Physiol. Psychol.* **88,** 917 (1975).
61. M. Nikoletseas and R. Lore, *Aggressive Behav.* **7,** 245 (1981).
62. N. Adams and R. Boice, *J. Comp. Physiol. Psychol.* **97,** 24 (1983).
63. R. Boice, *J. Comp. Physiol. Psychol.* **91,** 649 (1977).
64. R. Blanchard, K. Flannelly, and D. Blanchard, *Behav. Proc.* **11,** 209 (1985).
65. K. Flannelly, L. Flannelly, and R. Blanchard, *in* "Biological Perspectives on Aggression" (K. Flannelly, R. Blanchard, and D. Blanchard, eds.), p. 207. Alan R. Liss, New York, 1984.
66. K. Flannelly, L. Flannelly, and P. Pang, *Psychol. Rep.* **52,** 877 (1983).
67. R. Blanchard and D. Blanchard, *Behav. Neural. Biol.* **30,** 109 (1980).
68. V. Ryan and F. Wehmer, *Dev. Psychobiol.* **8,** 363 (1975).
69. J. Namikas and F. Wehmer, *Behav. Biol.* **23,** 219 (1978).
70. K. Flannelly and L. Flannelly, *Psychol. Rep.* **57,** 883 (1985).
71. D. Blanchard and R. Blanchard, *Neurosci. Biobehav. Rev.* **14,** 455 (1990).
72. R. Lore, M. Nikoletseas, and K. Flannelly, *Behav. Neural. Biol.* **28,** 243 (1980).
73. R. Blanchard and D. Blanchard, *J. Comp. Psychol.* **103,** 70 (1989).
74. R. Blanchard, personal communication (1992).
75. P. Brain, *Life Sci.* **16,** 187 (1975).
76. S. Parmigiani and P. Brain, *Behav. Proc.* **8,** 45 (1983).
77. S. Parmigiani, D. Mainardi, and A. Pasquali, *in* "The Biology of Aggression" (P. Brain and D. Benton, eds.), p. 563. Sijthoff & Noordhoff, Alphen aan den Rijn, The Netherlands, 1981.
78. K. Flannelly and L. Flannelly, *J. Comp. Psychol.* **101,** 101 (1987).
79. M. Haney, J. DeBold, and K. Miczek, *Aggressive Behav.* **15,** 443 (1989).
80. S. Parmigiani, P. Palanza, and P. Brain, *Ethol. Ecol. Evol.* **1,** 341 (1989).
81. F. vom Saal and L. Howard, *Science* **215,** 1270 (1982).
82. G. Perrigo, W. Bryant, L. Belvin, and F. vom Saal, *Anim. Behav.* **38,** 897 (1989).
83. R. Blanchard, D. Blanchard, J. Rodgers, and S. Weiss, *Neurosci. Biobehav. Rev.* **14,** 463 (1990).
84. J. Shepherd, T. Flores, R. Rodgers, R. Blanchard, and D. Blanchard, *Physiol. Behav.* **51,** 277 (1992).

# Section III

## Drug-Induced Behavior

# [9] Measurement of Cocaine-Induced Stereotyped Behavior in Response to Neuropeptides

Zoltan Sarnyai

## Introduction

The definition of a drug as a psychomotor stimulant is based on its effect on behavior. The behavioral profile of psychostimulants consists of a continuum of different behavioral elements, the occurrence of which depends on the dose, route of administration, treatment schedule, and the environmental cue. Apomorphine, amphetamine, and cocaine, all defined as psychomotor stimulants, elicit certain behavioral repertoires ranging from locomotor hyperactivity to stereotyped behaviors (highly repetitive, purposeless, compulsive, and restricted in variations). The behavioral repertoires induced by different psychomotor stimulants are similar but not identical. Locomotor hyperactivity, grooming, sniffing, signs of gnawing, biting, and licking, or, in some cases, dyskinetic behaviors and seizures are considered as the behavioral consequence of the administration of increasing doses of amphetamine and, at least in part, of cocaine (1). Stereotyped sniffing behavior has been demonstrated as the major component of the cocaine-induced behavioral repertoire in rats (2).

Two distinct neuroanatomical pathways have been postulated as the anatomical substrates of locomotor and stereotyped behavior induced by psychostimulants, namely, the mesolimbic dopaminergic system for the locomotor activity, on one hand, and the nigrostriatal dopaminergic system for the stereotyped movements, on the other. However, sniffing behavior seems to be mediated through the mesolimbic dopaminergic terminal structure, the nucleus accumbens. This is suggested by the lack of apomorphine- and amphetamine-induced sniffing after electrolytic or neurochemical lesion of the nucleus accumbens (3, 4).

In spite of the common neuroanatomical substrate of the three different psychomotor stimulants, the synaptic mechanisms mediating their actions have been considered to be different. Apomorphine acts as a direct dopamine agonist on the postsynaptic dopamine receptors (5); amphetamine acts as an indirect dopamine agonist, releasing dopamine from the nerve terminals (6, 7); and cocaine primarily acts on the dopamine transporter sites of presynaptic nerve endings, inhibiting the reuptake processes for dopamine (7, 8).

*Methods in Neurosciences, Volume 14*

The neurohypophyseal hormones oxytocin and vasopressin have a large number of behavioral actions mediated by the modulation of dopaminergic neurotransmission in the brain (9). Learning and memory processes, morphine and ethanol tolerance and dependence, and heroin self-administration are all related to changes of the brain catecholaminergic neurotransmission, and they seem to be affected by oxytocin and vasopressin in experimental animals (10). Neurohypophyseal hormone-induced modifications of the dopaminergic and noradrenergic neurotransmissions in discrete brain areas, especially in the mesolimbic basal forebrain structures, have also been evaluated (11).

Particular interest in the processes mediated through the mesolimbic dopaminergic system has emerged because both the rewarding effects of drugs of abuse and the effects of antipsychotic agents might be related to this neuroanatomical structure (12). The measurement of stereotyped behavior as an index of the functional activity of dopaminergic system has become an important strategy of behavioral pharmacology. A number of different techniques have been used to measure the hyperactivity and stereotyped behavior exhibited in response to psychostimulants. In rats these range from a simple determination of whether some signs of the response are present or absent to complex phenomenological descriptions. The widely used tests to measure the behavioral profiles of psychostimulants employ a rating scale to describe the different behavioral elements from inactivity through low-intensity components (such as grooming and episodic sniffing) and high-intensity components (such as continuous sniffing in one location) to the dramatic dyskinetic movements (jumping and convulsions). These scales have been based on the observation that this behavioral profile is on a continuum related to the doses of the drug (1, 3, 13, 14). Our current knowledge of the behavioral effects and neuroanatomical/neurochemical substrates of psychomotor stimulants originated from these excellent qualitative and semiquantitative descriptions.

Because stereotyped sniffing is the major component of the behavioral repertoire induced by moderate doses of cocaine, we aimed to develop a sensitive, quantitative rating scale, which would be easy to use, be readily adapted for use by other laboratories, and have a high intra- and interrater reliability. Basic requirements for such a scale would be high sensitivity to measure the effects of neuromodulatory neuropeptides, which probably do not have such a robust effect, like transmitter antagonists, on the action of cocaine in the brain. Additionally, the rating scale should reflect more detailed documentation of the effects of neuropeptides on different basal forebrain structures to elucidate the sites of potential interactions between peptides and the highly abused drug cocaine in rats (15).

To investigate the reliability of the scale, the first study utilized a dose–response and a time–effect relationship of the cocaine on two forms of stereotyped behavior. Cocaine was injected both subcutaneously and into the different dopaminergic structures in the brain. The second study demonstrated that the scale accurately describes the pharmacological interactions. The effect of cocaine was dose-dependently inhibited when the neurohypophyseal hormone oxytocin was administered subcutaneously, intracerebroventricularly, and locally into the basal forebrain structures containing oxytocin receptors. In the third study, the selectivity of the action of oxytocin on cocaine was demonstrated by the lack of effect of the neuropeptide on sniffing behavior induced by other pharmacologically related motor stimulants (i.e., amphetamine and apomorphine), suggesting the great sensitivity of this behavioral rating.

## Methods

### Animals and Surgery

Male rats of the Wistar strain (LATI, Gödöllö, Hungary) weighing 150–180 g are used. Five animals are housed per cage and kept at room temperature under a constant light/dark cycle (lights on between 6:00 a.m. and 6:00 p.m.). Five days prior to the experimental session, the animals are subjected to the surgical procedure. For the intracerebroventricular and local cerebral microinjection, animals are operated on with sodium pentobarbital (Nembutal, CEVA, Paris, France, 40 mg/kg i.p.) anesthesia using a stereotaxic apparatus. A 23-gauge stainless steel guide cannula is implanted unilaterally into the right cerebral ventricle and bilaterally into the nucleus accumbens, tuberculum olfactorium, nucleus olfactorius, central amygdaloid nucleus, and nucleus caudatus according to the coordinates of Pellegrino et al. (16). Thus the tip of the cannula rests 1 mm above the intended site of injection. The cannulae are cemented to the skull by dental acrylic cement. At the end of the experiment, the location of the cannula is checked histologically. The animals are allowed at least 7 days to recover from surgery before the behavioral testing is started. Drugs are infused in a volume of 2 μl by a Hamilton microsyringe for the intracerebroventricular and intracerebral treatments.

### Behavioral Testing

During the period of recovery the animals are subjected to a repeated handling procedure (5 min/day). On the day of experiment the rats are carried to the experimental room for an adaptation period at least 2 hr prior to the testing session. For the behavioral experiment the animals are placed individually into transparent plastic cages (20 × 6 × 34 cm) in a low-noise environment immediately after cocaine or saline injection and are left for 30 min prior to testing. During the 30 min, intense sniffing behavior develops and the unspecific stress-related behavioral hyperactivity due to the injection procedure and the new environment is extinguished.

The behavioral effect of cocaine is judged to be stereotyped if the behavioral repertoire is strongly restricted in variations and consists of continous repetition of one or a few items of behavior. According to Scheel-Krüger (13), ". . . sniffing behavior is considered to be stereotyped if the rats sit in a crouched posture and behavioral stimulation is strongly restricted to a long lasting performance of continous sniffing, which covers only a small area of the cage. The sniffing activity is performed by the rats on either the lower part of the wall or on the floor with head movement from side to side." Increased rearing activity is present only after the administration of the lower doses of cocaine. The occurrence of the following activities is recorded as grooming behavior: face washing, body stroking, scratching, paw licking, and genital washing.

The rating is started 30 min after the injection of saline or the psychomotor stimulants. The behavior (continous sniffing = 1, no sniffing = 0) is recorded visually every third minute (every minute after the intracerebral cocaine/saline administration) for 120 min; the maximum sniffing score is therefore 40 (120 after the intracerebral treatment). Ten seconds is allowed to record the score of each animal. To assess interrater reliability, 2 observers are used. Neither knows the drug or dose the animals have received. Ratings are not compared until completion of the entire test session.

## Drugs

Peptides are administered 1 hr prior to cocaine (or amphetamine and apomorphine) treatment. Peptide solutions are freshly prepared before administration. Oxytocin (Gedeon Richter Pharmaceutical Works, Budapest, Hungary) is dissolved in 0.9% NaCl for subcutaneous administration or in an artificial cerebrospinal fluid (CSF) for intracerebroventricular and intracerebral administration. Arginine-vasopressin (Sigma, St. Louis, MO) and lysine-vasopressin (Sandoz, Basle, Switzerland) are also dissolved in saline. An oxytocin receptor antagonist, $N^\alpha$-(2-O-methyltyrosine)-oxytocin (kindly donated by T. Barth, Prague, Czechoslovakia), is dissolved in CSF. Each animal is injected with 0.2 ml solution for subcutaneous and with 2 μl for intracerebroventricular and intracerebral administration. Control animals receive saline or CSF in the same volume. Cocaine (cocaine hydrochloride, E. Merck, Darmstadt, Germany), apomorphine (apomorphine hydrochloride, (Sigma), and amphetamine (Aktedron, Gedeon Richter Pharmaceutical Works) are dissolved in saline.

## Statistical Analysis

Statistical analysis of the data is performed by one-way analysis of variance (ANOVA) followed by Tukey's and Dunnett's tests for multiple comparison and by Student's $t$-test. A probability level of 0.05 is accepted as indicating a significant difference.

## Results and Discussion

### Experiment 1

The effects of different doses of cocaine injected subcutaneously or intracerebrally on stereotyped sniffing behavior were evaluated in Experiment 1. Peripheral administration of increasing doses of cocaine produced an intense, dose-dependent stereotyped sniffing, which lasted more than 2 hr (Fig. 1). The grooming behavior was elicited only after treatment with a low dose (3.25 mg/kg) of cocaine, because it might be overlapped by sniffing induced by the higher doses of cocaine. The reliability of the scale is acceptable from 30 min postinjection on. The rather poor reliability of the first 30 min after the injection reflects the rapidly changing nature of the behavior at this time owing to the stressful effects of the injection and the new environment.

Sniffing behavior was induced by the microinjection of cocaine into the mesolimbic terminal nucleus accumbens, the tuberculum olfactorium, and the nigrostriatal terminal caudate nucleus (Fig. 2). Grooming behavior was elicited by the microinjection of cocaine into the caudate nucleus, but not into the mesolimbic terminal structures. The registration of behavioral signs every minute was required by the weaker effect and shorter duration of action of cocaine injected intracerebrally. The changes in frequency and duration of the effect of cocaine could be explained by assuming that not only one structure mediates the behavioral effect of cocaine, or that the unspecific adverse effects of intracerebral implantation of the cannula decreased the general behavioral activity of the animals. The later proposition could be excluded because the intraaccumbens injection of cocaine produced a high rate of sniffing, which was similar to that described after subcutaneous treatment.

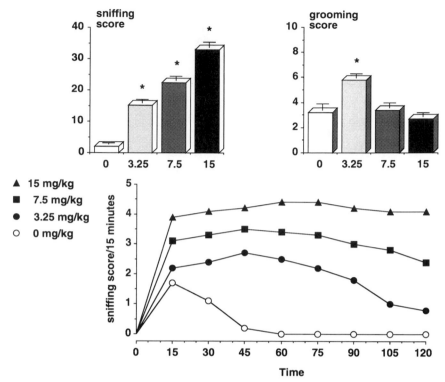

FIG. 1   Effects on sniffing and grooming activity in rats of cocaine administered subcutaneously. Bar graphs represent the behavioral activity in response to different doses of cocaine measured for 120 min. The bottom graph shows the time–effect and dose–response relationships of cocaine on stereotyped sniffing behavior in rats ($N$ = 15–20 rats/group). Asterisks denote $p < 0.05$ versus control (0 mg/kg).

These results showed that the stereotyped behavioral activity induced by cocaine is not a homogeneous entity in its neuroanatomical substrate. Cocaine-induced grooming activity was localized in the nigrostriatal terminal caudate nucleus, in agreement with some previous observations (17). Micro-injection of cocaine into the mesolimbic terminal structures elicited an intense sniffing activity, suggesting that the sniffing induced by cocaine, as a unique form of stereotyped behavior, could not be associated with the nigrostriatal dopaminergic system alone. Similar or more pronounced involvement of mesolimbic dopaminergic structures in the mediation of cocaine-induced sniffing behavior is suggested. This result has been supported by findings that electrolytic lesions of the nucleus accumbens and the tuberculum olfact-orium, but not the caudate putamen, blocked the apomorphine-induced snif-

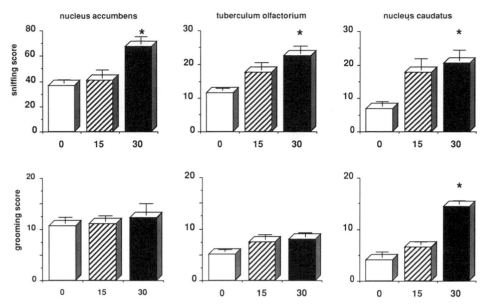

FIG. 2   Sniffing and grooming behavior induced by local microinjection of cocaine into the mesolimbic and nigrostriatal terminal structures. Cocaine was injected bilaterally in doses of 15 and 30 $\mu g/2\mu l$. Asterisks denote $p < 0.05$ versus control $(0\mu g/2\mu l)$.

fing (3). In addition, local microinjections of apomorphine into the tuberculum olfactorium and into the nucleus accumbens, but not the caudate putamen, elicited sniffing behavior in rats (3). Neurochemical lesion of the nucleus accumbens by 6-hydroxydopamine decreased significantly the sniffing component of the amphetamine-induced stereotyped behavioral response (18). It has been suggested that sniffing is an exploratory act of the behavioral repertoire which could be closely coupled to locomotor behavior in the environment. It seems that in selecting sniffing behavior as a measure of stereotypy, we may have been studying a function of the mesolimbic rather than the striatal dopamine system.

The behavioral rating scale used was very sensitive to the effects of different doses and routes of administration of cocaine. The scale was able to discriminate two components of cocaine-induced behavior as a function of dose.

## Experiment 2

Testing the sensitivity of the rating scale to the effects of neurohypophyseal peptides on cocaine-induced stereotyped behavior was the critical step in the evaluation of its reliability because the neuropeptides, which probably

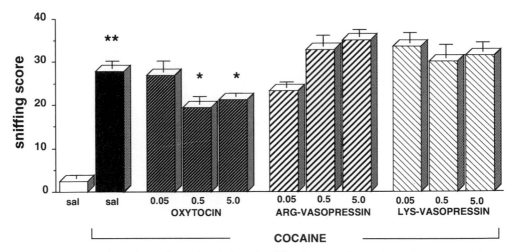

FIG. 3    Alterations of cocaine-induced sniffing by neurohypophyseal hormones. Oxy-tocin and Arg- and Lys-vasopressin (0.05–5.0 µg/animal, s.c.) were administered 60 min prior to cocaine treatment (15 mg/kg, s.c.). Single asterisks denote $p < 0.05$ versus sal (0.9% NaCl) plus cocaine-treated control; double asterisk denotes $p < 0.005$ versus sal-treated control.

act as neuromodulators, do not have a robust, well-defined action like the neurotransmitter antagonists. To determine the sensitivity of our rating scale to the neuropeptide–cocaine interaction, a sequence of studies was designed to investigate the effects of oxytocin and Arg- and Lys-vasopressin on co-caine-induced sniffing behavior. Peripheral (subcutaneous) administration of oxytocin, but not Arg- or Lys-vasopressin, was able to inhibit the effect of cocaine dose-dependently (Fig. 3). If an oxytocin receptor antagonist was injected into the lateral cerebral ventricle prior to subcutaneous administra-tion of oxytocin, the effect of peripherally administered oxytocin disap-peared, suggesting a target site for the action of oxytocin in the brain (data not shown here; see Ref. 15). Intracerebroventricular application of oxytocin in nanogram quantities inhibited the cocaine-induced sniffing behavior (Fig. 4). To determine the site(s) of action of oxytocin, the peptide was adminis-tered topically into selected brain areas. The cocaine-induced sniffing behav-ior decreased when a 100-pg dose of oxytocin was injected into the nucleus accumbens or tuberculum olfactorium, but not when oxytocin was injected into the nucleus olfactorius, central amygdaloid nucleus, or nucleus cauda-tus (Fig. 5). Acute tolerance to morphine (19) and intravenous heroin self-administration (20) have also been blocked by the local application of oxyto-cin into the limbic forebrain areas containing oxytocin receptors, suggesting

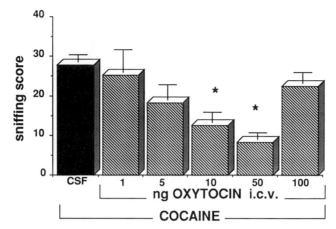

FIG. 4   Effect of intracerebroventricularly administered oxytocin on cocaine-induced (15 mg/kg) sniffing behavior. Asterisks denote $p < 0.05$ versus CSF (artificial cerebrospinal fluid) plus cocaine-treated control.

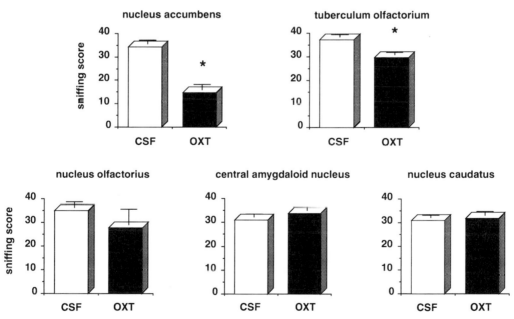

FIG. 5   Effect of oxytocin (OXT, 100 pg) administered into different basal forebrain structures on cocaine-induced (15 mg/kg) sniffing behavior. Asterisks denote $p < 0.05$ versus CSF (artificial cerebrospinal fluid) plus cocaine treated control.

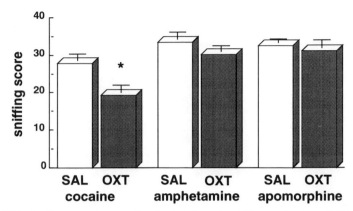

FIG. 6   Effect of oxytocin on stimulant-induced sniffing behavior. SAL refers to control administration of 0.9% NaCl. Oxytocin (OXT) was given at 0.5 $\mu$g/animal. Cocaine (15 mg/kg), amphetamine (5 mg/kg), and apomorphine (250 $\mu$g/kg) were administered 60 min after the SAL or OXT. Asterisk denotes $p < 0.05$ versus SAL plus cocaine.

the regulatory role of oxytocin in the drug-induced behavior. Oxytocin receptors located in the nucleus accumbens mediate the stimulatory effect of oxytocin on the novelty-induced grooming behavior (21).

These data indicate that oxytocin selectively inhibits cocaine-induced sniffing behavior through the basal forebrain target sites. However, the effects of structurally related neuropeptides and the sensitivity of different brain structures to oxytocin were effectively discriminated by the behavioral rating scale.

## Experiment 3

As a further step to validate the behavioral rating scale, the effect of oxytocin on the behavioral action of different psychomotor stimulants was investigated (Fig. 6). Cocaine, amphetamine, and apomorphine elicit a similar behavioral profile via the activation of the same neuroanatomical structures by modifying different sites of the dopaminergic nerve terminals and receptors. Functionally equivalent doses of cocaine (15 mg/kg), amphetamine (5 mg/kg), and apomorphine (250 $\mu$g/kg) were chosen. A similar behavioral repertoire was induced by these stimulants, with the episodic appearance of licking activity after amphetamine and apomorphine treatment. Oxytocin pretreatment decreased the sniffing activity induced by cocaine without any effect on amphetamine- and apomorphine-induced sniffing at the dose investigated. Because the proposed site of action of cocaine is the presynaptic dopamine transporter (reuptake) site, whereas amphetamine acts on dopamine release from the presynaptic transmitter-containing vesicles and apomorphine acts as a direct dopamine agonist on the postsynaptic dopamine receptors, it is possible that oxytocin interacts selectively with the dopamine reuptake sites and/or cocaine-receptors in the mesolimbic dopaminergic structures.

This result demonstrates that the rating scale is sensitive enough to compare the effect of oxytocin on the behavior induced by functionally related stimulants. In addition, not only could the important brain structures mediating the effect of cocaine be identified by these behavioral methods, but a putative synaptic site of action of oxytocin has also been proposed.

## Conclusions

We have tried to develop a sensitive behavioral test to measure the cocaine-induced sniffing activity in response to neuropeptides. Sniffing behavior was chosen because this element of the cocaine-induced behavioral profile can be elicited by moderate doses of cocaine and also because this behavior is thought to be mediated by mesolimbic dopaminergic structures. These structures are critically important in the mediation of the rewarding effect of abused drugs, and they might also be potential targets for the actions of neurohypophyseal peptides in modulating behavioral changes.

A quantitative rating scale has been developed which can measure the frequency of a certain behavior. The rating scale was tested by a sequence of experiments to evaluate its sensitivity, specificity, and intra- and interrater reliability. According to the results, this behavioral test was very sensitive in measuring the effects of cocaine, injected peripherally or locally into the brain, on behavior. The rating scale appeared to be sensitive enough to detect the effect of neuropeptides on cocaine-induced sniffing and to discriminate the effects of structurally related peptides. The neuroanatomical mediation of the effect of oxytocin on cocaine-induced behavior has also been evaluated by this procedure. Additionally, a presynaptic location of the action of oxytocin on dopamine uptake sites has also been suggested based on the selective inhibition of cocaine-induced (but not amphetamine- or apomorphine-induced) sniffing by oxytocin.

The results gained by using this behavioral rating scale suggest that cocaine-induced sniffing behavior might be mediated primarily through the mesolimbic dopaminergic terminal areas and can be affected by oxytocin introduced into the basal forebrain regions containing oxytocin receptors. Our data illustrate the usefulness of the scale for sensitive and selective quantitative determination of the interactions between psychostimulant cocaine and certain neuropeptides. This simple behavioral method could be useful in evaluating the effects of chronic cocaine treatment on behavior and could reflect the possible interactions with various neuropeptides as potential modulators for cocaine addiction.

## Acknowledgments

These studies were performed at the Department of Pathophysiology (Head: Prof. G. Telegdy), Albert Szent-Györgyi Medical University, Szeged, Hungary. The author wishes to thank Prof. G. L. Kovács (Szombathely, Hungary) for stimulating discussions during the experiments. The important contribution of Dr. E. Babarczy, E. Pollák, and I. Baczkó in these studies is gratefully acknowledged. The author thanks Wallis Scholar for assistance in graphic displays and Lynne G. Wighton for editorial correction of the manuscript.

## References

1. E. H. Ellinwood, Jr., and R. L. Balster, *Eur. J. Pharmacol.* **28,** 35 (1975).
2. J. Scheel-Krüger, C. Braestrup, M. Nielson, K. Golembiowska, and E. Mogilnicka, *in* "Cocaine and Other Stimulants" (E. H. Ellinwood, Jr., and M. M. Kilbey, eds.), p. 373. Plenum, New York and London, 1975.
3. B. Costall, R. J. Naylor, and J. Neumeyer, *Eur. J. Pharmacol.* **31,** 1 (1975).
4. P. H. Kelly and S. D. Iversen, *Eur. J. Pharmacol.* **40,** 45 (1976).
5. A. M. Ernest, *Psychopharmacologia* **10,** 316 (1967).
6. A. Randrup and I. Munkvad, *Nature (London)* **211,** 540 (1966).
7. R. Kuczenski, *in* "Stimulants: Neurochemical, Behavioral, and Clinical Perspectives" (I. Creese, ed.), p. 31. Raven, New York, 1983.
8. M. C. Ritz, R. J. Lamb, S. R. Goldberg, and M. J. Kuhar, *Science* **237,** 1219 (1987).
9. G. L. Kovács, *Curr. Top. Neuroendocrinol.* **6,** 91 (1986).
10. G. L. Kovács and G. Teledgy, *in* "Frontiers in Hormone Research, Volume 15: Neuropeptides and Brain Function" (G. Telegdy, ed.), p. 138. Karger, Basel, 1987.
11. D. H. G. Versteeg, *Pharmacol. Ther.* **19,** 297 (1983).
12. G. F. Koob, F. Vaccarino, M. Amalric, and F. E. Bloom, *in* "Brain Reward Systems and Abuse" (J. Engel and L. Oreland, eds.), p. 35. Raven, New York, 1987.
13. J. Scheel-Krüger, *Eur. J. Pharmacol.* **18,** 63 (1972).
14. I. Creese and S. D. Iversen, *Brain Res.* **55,** 369 (1973).
15. Z. Sarnyai, E. Babarczy, M. Kriván, G. Szabó, G. L. Kovács, and G. Telegdy, *Neuropeptides* **19,** 51 (1991).

16. L. J. Pellegrino, A. S. Pellegrino, and A. J. Cushman, "A Stereotaxic Atlas of the Rat Brain." Plenum, New York, 1979.
17. I. Creese and S. D. Iversen, *Brain Res.* **83,** 419 (1975).
18. P. H. Kelly, P. W. Seviour, and S. D. Iversen, *Brain Res.* **94,** 507 (1975)
19. Z. Sarnyai, S. Viski, M. Kriván, Szabó, G. L. Kovács, and G. Telegdy, *Brain Res.* **463,** 284 (1988).
20. R. Ibragimov, G. L. Kovács, and G. Telegdy, *Life Sci.* **41,** 1265 (1987).
21. F. Drago, Z. Sarnyai, and V. D'Agata, *Physiol. Behav.* **50,** 533 (1991).

# [10] The Circling Training: A Behavioral Paradigm for Functional Teratology Testing

Juan L. Brusés and Julio M. Azcurra

## Introduction

Prenatal exposure to certain drugs or chemicals can affect the embryonic development of the central nervous system (CNS), resulting in structural malformations or teratogenesis (*terato* = monster). Nonteratogenic doses of these compounds can still alter the normal development of the CNS and induce subteratogenic functional defects, represented in abnormal behavioral patterns of the offspring. Moreover, drugs or environmental agents that lack any teratogenicity can also cause subteratogenic functional deficits in the CNS. These subteratogenic alterations are called behavioral or functional teratology, because brain functions are affected but morphological damage cannot be detected (1). Although any anatomical malformation results in functional teratology, there might be a neurophysiological or neurochemical impairment underlining the functional deficit (2). It is worth noting that the vulnerability of the CNS to subteratogenic effects extends to the postnatal period because development of the CNS continues for weeks or years after birth depending on the species.

The subteratogenic effect of drugs and chemicals on humans is difficult to assess, because CNS functional deficits become more evident when high brain functions such as learning or complex behaviors are performed, as is the case with children of elementary school age. By this time the correlation between a functional deficit and the administration of a drug during prenatal or early postnatal development is difficult to establish.

There are two main approaches for detecting functional teratology. One is to perform longitudinal studies of behavioral patterns and intellectual skills on humans after the administration of a drug during pregnancy or during the first years of life. This necessary approach is the only way to confirm that certain substances induce functional teratogenicity in humans. However, these studies are difficult and expensive to perform because the subject must be followed for many years, and also the damage induced to the subject is irreparable. In contrast, the aim of the other approach is to predict the undesirable effects of a drug before humans are exposed to the substance, and thus the deleterious effects of drugs are evaluated with studies on experimental animals. Because behavior is the final common

*Methods in Neurosciences, Volume 14*

pathway of several brain functions, an anomalous behavioral pattern reflects a CNS functional impairment. Moreover, behavioral indices of neurotoxicity are reliable and sensitive for detecting functional teratology (3); thus, behavioral analysis is the first step in detecting subteratogenic drug-induced effects.

Many behavioral tests are used to screen the functional teratogenicity of drugs and chemicals in laboratory animals. These tests have been categorized according to the brain function most involved in that particular task, such as learning or neuromotor abilities, sensory function, level of activity, and pharmacological responsiveness (4). On the other hand, behavioral teratology tests can be classified into apical and nonapical tests (3). Nonapical tests focus on the evaluation of a specific behavior that reflects a particular brain function. In contrast, in an apical test, several brain functions, such as learning skills, sensory functions, and motor abilities, are involved in the performance of the task and must be correctly integrated by the CNS to complete the test successfully. Therefore, it is convenient for screening methods aimed at detecting functional teratology to start with apical tests, which are sensitive to changes in any of several brain functions and thus increase the probability of detecting a functional impairment. Some requirements for these tests are as follows: (a) the measured end point(s) derived from the test should be objective and simple to define; (b) the test has to be cost-effective and objective, and, hence, automated procedures for experimental performance and data collection are desirable; (c) the test has to be amenable to longitudinal testing of the same animal, and, therefore, apical tests should be noninvasive in nature; and (d) it is preferable that the quantifiable end point(s) be sensitive to changes of control levels in either direction (5).

Another valuable characteristic of these tests is that the behavioral response induces measurable neurophysiological activity; thus, behavioral performance can be correlated with a neurochemical or neurophysiological parameter. This association between a behavioral response and a neurophysiological change improves the study of functional teratology because biochemical measurements can be used as indicators of functional impairments.

The circling behavior in animals has been intensively studied since it was shown to be the result of an imbalance of dopaminergic activity in motor brain nuclei. This imbalance between the left and right sides of the brain can be induced by specific lesions and pharmacological manipulations in the motor circuit of the CNS (6). In addition, conditioned circling behavior (without any lesion) induces changes in the caudate dopamine (DA) metabolism (7), activation of the caudate tyrosine hydroxylase (8), and firing of substantia nigra DA neurons (9). Thus, the circling behavior is an extensively used biological model to study the neurophysiology and neurochemistry of motor functions.

The circling training model is an operant conditioning behavioral paradigm in which water-deprived rats are trained to turn in a circle trail for a reward during eight consecutive daily sessions. The time required by the animal to perform a defined number of turns is measured in each session, and then the average speed in turns per minute is calculated. The speed achieved by the animal increases in each consecutive session until a plateau is reached, and the time course of the increment in speed is sensitive to changes arising from the prenatal administration of behavioral teratogens (10). Consequently, the time course of the rate in turns per minute achieved by the animal during the training period can be used as an indicator of functional teratology. Two elements are involved in the task, namely, the learning skills and motor abilities of the animals, both of which can be independently evaluated following a mathematical analysis (10). These are some reasons why the circling training and the neostriatum are a useful system for behavioral and neurophysiological studies of functional teratology. According to the classification criteria developed by Tilson for behavioral toxicology tests (11), the circling training can be classified as follows: (1) type of behavior: instrumentally conditioned; (2) quantifiable end points: motor, on-going performance, and cognitive, associative learning; and (3) behavioral testing scheme: apical test.

## Apparatus

The apparatus is built with two steel (1.8 mm thick) cylinders, 24 and 40 cm in diameter and 31 cm high, mounted concentrically on a wooden base. A schematic diagram of this apparatus is shown in Fig. 1. The area between the two cylinders, the testing trail, is 8 cm wide, 100 cm long, and is covered with nonslip rubber. The inner walls of the cylinder and the floor rubber are black and without any irregularities to create a homogeneous and undistracting testing environment. The testing device has a plastic cover with ventilation ports and an orifice in the middle of the pathway for delivering the reward. The height of the chamber is 31 cm to prevent the animal from reaching the opening through which the reward is delivered. The trail is homogeneously illuminated by external soft white light.

A liquid dispenser that delivers the reward is connected to the testing device and is filled with a 10% sucrose/water solution. The reward consists of a 50-$\mu$l drop of the sucrose/water solution. When the animal has to be rewarded, the liquid dispenser discharges a drop of the solution which always falls in the same place on the floor of the pathway (point A in Fig. 1).

Four infrared light-emitting diodes (E) are placed in the inner cylinder, and four infrared detectors (D) in the outer one are aligned with the emitter. Each pair (emitter–detector) is located 2 cm above the floor, separated by 90°, and numbered in a clockwise (right-hand) direction. The detectors and the liquid dispenser are connected to an electronic device with a timer that detects the interruption of the infrared beams, counts the number of turns, and activates the reward dispenser. The infrared beams have to be interrupted in a defined sequence to activate the reward dispenser, and the operator can select the interruption sequence to assign different rewarded turning directions (left- or right-hand). If the right-hand turn is rewarded, the reward is delivered right over beam number 4 after interrupting beams 1, 2, and 3. When beam 3 is interrupted (1/4 of the circle away from the place where the reward is delivered) the liquid dispenser is activated, and a drop falls on the floor at the rewarding place (A). When the animal reaches the reward, beam 4 is interrupted; as a result one turn is counted, and the system is reset and ready to detect the next turn. If a left-hand turn is rewarded the infrared beams have to be interrupted in the sequence 3, 2, and 1, and the reward is delivered over beam 4 after the interruption of beam 1.

The dimensions of the testing device have to be changed according to requirements of the experiment. If larger animals are tested, increasing the height of the walls is recommended to prevent the animal from reaching the roof orifice from which the reward is delivered. Also, the width of the pathway should be increased if larger animals are tested to allow them to move freely in either direction. The animal has to be free to move in any turning direction in order for learning skills and spatial discrimination to be evaluated. The width of the training pathway can be increased by enlarging the diameter of the outer cylinder; thus, the arc of the animal's body is not modified during the turning (see Speed and Posture section).

## Experimental Subjects

The rat is a good subject for functional teratology testing procedures because it is regularly used for teratological and toxicological tests, and information about the effect of a wide range of compounds is available for comparing the results. There is also the advantage that the anatomy, neurochemistry, and development of motor circuits in the rat have been well characterized. Moreover, most of the neurophysiological, neurochemical, and pharmacological studies of the motor circuit during turning behavior have previously been done in the rat.

A
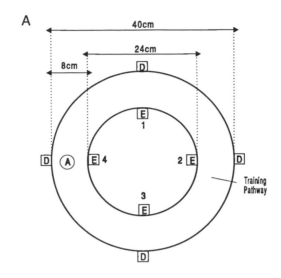

B
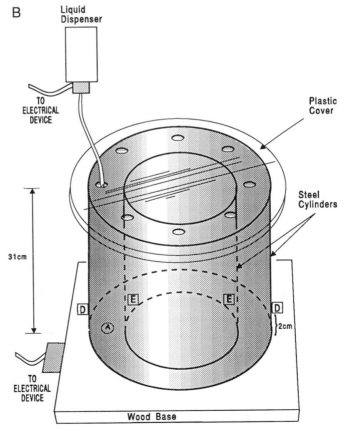

Prior to starting the test, the animal has to be weaned, and the nervous system has to be mature enough to perform learning and motor activities. The youngest rat tested was 29 days old. The rat is usually weaned at 21 days of age, and at day 29 it is mature enough to perform motor and learning tasks and to tolerate the stress of water deprivation without any detectable impairment in long-term development (10). There is no upper age limit to start the test.

The mouse can also be used as an experimental subject. The mouse has some advantages compared with the rat, such as reduced housing and feeding costs and small variation in body size during development. Many toxicological studies also use the mouse as an experimental subject.

## Procedure

### Training

The training is conducted for eight consecutive days. The animal is water deprived 24 hr before the first training session, and during the testing period water is provided only at the training session. Food is provided *ad libitum* in the home cage. The rewarded turning direction is randomly assigned to each animal before the test is started. The first training session (S0) is used to habituate the animal to the testing environment and to show the animal the rewarded turning side. This session takes 30 min and consists of rewarding the animal by successive approximation to a full turn in the assigned direction. The reward is delivered by hand using a Pasteur pipette at any location of the training pathway. During the first 10 min of the session a quarter-turn is rewarded, the next 10 min a half-turn, and the last 10 min a full turn in the direction assigned is rewarded. No data are collected from this session. After that, the animal is trained daily, one session a day, for an additional 7 days (S1 to S7). The animal is required to perform 100 complete turns in each of the first three sessions (S1 to S3) and 150 complete turns in each of the four

FIG. 1 Schematic diagram of the testing apparatus. (A) Top view of the training pathway. The dimensions of the pathway and the locations of infrared light-emitting diodes (E), infrared detectors (D), and reward place (A) are indicated. (B) Frontal view of the apparatus. The liquid dispenser and the four infrared detectors are connected to an electronic device that controls the delivery of the reward and records the turns performed in the assigned direction. Only two of the four emitter–detector pairs are shown; they are located 2 cm above the floor.

subsequent sessions (S4 to S7). When the number of turns required is reached, the animal is taken from the testing apparatus and returned to its cage. If the animal does not complete the number of turns required in a session within 30 min the session is ended, but this is very rare after session 2 or 3. If the animal does not reach the number of turns required by session 3, it is eliminated from the testing group. Records have to be kept, however, because this probably indicates severe functional impairment due to the drug treatment.

## Data Acquisition and Behavioral Analysis

In each session the data obtained comprise the total turns performed and the time employed for the session. These values are transformed into turns per minute, dividing total turns by the time employed. If spatial discrimination is evaluated, the errors made during the session are recorded (see section on Spatial Discrimination).

During the first three or four testing sessions the performance of the animal in terms of turns per minute increases very rapidly, then the increment in speed starts to decrease and eventually reaches a plateau. This is easily seen by plotting sessions against turns per minute. The first step in analyzing the data is the statistical comparison between saline- and drug-treated animals. A typical experiment is as follows. Three groups of pregnant rats are treated daily with saline [200 $\mu$l from embryonic day (E) 5 to E20], vitamin A (80,000 IU/kg/day from E11 to E14), and haloperidol (2.5 mg/kg/day from E5 to E20) (10). The date of birth of resulting offspring is considered day 0, and the test is started at 29 days. The results shown in Table I are the values (means $\pm$ SD) of turns per minute performed by male animals in each session. The statistical analyses are done with the analysis of variance (ANOVA) test using the split-plot repeated measure design (12). The results of the statistical comparisons of the entire testing period (group main effect) and of each individual session (simple main effect) between saline and drug-treated animals are indicated.

Further analyses of the circling behavior are based on the fact that the circling training is an instrumentally conditioned behavior with a motor performance as the quantifiable end point (turns per minute). However, two elements are involved in the animal's behavior, namely, the skill to learn the task and the motor capability of the animal. These two components, the associative learning process and the motor ability, can be differentiated by following a mathematical analysis based on a derivation of Herrnstein's matching law equation (13).

TABLE I   Circling Training[a]

| Session | Age (days) | Turns/min[b] C ($n = 7$) | VA ($n = 9$) | HP ($n = 9$) |
|---------|-----------|--------------------------|--------------|--------------|
| 1 | 30 | $1.1 \pm 0.4$ | $0.4 \pm 0.3$ | $0.9 \pm 0.7$ |
| 2 | 31 | $5.2 \pm 1.2$ | $1.8 \pm 1.2^d$ | $2.9 \pm 1.7$ |
| 3 | 32 | $8.4 \pm 2.4$ | $4.5 \pm 2.0^d$ | $5.7 \pm 2.2^c$ |
| 4 | 33 | $9.0 \pm 2.9$ | $6.4 \pm 1.8^c$ | $6.5 \pm 1.7^c$ |
| 5 | 34 | $10.7 \pm 4.1$ | $6.7 \pm 2.0^d$ | $7.9 \pm 1.1^c$ |
| 6 | 35 | $12.3 \pm 3.2$ | $9.2 \pm 2.8^d$ | $9.4 \pm 2.1^d$ |
| 7 | 36 | $13.5 \pm 2.9$ | $9.1 \pm 2.2^d$ | $9.2 \pm 2.2^d$ |

[a] Reprinted from J. L. Brusés, P. M. Berninsone, S. I. Ojea, and J. M. Azcurra, The circling training rat model as a behavioral teratology test, *Pharmacol. Biochem. Behav.* **38** (1989), with permission from Pergamon Press plc.

[b] Data are expressed as means ± *SD*. Statistical analysis was done with the ANOVA test using a split-plot design. Group main effect, $F(2,22) = 9.51$, $p < 0.01$. Pairwise comparisons of differences among means compared to the control group by *a posteriori* Tukey's ratio were as follows: VA, $p < 0.01$; HP, $p < 0.01$. The simple main effect was analyzed with the ANOVA test, and differences among means were tested by pairwise comparison with respect to the control group by *a posteriori* Tukey's ratio. C, Saline-treated group; VA, vitamin A-treated group; HP, haloperidol-treated group; see text for details.

[c] $p < 0.05$.

[d] $p < 0.01$.

The number of turns completed by the animal in each session is known; therefore, at any time during the testing period the amount of reinforcers administered is also known because the animal is rewarded each time it completes a turn in the assigned direction. Thus, the response rate in turns per minute ($V$) can be plotted versus the amount of reinforcers received up to that moment (accumulated reinforcers, $Ra$) (Fig. 2A). The relationship between the two components follows the equation $V = V_m \times Ra/(L + Ra)$, where $V$ is the response rate, $V_m$ is the asymptotic response rate or maximal velocity in turns per minute, $Ra$ is the accumulated reinforcers, and $L$ is the amount of reinforcers necessary to reach one-half of the maximal velocity ($V_m$).

Figure 2A shows the data from Table I plotted as the rate of response in turns per minute ($V$) versus the accumulated reinforcers ($Ra$). The symbols mark the turns per minute performed at that amount of reinforcers, and the curves were obtained by fitting the equation $V = V_m \times Ra/(L + Ra)$ to the experimental data. This equation describes a rectangular hyperbola; thus, with the known parameters ($V$ and $Ra$), the unknown parameters $L$ and $V_m$

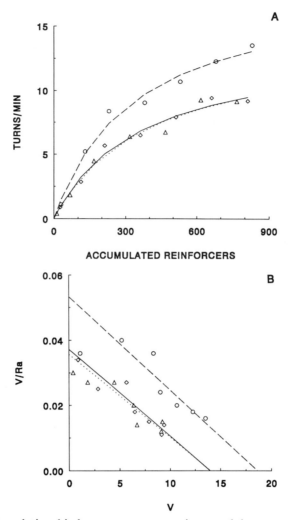

FIG. 2 (A) The relationship between turns per minute and the amount of reinforcers administered were calculated from the data of Table I and plotted (symbols). The curves were obtained by fitting the data with the equation $V = V_m \times Ra/(L + Ra)$, where $V$ is the response rate, $V_m$ is the asymptotic response rate, $Ra$ is the accumulated reinforcers, and $L$ is the accumulated reinforcers to reach one-half of the asymptotic response rate. (B) Scatchard plot of the data shown in (A). The slope of the curves reflects the magnitude of $L$, and the $x$ axis values of each group at $y = 0$ are equal to $V_m$. Symbols and lines: circles and dashed lines, saline-treated group; triangles and dotted line, vitamin A-treated group; diamonds and solid line, haloperidol-treated group (see text for details). [Reprinted from J. L. Brusés, P. M. Berninsone, S. I. Ojea, and J. M. Azcurra, The circling training rat model as a behavioral teratology test, *Pharmacol. Biochem. Behav.* **38** (1989), with permission of Pergamon Press plc.]

can be calculated.* Moreover, several data transformations can be applied to make the results graphically comparable, one of the most commonly used being the Scatchard plot (14). Figure 2B shows the Scatchard plot of the data from Fig. 2A, where the $y$ axis is the ratio of the animal response in turns per minute ($V$) to the accumulated reinforcers ($Ra$), and the $x$ axis is the animal response rate ($V$). The merit of the application of the Scatchard tranformation is that the two components of behavioral response that were difficult to distinguish in Fig. 2A are represented in two components of the straight line: (a) the theoretical maximal response of the animal ($V_m$), which is the $x$ axis value of the curve at $y = 0$, and (b) the slope ($L$), which reflects the number of reinforcers necessary to reach one-half of $V_m$. Consequently, $V_m$ represents the maximal motor capability of the animal, and $L$ is considered an indicator of the associative learning component of the behavioral response.

The usefulness of this data analysis and transformation is clearly shown in this experiment. If Fig. 2A the performance of both drug-treated groups was clearly impaired, but it is almost impossible to distinguish which component of the behavioral response (learning or motor performance) is responsible for the behavioral impairment. However, in the Scatchard plot (Fig. 2B) the slopes of the lines in the three groups have not changed. Therefore, learning skills of drug-treated animals compared with control animals have not been greatly affected, or this parameter is not sensitive enough to detect the behavioral impairment. On the other hand, $V_m$ values are much lower than control values for both drug-treated groups, indicating that in this case the most affected component of the behavior was motor performance. $V_m$ and $L$ values obtained and their asymptotic standard error (given in parentheses) for each group were as follows: control: $V_m = 18.4(1.6)$, $L = 332.3(71.7)$; vitamin A: $V_m = 12.2(1.5)$, $L = 388.8(95.0)$; and haloperidol: $V_m = 12.1(1.3)$, $L = 364.6(80.0)$. The validation of this analysis has been corroborated by testing the animals with other specific tests that have motor and learning end points such as T-maze, negative geotaxis, and surface righting (10).

## Spatial Discrimination

Spatial discrimination can be directly evaluated by recording the number of times the animal chooses the opposite turning direction to the one assigned. The operator can record this by hand, or automatic recording may be used.

---

* The form of the equation is the same as the Michaelis–Menten equation for enzyme kinetics. Thus, any computer program for enzyme kinetics is useful for the calculation of $V_m$ and $L$. In this case, $Ra$ is considered to be the substrate concentration and $V$ the velocity of the enzyme.

Two lines are marked on the floor dividing the circular trail in two semicircles to help in the evaluation of spatial discrimination. It is not frequently that the animal completes an entire turn in the wrong direction after session 2 or 3; therefore, one error is recorded each time the animal walks through one of the two lines marked on the floor in the opposite direction to the one assigned. Spatial discrimination can be tested in this behavioral paradigm because nothing is attached to the body of the animal to record turning behavior, and, therefore the animal is completely free to select the turning direction. The spatial discrimination evaluated in the circling training shows many similarities to the spatial discrimination observed in a T-maze. In this particular case, where an impairment in the animal's skill due to the effect of a drug treatment is screened, the spatial discrimination did not show high sensitivity compared with the other measurements obtained by the circling training (10).

## Extended Training

The animals reach a plateau speed in 8 consecutive days, the time period selected for the training. Although there is no limitation for extending the training period to more sessions, the weight of the animals has to be carefully controlled because physical impairments could occur owing to the water deprivation, especially when young animals are tested. In the experiment shown here, where training was started at 29 days of age and conducted for eight consecutive days, the animals recovered the weight of nondeprived animals 3 or 4 weeks after the training was ended (10). Providing water in the animal's home cage is not recommended in order to avoid the animal's weight loss. However, it is offered for defined periods, because an extra supply of water strongly affects the animal's performance at the task. Moreover, the amount of ingestion can vary from one animal to another, affecting their motivational levels.

The turning behavior remains stable for weeks after the training period is ended. Animals trained at the age of 29 to 36 days were kept in their cages with *free* access to water and food, then retrained at 60 days of age after 24 hr of water deprivation. The animals easily remember the rewarded turning direction they were assigned, and in 2 or 3 sessions they attain the same maximal speed that they reached at the end of the first training period (15). This characteristic of the circling training is particularly useful for the study of acute responses of drugs on animals treated during the perinatal period of life and trained at 30 days of age. Moreover, the long-term effects of early drug treatment can be followed.

## Speed and Posture

There is a close relationship between the arc of the animal's body during turning behavior and the increment of DA and 3,4-dihydroxyphenylacetic acid (DOPAC) concentrations in the neostriatum, contralateral to the turning direction (16). The diameter of the inner cylinder of the apparatus sets the body arc during turning. A diameter of 24 cm leads to a body arc of about 120°. If the diameter is reduced to 12 cm, the body arc is about 300°, and a maximal difference in the DA and DOPAC concentration between ipsi- and contralateral caudate nucleus and nucleus accumbens is obtained (16). The increment of DA and DOPAC in the motor circuit is also affected by the speed of the animal during rotation. The lowest speed required to induce a detectable neurophysiological change in the caudate nucleus or nucleus accumbens is 5 revolutions/min (rpm) and has to be maintained for 20 min. However, 10 rpm is an optimal speed for the detection of neurochemical concentration changes (16). This speed is reached by the animal during the last sessions of the circling training (Table I). It is worth noting that in the circling training the animal is completely free to move because nothing is attached to its body for recording turning behavior as is done in other testing devices.

## Detection Sensitivity

The ability of a behavioral test to detect variations in behavioral response as a consequence of drug treatment is referred to as the sensitivity of that test. One parameter commonly used in the evaluation of sensitivity detection is the coefficient of detection ($CD$) that estimates the percentage of change needed in the behavioral response to detect the effect of a drug (17): $CD = (t_\alpha/n^{1/2})(SD/X)100$, where $t$ is a constant at a given $\alpha$ that is derived from a two-tailed distribution table for a known group size, $n$ is the number of subjects in each experimental group, and $SD$ and $X$ are the standard deviations and means of the results of the experiments, respectively (3). To assess the behavioral sensitivity of the circling training, control animals from 10 independent experiments were analyzed to calculate the $CD$. Because the lowest $n$ used in the set of experiments evaluated was 6, we used five degrees of freedom. The resulting $CD$ for each component of the behavioral response obtained in the circling training was 25.8 and 46.6% for $V_m$ and $L$, respectively (10). Considering the low $n$ used, the $V_m$ $CD$ indicates a high sensitivity of the circling training in detection of behavioral variations.

# The Circling Training and the Neostriatum in the Study of Functional Teratology

Behavioral response is the final common pathway for the action of many brain functions. Thus, behavioral analyses are important for screening drug-induced functional brain damage because impairments in any of several brain activities can be detected through a variation of behavioral responses. In addition, since neurochemical modifications underlie the neurophysiological impairment, biochemical brain analysis can be very useful in assessing functional teratology. However, it is difficult to determine when a biochemical deviation reflects a brain function impairment. Perinatal drug treatment can affect several brain biochemical measurements such as metabolites, receptors, or neurotransmitter concentrations, but it does not mean that brain functions are damaged. In contrast, the alteration of the biochemical correlate of a physiological brain activity can more readily express the pathological basis of the biochemical modification. In this framework, the study of a specific neurochemical parameter related to a neurophysiological activity that is triggered by a behavioral response could be useful for detecting functional CNS damage.

The circling training is a behavioral conditioned test advantageous for detecting functional teratology. As an apical test for screening procedures, it is easy to perform, it has a short defined testing period (8 days), each session is no longer than 30 min, it provides objective measures, and learning and motor components of the behavioral response can be discerned.

The relationship between circling behavior and physiological activity in the caudate nucleus and nucleus accumbens makes this behavioral paradigm a useful system for studying behavioral and neurophysiological components of functional teratology. The neostriatum is an appropriate brain structure for studying physiopathological consequences of perinatal drug treatment because (1) the afferent and efferent connections of the rat neostriatum are formed during the last week of gestation (18), but the functional maturation of the synaptic contacts occur during the first weeks after birth (19) and, therefore, is vulnerable to the action of drugs during pre- and postnatal developmental periods; (2) the neostriatum is a well-defined anatomical area, it is easy to dissect, and its size provides enough tissue samples for biochemical analysis; (3) dopaminergic, cholinergic, and GABAergic systems are involved in the inputs, interconnections, and outputs of the neostriatum, and, hence, it is the target for a wide range of drugs; and finally (4) the activation of a dopaminergic system due to turning behavior appears to be a reliable marker for the function of the neostriatum and may offer a neurochemical correlation for a specific brain function.

# Acknowledgments

This work was supported by research grants from Comisión de Investigaciones Científicas de la Provincia de Buenos Aires (CIC) and Universidad de Buenos Aires. J. L. Brusés acknowledges CIC for the fellowship award. We especially thank Dr. G. R. Pilar for support during preparation of the manuscript. We also thank E. M. Zaritzky for collaboration in the design and construction of the testing device, S. I. Ojea for assistance in the mathematical analysis, K. Sommer for editorial assistance, and M. J. Spring for the design of illustrations.

# References

1. J. Werboff and J. S. Gottlieb, *Obstet. Gynecol. Surv.* **18,** 420 (1963).
2. D. F. Swaab, G. J. Boer, and M. G. P. Feenstra, *Prog. Brain Res.* **73,** 3 (1988).
3. C. V. Vorhees, *Neurotoxicol. Teratol.* **9,** 445 (1987).
4. J. Adams, *in* "Handbook of Behavioral Teratology" (E. P. Riley and C. V. Vorhees, eds.), p. 67. Plenum, New York, 1986.
5. M. A. Geyer and L. W. Reiter, *Neurobehav. Toxicol. Teratol.* **7,** 661 (1985).
6. C. J. Pycock, *Neuroscience (Oxford)* **5,** 461 (1980).
7. B. K. Yamamoto and C. R. Freed, *Nature (London)* **298,** 467 (1982).
8. M. E. Morgan, B. K. Yamamoto, and C. R. Freed, *J Neurochem.* **43,** 737 (1984).
9. M. Diana, M. Garcia-Munoz, J. Richards, and C. R. Freed, *Exp. Brain Res.* **74,** 625 (1989).
10. J. L. Brusés, P. M. Berninsone, S. I. Ojea, and J. M. Azcurra, *Pharmacol. Biochem. Behav.* **38,** 739 (1991).
11. H. A. Tilson, *Neurotoxicol. Teratol.* **9,** 427 (1987).
12. R. E. Kirk, "Experimental Design: Procedures for the Behavioral Sciences" Wadsworth, Belmont, California, 1968.
13. P. A. de Villares and R. J. Herrnstein, *Psychol. Bull.* **83,** 1131 (1976).
14. G. Scatchard, *Ann. N.Y. Acad. Sci.* **51,** 660 (1949).
15. J. L. Brusés, unpublished observations (1988).
16. C. R. Freed and B. K. Yamamoto, *Science* **229,** 62 (1985).
17. J. Buelke-Sam, C. A. Kimmel, J. Adams, C. J. Nelson, C. V. Vorhees, D. C. Wright, V. St. Omer, B. A. Korol, R. E. Butcher, M. A. Geyer, J. F. Holson, C. L. Kutscher, and M. J. Wayner, *Neurobehav. Toxicol. Teratol.* **7,** 591 (1985).
18. J. M. Lauder and F. E. Bloom, *J. Comp. Neurol.* **155,** 469 (1974).
19. J. T. Coyle and P. Campochiaro, *J. Neurochem.* **27,** 673 (1976).

## [11] Intracerebral Administration of Neuropeptides: An Assessment of Behavioral Change

Sergio Mora and Gabriela Díaz-Véliz

## Introduction

Research on the interaction of neuropeptides and the brain has grown explosively in the last 20 years, and it is widely accepted that these substances have significant influences on the regulation of mammalian behavior. The studies performed by de Wied and collaborators in the early 1970s showed, for the first time, that neuropeptides, such as adrenocorticotropin (ACTH) and related fragments, exert profound behavioral effects on animals and humans (1). Most of these studies included acquisition and extinction of both aversive and appetitive behavioral responses. Much attention in this field was focused on two important findings. The first was the demonstration of the existence of several hypothalamic releasing and inhibiting hormones in regions of the brain outside the hypothalamus (2). The second was the discovery of the enkephalins (3). Since then many neuropeptides, both natural and synthetic, have been studied in a variety of behavioral paradigms including avoidance conditioning (4), approach behavior (5), reversal learning of discrimination tasks (6), memory retrieval (7), and sexually motivated behavior (8).

## Peripheral versus Intracerebral Administration

The study of the behavioral effects of neuropeptides presents some methodological problems. One problem is the route of administration. No doubt exists that peripherally administered peptides are able to exert significant alterations in the central nervous system (CNS), especially in relation to learning and memory of some avoidance behaviors (9). For instance, Leu-enkephalin and $\beta$-endorphin administered intraperitoneally cause complete retrograde amnesia for a shuttle avoidance behavior (10); enkephalins and $ACTH_{4-10}$ administered intraperitoneally impair acquisition of a one-way active avoidance response (11) in rats; peripherally administered vasopressin facilitates retention of active and passive avoidance responses

*Methods in Neurosciences, Volume 14*

(12); gonadotropin-releasing hormone (GnRH), after subcutaneous injection, impairs the acquisition of an active avoidance response (13) but improves the retention of both active and passive conditioning (14). Nevertheless, experiments of this nature must be interpreted with caution for at least two reasons. First, it is difficult to differentiate between the direct central effects and the multiple peripheral actions of a peptide that can secondarily affect brain function by stimulation of peripheral receptors and/or metabolic peripheral changes. Second, only a fraction of systemically injected peptides reaches the brain by virtue of their rapid peripheral metabolism; in addition, the free passage of peptides across the blood–brain barrier (BBB) is restricted.

The enzymes of living organisms metabolize peptides effectively and quickly. For example, after intravenous administration to rats, ACTH has a half-life ($t_{1/2}$) of less than 1.5 min, whereas $ACTH_{4-9}$ has a $t_{1/2}$ of 5 to 10 min (15, 16) and GnRH a $t_{1/2}$ of 4 min (17). The fundamental problem of whether peptides can penetrate the BBB has been extensively studied. Moreover, some studies show conflicting data. A study of the penetration of $ACTH_{4-9}$ into the brain showed that 5 min after intravenous administration approximately two-thousandths of the dose was present in the whole brain of a rat (18). Other studies could not demonstrate radioactivity inside the BBB after intracarotid injection of $^{125}I$-labeled lysine-vasopressin, whereas relatively high doses of tritiated oxytocin (100 $\mu$g/rat) were needed to observe radioactive label inside the BBB (19). Penetration of radioactive enkephalins and other peptides measured after intracarotid injection showed that their extraction by the brain was very low (2–3%) and similar to that of neurotransmitters, such as monoamines and acetylcholine (20). In contrast, other investigators have reported brain extractions of 15 and 20% for Met-enkephalin (21) and $\beta$-endorphin (22), respectively, concluding that they readily cross the BBB.

Central application of neuropeptides circumvents some of these problems. Only by this means could the actual potency and sites of action of the neuropeptides be determined. It is clear that peptides must be extremely active in order for behavioral effects to be seen after peripheral administration; however, the conditions for showing such effects are not always favorable and are often difficult to reproduce. The ability to mimic, with much lower doses, by intracerebral application the behavioral effects produced by peripheral administration argues strongly for a direct CNS effect and against an indirect effect via peripherally evoked reflexes. The dosage must be of the order of what can be expected to reach the brain after systemic administration. When the effect is seen only after application of the substance to a particular brain region and not to other areas, it suggests that the activity is localized to that area. Like other substances, neuropeptides have reasonable specificity when given in low doses, whereas higher doses may exert nonspecific actions on other receptors. For that reason it is important to investigate a wide dose range and to control the experimental conditions rigidly. With careful experimental technique, dose–response relationships can be determined. Hypotheses about the putative physiological functions of neuropeptides should be based only on effects seen at doses consistent with physiological concentrations.

## Behavioral Effects of Neuropeptides

The microinjection technique has been found particularly useful to study the involvement of neuropeptides in producing a given behavioral response. For example, applying neuropeptides into the cerebrospinal fluid (CSF) or directly into a brain region can elicit a behavioral response. ACTH, after intracerebroventricular injection, induces excessive grooming, stretching, and yawning (23). Intracerebroventricular application of GnRH (24) induces hypermotility and increases head shakes and grooming. These effects are not evident when similar doses of GnRH are injected into the striatum or hippocampus. Nevertheless, grooming, gnawing, and head shakes are increased when GnRH is injected into the mesencephalic periaqueductal gray substance (25).

Neuropeptides can also modulate a response evoked by environmental stimuli: β-endorphins injected into the CSF or periaqueductal gray matter results in antinociception (26); injections into the nucleus accumbens of γ-endorphins attenuates passive avoidance behavior (27); thyrotropin-releasing hormome (TRH) (28) and GnRH (24), administered intraventricularly in the rat, induce opposite effects on the acquisition of conditioned avoidance responses. Studies such as these can reveal the site of action of neuropeptides in the brain for eliciting a certain behavioral effect, or they may allow the different behavioral effects of a given neuropeptide to be understood.

The apparatus and procedures described below were used in our laboratory for the study of the behavioral effects of GnRH after its direct infusion into the rat brain.

## Subjects

The studies have been carried out using adult male rats (Sprague-Dawley, Wistar, or similar strain), weighing $200 \pm 20$ g at the time of surgery. Rats are housed in groups of less than 6 animals per cage under standard conditions in a temperature-controlled (22°C) vivarium on a 12-hr light/dark schedule with dark onset at 2000 hr; the rats receive *ad libitum* access to food and tap water.

## Surgical and Stereotaxic Methods

Animals can be anesthesized intraperitoneally with sodium pentobarbital (50 mg/kg) or ketamine (100 mg/kg). Although the entire surgical procedure should be aseptic, it is advisable to administer a prophylactic dose of an antimicrobial agent.

The external guide cannulae are constructed from 23-gauge stainless steel tubing and are permanently implanted in the brain by immobilizing the head of the rat in an appropriate stereotaxic apparatus. Each cannula is introduced into a certain brain structure, the position of which is determined using a system of three coordinates indicating its distance from the point of intersection of the three zero planes or from a certain point on the skull (bregma, lambda). In the stereotaxic atlases for the rat (29) the basal plane is defined as that perpendicular to the sagittal plane and passing through the bregma and a point 1 mm above lambda. The sagittal plane passes through the head medially in the region of the sagittal suture; the frontal plane is given by the line joining both auditory meati.

The head of the animal is fixed in the stereotaxic head holder by bars introduced into the external meati and by bars or clamps for the upper jaw, placing the head symmetrically and firmly in a position corresponding to the basal plane. Coordinates of the zero point are established after cutting the skin and cleaning the skull bones, and the sites of trepanation are marked on the skull surface with reference to that point. Trephine openings are drilled with a dental drill or a hand trephine at the sites marked, the dura mater is pierced with a sharp needle, and the cannula is introduced and fixed to the skull with dental acrylate. Stainless steel stylets, made from closed 30-gauge tubing, protect the guide cannulae when the animals are not being injected intracerebrally.

## Intracerebral Administration

A recovery period of at least 7 days after surgery is needed before behavioral testing begins. The intracerebral application is made by injecting the drug solution through an internal cannula constructed using 30-gauge stainless steel tubing which is inserted with its tip terminating 1 mm below the guide cannulae. The internal cannula is connected via polyethylene tubing to a Hamilton microsyringe mounted on a syringe pump. Neuropeptides are dissolved in sterile saline and injected in a volume of 0.5 $\mu$l at a constant rate of 1 $\mu$l/min, and the cannula is left in place for an additional 30 sec after completion of the injection to allow for diffusion away from the cannula tip.

## Histology

If a guide cannula is to be implanted into a certain brain structure, it is of advantage to use the stereotaxic atlas only as a first approximation and to determine its precise localization by subsequent histological examination. After behavioral experiments, the rat receives a volume of methylene blue or other appropriate stain infused through the cannula. Then the animal is administered an overdose of sodium pentobarbital, is perfused intracardially with saline followed by 10% formalin, and is decapitated. The brain is carefully removed from the skull and placed into 10% formalin for at least 24 hr. The location of the cannula tip is verified in serial sections with respect to the known stereotaxic coordinates.

## Behavioral Testing

There are various methods to assess the behavioral activity of neuropeptides, and they are not different from those used to evaluate other psychotropic drugs. Assay methods which give a graded dose–response relationship, allowing for the estimation of an ED50, are necessary. The evaluation should not be based on only one test, but on a battery of pharmacological tests, including spontaneous and conditioned behaviors.

### Spontaneous Motor Activity

The subjective observation of drug-induced changes in the spontaneous activity of laboratory animals has long been a preliminary procedure in drug testing. Because the observable behavioral patterns are under the influence of innumerable internal and external factors, the experimental conditions must be carefully controlled. There are many variables that must be regulated before the observational test can be used efficiently. Perhaps most important is the subject of the experiment. The health of the animals should be a primary concern, and the experiment should be designed to ensure a minimum of stress. The investigator should determine the optimum set of experimental conditions for the laboratory, since external influences will vary from one laboratory to another. The initial standardization of the environment of the observation area (temperature, light, humidity, etc.) should not be altered during the experiment in order to prevent behavioral changes from occurring. Also, the influence of unexpected external stimuli, such as noises, that can disturb the effects of drugs on behavior must be considered. The observer must be a highly trained, attentive, and patient individual, and must be fully cognizant of the animal's normal behavior and totally aware of the possible changes in behavior that a drug can produce.

The number of observational test procedures involving rats varies greatly depending on the need of each laboratory. A simple measure of motor activity that has been used to test emotionality is the open field. Classically, the animals are placed at the center of a circular or rectangular open area, which has been suitably marked off, and the number of times that an animal moves across the zone lines is recorded. In addition, other behavioral signs such as grooming, sniffing, lying, and rearing can be checked simultaneously. The frequency and duration of some behavioral signs are easily measured; others such as stereotypy, tremor, piloerection, and posture must be evaluated by means of arbitrary score tables.

## Apparatus

Spontaneous motor activity can be recorded automatically by using "activity monitors." Many of these devices have been used to study the effects of psychoactive drugs, including neuropeptides. They are composed of an activity recording unit connected to an appropriate digital counter, which operates by means of photocells (30), piezoelectric devices (31), or tilt cages (32). Our experiments are performed using an activity platform (Lafayette Instrument Co., Lafayette, Indiana, Model 86010) which operates on a floor displacement principle. The transducer consists of a permanent magnet suspended within a coil that transduces normal floor vibrations which occur during motor activity into electrical signals. The amplifier has a relay output for an impulse counter, and it is provided with a threshold control for varying the sensitivity of the platform.

## Procedure

The animals are individually placed in a plexiglass cage (30 × 30 × 30 cm) supported on the platform (Fig. 1). To avoid disturbing noises, the activity monitor is placed into a wooden soundproof chamber (60 × 60 × 60 cm), and behaviors are monitored continuously until the end of the session by means of a video camera connected to a tape recorder. Scores are made from the live observations, and the video sequences are used for subsequent reanalyses. Spontaneous motor activity is recorded every 5 min for a period of 15 or 30 min, and, simultaneously, the following responses are also registered:

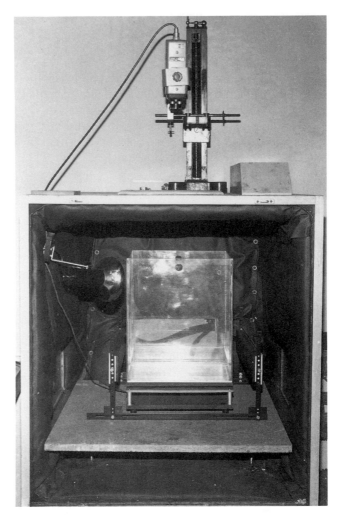

FIG. 1   Activity monitor. The system is placed into a soundproof chamber, and the behavioral observations are made via a video camera.

number of times each animal rears, number of head shakes, and the time (seconds) spent in grooming behavior. By means of this method we have demonstrated that GnRH infused directly into one of the lateral ventricles of the rat brain induces significant enhancement in the four motor behaviors studied (Fig. 2).

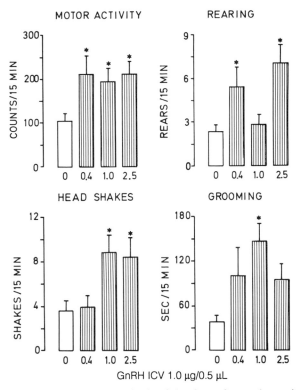

FIG. 2   Effects of intracerebroventricular injection of gonadotropin-releasing hormone (GnRH) on rat spontaneous motor activity. Each rat received either saline (0.5 μl) or GnRH (0.4, 1.0, or 2.5 μg/0.5 μl). Behavioral testing was initiated immediately after injection. Each bar represents the mean ± SEM of motility counts, number of rears, number of head shakes, or the time in seconds spent in grooming during a 15-min period of observation. Statistical comparisons were made by using one-way analysis of variance (ANOVA) followed by Dunnett's test (asterisks denote significant differences from the saline group, $p < 0.05$). Eight or nine animals were in each group. GnRH increased the four motor behaviors studied, but no dose–response relationship was observed. (Reprinted with permission from Ref. 24, Pergamon Press Ltd., Oxford, England.)

## Conditioned Behavior

Acquisition and extinction of conditioned avoidance behaviors (CARs) have proved to be very useful for the study of CNS effects of neuropeptides. Evidence indicates that the training and testing of animals for this kind of behavior is accompanied by the release of peptide hormones such as ACTH

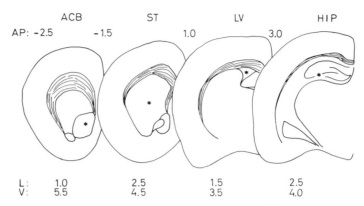

FIG. 3   Locations of cannula tips for the infusion of GnRH into the nucleus accumbens (ACB), striatum (ST), lateral ventricle (LV), or hippocampus (HIP). The coordinates used were derived from Fifková and Marsala (29).

and vasopressin, (33, 34). These hormones and their fragments may modulate the acquisition or maintenance of the behavioral response.

### Apparatus

In our studies on the influence of GnRH on avoidance behavior, a two-compartment conditioning box (shuttle box) has been used. Rats have to shuttle from one compartment to the other by crossing a door in response to a buzzer. When the rat does not cross in time, it receives an electric foot shock through the grid floor. The apparatus (Lafayette Instrument Co., Model 85103) is composed of two stainless steel modular testing units. Each unit is equipped with a one-way mirror, an 18-bar insulated shock grid floor, two 28 V dc lights, and a tone generator (Mallory Sonalert, 2800 Hz). Electric foot shocks are provided to the grid floor by a Master Shock Supply (Lafayette Instrument Co., Model 82400).

### Procedure

The guide cannulae are stereotaxically implanted into the lateral ventricle, following the coordinates derived from Fifková and Marsala (29) (+1 mm with respect to bregma; 1.5 mm lateral to bregma; 3.5 mm below the surface of the skull), into the right striatum (AP −1.5, L +2.5, V −4.5 mm), into the right hippocampus (AP +3, L +2.5, V −4.0 mm), or into the right nucleus accumbens (AP −2.5, L −1.0, V −5.5 mm). Cannula tip placements used in the experiments are shown in Fig. 3. After intracerebral neuropeptide

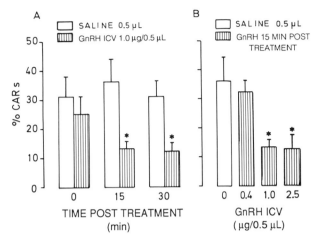

FIG. 4 Effects of intracerebroventricular injection of GnRH on the acquisition of conditioned avoidance responses (CARs): influence of the time of treatment (A) and dose (B). (A) Each rat received either saline (0.5 $\mu$l) or GnRH (1.0 $\mu$g/0.5 $\mu$l). The behavioral test was applied at 0, 15, or 30 min after injection. Each rat received saline or GnRH (0.4, 1.0, or 2.5 $\mu$g/$\mu$l), and the behavioral test was performed 15 min later. Bars represent the means ± SEM of the percentage of CARs out of 50 trials. Statistical comparisons were made by using Student's $t$-test (A) and one-way ANOVA followed by Dunnett's test (B) (asterisks denote significant differences from the saline group, $p < 0.01$). Eight or nine animals were in each group. GnRH impaired the response when injected at least 15 min before the beginning of the test. There is no dose–response relationship. (Reprinted with permission from Ref. 24, Pergamon Press Ltd., Oxford, England.)

infusion, the animals are individually placed in the shuttle box; after a 5-min period of habituation, they are trained over 50 trials. Each trial consists of the presentation of a tone which after 5 sec is overlapped with a 0.20–0.30 mA foot shock until the animal escapes to the opposite chamber. The intertone interval is 30 sec. A conditioned avoidance response is defined as a crossing within the first 5 sec.

The effects of the intracerebroventricular administration of GnRH on the acquisition of CARs is shown in Fig. 4. GnRH (1 and 2.5 $\mu$g/rat) impairs acquisition of CARs. The effect of GnRH (1 $\mu$g/rat) is significant when it is administered either 15 or 30 min before the beginning of the test, but not when it is administered immediately before the test. Figure 5 shows that 15 min after the administration of GnRH (1 $\mu$g/rat) into the striatum, hippocampus, or nucleus accumbens there is a significant impairment in conditioning. Only the infusion into the striatum is able to induce an immediate effect on the acquisition of CARs, leading to the

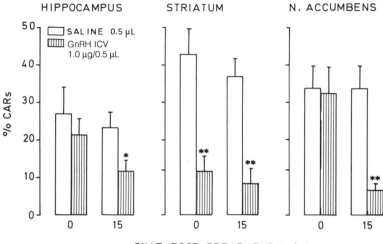

FIG. 5   Effects of GnRH infusion into hippocampus, striatum, and nucleus accumbens on the acquisition of CARs. Each rat received either saline (0.5 μl) or GnRH (1 μg/0.5 μl), and behavioral testings were applied immediately or 15 min after injection. Bars represent the means ± SEM of the percentage of CARs out of 50 trials. Statistical comparisons were made by using Student's *t*-test for independent groups (single asterisk, $p < 0.025$; double asterisk, $p < 0.0005$). Eight to ten animals were in each group. Only infusion into the striatum induced an immediate effect on conditioning. (Reprinted with permission from Ref. 24, Pergamon Press Ltd., Oxford, England.)

suggestion that this nucleus could be one of the primary sites of GnRH action in the CNS.

We decided to study the effects of a wider range of doses injected into the striatum in order to establish a reliable dose–response relationship and to determine the real potency of the neuropeptide. The results of this study are shown in Fig. 6. There is a good correlation with the dose in the range of 3.9 to 62.5 ng/rat; the ED50 is estimated to be approximately 23.5 ng/rat, and the maximal effect is observed after 62.5 ng/rat. We have reported previously (35) that GnRH (20 μg/rat s.c., a dose 320 times higher) is needed to obtain a similar behavioral effect. In conclusion, this evidence indicates that GnRH exerts a very potent effect on the acquisition of CARs, that the striatum could be the site of this action, and that the blood–brain barrier inhibits the effect of peripherally administered doses of this neuropeptide. Additional research is needed to establish the mechanism of action of GnRH on behavior and the physiological role of this action.

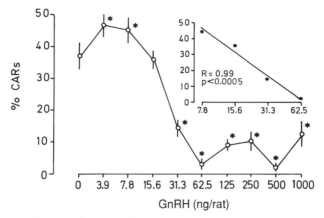

FIG. 6   Effects of intrastriatal administration of GnRH on the acquisition of CARs: dose–response relationship. Each animal received an intrastriatal infusion of saline (0.5 μl) or one of the following doses of GnRH: 3.9, 7.8, 15.6, 31.3, 62.5, 125, 250, 500, and 1000 ng/0.5 μl. The behavioral test began immediately after treatment. Each point of the principal curve represents the mean ± SEM of the percentages of CARs out of 50 trials. Comparisons were made by using one-way ANOVA followed by Dunnett's test (asterisks denote significant differences from the saline group, $p < 0.05$). The inset represents the dose–response analysis in a selected range of doses, assessed by Pearson's correlation test. Nine to eleven animals were in each group. (Reprinted with permission from Ref. 24, Pergamon Press Ltd., Oxford, England.)

## Acknowledgments

This work was supported by Grant 1084/89 from FONDECYT, Chile.

## References

1. D. de Wied, in "Frontiers in Neuroendocrinology" (W. F. Ganong and L. Martini, eds.), p. 97. Oxford Univ. Press, New York, 1969.
2. M. J. Brownstein, in "Peptides in Neurobiology" (H. Gainer, ed.), p. 295. Plenum, New York, 1977.
3. J. Hughes, T. W. Smith, H. W. Kosterlitz, L. A. Fothergill, B. A. Morgan, and H. R. Morris, Nature (London) 258, 577 (1975).
4. Tj. B. van Wimersma Greidanus, in "Central Nervous System Effects of Hypothalamic Hormones and Other Peptides" (R. Collu, A. Barbeau, J. R. Ducharme, and J. G. Rochefort, eds.), p. 177. Raven, New York, 1979.
5. J. T. Martin, in "Endogenous Peptides and Learning and Memory Processes"

(J. L. Martinez, Jr., R. A. Jensen, R. B. Messing, H. Rigter, and J. L. McGaugh, eds.), p. 99. Academic Press, New York, 1981.

6. C. A. Sandman, W. D. Alexander, and A. J. Kastin, *Physiol. Behav.* **11,** 613 (1973).

7. H. Rigter, H. van Riezen, and D. de Wied, *Physiol. Behav.* **13,** 381 (1974).

8. B. Bohus, H. H. L. Hendrick, A. A. van Kolfschoten, and T. G. Krediet, *J. Endocrinol.* **64,** 37p (1975).

9. J. W. van Nispen and H. M. Greven, *Pharmacol. Ther.* **16,** 67 (1982).

10. I. Izquierdo, A. C. M. Paiva, and E. Elisabetsky, *Behav. Neural. Biol.* **28,** 246 (1980).

11. H. Rigter, T. J. Hannan, R. B. Messing, J. L. Martinez, Jr., J. L. Vasquez, R. A. Jensen, J. Veliquette, and J. L. McGaugh, *Life Sci.* **26,** 337 (1980).

12. G. L. Kovacs, B. Bohus, and D. H. G. Versteeg, *Neuroscience (Oxford)* **4,** 1529 (1979).

13. S. Mora, A. G. Nasello, M. Mandelli-Lopes, and G. Díaz-Véliz, *Physiol. Behav.* **30,** 19 (1983).

14. S. Mora and G. Díaz-Véliz, *Psychopharmacology (Berlin)* **85,** 315 (1985).

15. J. Verhoef and A. Witter, *Proc. 17th Dutch Fed. Meet. Med. Biol. Soc., Abstr. Commun.,* p. 397 (1976).

16. A. Witter, H. M. Greven, and D. de Wied, *J. Pharmacol. Exp. Ther.* **5,** 853 (1975).

17. M. D. Mauk, G. A. Olson, A. J. Kastin, and R. D. Olson, *Neurosci. Biobehav. Rev.* **4,** 1 (1980).

18. H. van Riezen, H. Rigter, and H. M. Greven, *in* "Neuropeptide Influences on the Brain and Behavior" (L. H. Miller, C. A. Sandman, and A. J. Kastin, eds.), p. 11. Raven, New York, 1977.

19. H. J. Ruhle and A. Ermisch, *in* "Neurosecretion and Neuroendocrine Activity" (W. Bargmann, A. Oksche, A. Polenov, and B. Scharrer, eds.), p. 275. Springer-Verlag, Berlin, 1978.

20. E. M. Cornford, L. D. Braun, P. D. Crane, and W. H. Oldendorf, *Endocrinology (Baltimore)* **103,** 1297 (1978).

21. A. J. Kastin, C. Nissen, A. V. Schally, and D. H. Coy, *Brain Res. Bull.* **1,** 583 (1976).

22. I. Izquierdo, D. O. Souza, M. A. Carrasco, R. D. Dias, M. L. Perry, S. Eisinger, E. Elisabetsky, and D. A. Vendite, *Psychopharmacology (Berlin)* **70,** 173 (1980).

23. W. H. Gispen and R. L. Isaacson, *Pharmacol. Ther.* **12,** 209 (1981).

24. S. Mora, A. Afani, R. Kusanovic, C. Tapia, and G. Díaz-Véliz, *Pharmacol. Biochem. Behav.* **38,** 705 (1991).

25. P. A. Gargiulo and A. O. Donoso, *Pharmacol. Biochem. Behav.* **32,** 853 (1989).

26. L. F. Tseng, E. T. Wei, H. H. Loh, and C. H. Li, *J. Pharmacol. Exp. Ther.* **214,** 328 (1980).

27. G. L. Kovacs, G. Telegdy, and D. de Wied, *Neuropharmacology* **21,** 451 (1982).

28. S. Mora, A. G. Nasello, and L. Fieschi, *Pharmacol. Biochem. Behav.* **13,** 137 (1980).

29. E. Fifková and J. Marsala, *in* "Electrophysiological Methods in Biological Re-

search'' (J. Bures, M. Petrán, and J. Zachar, eds.), p. 653. Academic Press, New York, 1967.

30. L. S. Wright, Jr., H. J. Horn, and G. Woodard, *Fed. Proc.* **21,** 420 (1962).
31. E. Frommel, *Arch. Int. Pharmacodyn.* **154,** 231 (1965).
32. J. W. Kissel, *Science* **139,** 1224 (1963).
33. F. Laczi, O. Gaffori, E. R. de Kloet, and D. de Wied, *Brain. Res.* **280,** 309 (1983).
34. P. E. Gold and R. L. Delanoy, *in* "Endogenous Peptides and Learning and Memory Processes" (J. Martinez, R. A. Jensen, R. B. Messing, H. Rigter, and J. L. McGaugh, eds.), p. 79. Academic Press, New York, 1981.
35. S. Mora, F. Caro, M. G. Cárdenas, M. Espinoza, and G. Díaz-Véliz, *IRCS Med. Sci.* **11,** 1108 (1983).

# [12] Measurement of the Effects of Intrathecal Administration of Neuropeptides on Motor Behavior

Kevin C. F. Fone

## I. Introduction

Whether neuropeptides are neurotransmitters or neuromodulators has been the subject of considerable research effort. Certain prerequisite criteria for a neurotransmitter, such as the presence in specific neurons of precursors, synthetic enzymes, and the neurotransmitter itself, of specific binding sites, and mechanisms for deactivation or removal, have been verified for a number of neuropeptides. Furthermore, the use of intracranial dialysis (1, 2) and antibody microprobes (3) is beginning to provide evidence for the release of several neuropeptides in specific brain and spinal cord regions. However, obtaining definitive proof of the function of neuropeptides in normal behavior is proving to be a more elusive problem. The identification of neuropeptide-containing neurons within specific neuronal pathways using immunohisto-chemistry and the qualitative distribution of corresponding ligand binding sites usually provide the first indication of which behavioral functions may be modulated by a particular neuropeptide (4). Many neuropeptides are known to coexist with a more "classical" neurotransmitter, but they are differentially stored in large and not small vesicles. The large, dense-core vesicles may only be released from nerve terminals following high-frequency neuronal activation and/or in conditions such as neuronal damage or target organ injury, to alter the effectiveness of the "classical" neurotransmitter. In such situations any resultant response might be expected to be slow in onset but prolonged in nature (5). These theoretical aspects of neuropeptide modulation should be borne in mind when designing behavioral (or other) paradigms to investigate functional roles.

Because many neuropeptides (a) poorly cross the blood–brain barrier, (b) are rapidly metabolized by plasma enzymes into both active and inactive metabolites, and (c) would be prohibitively expensive to deliver by a systemic route, direct intracerebroventricular or intrathecal administration, which also offers the advantage of allowing a more discrete locus of action to be determined, is particularly advantageous. This chapter illustrates how the intrathecal administration technique can be utilized to examine the spinal motor

*Methods in Neurosciences, Volume 14*

behavioral function of neuropeptides, although the same approach can be, and has been, used to assess autonomic and sensory roles of neuropeptides. The term motor is used here in the broadest sense to include all unconditioned movements and not just locomotion.

## II. Intrathecal Injection Techniques

### A. *Methods of Intrathecal Injection*

Intrathecal injection of drugs can be performed either directly by lumbar puncture or with indwelling intrathecal cannula. These techniques have been utilized most frequently in adult mice and rats but have also been documented in several other experimental species including rabbits (6), guinea pigs (7), cats (8), and sheep (9). Intrathecal injection by transcutaneous lumbar puncture into the L4–5 region has been reported in mice and rats either anesthetized with ether or unanesthetized (10). Lumbar puncture offers the advantage of delivering small drug volumes to a very localized region of the cord, but behavioral observations are severely compromised by the use of anesthetic and/or the stress associated with the technique. Papir-Kricheli *et al.* (11) have recently reported, however, that the behavioral effect of two substance P analogs was comparable irrespective of whether lumbar puncture or indwelling cannulae were used for drug delivery, although the first 5 min of the behavioral response was not monitored owing to the effects of the anesthetic. Indwelling intrathecal cannulae can be readily implanted in adult rats (250 to 350 g) in a period of 20 min using a suitable anesthetic, such as sodium methohexitone (60 mg/kg i.p.) or halothane, followed by treatment with a postoperative analgesic (e.g., buprenorphine 30 $\mu$g i.m.). Intrathecal implantation has been described in some detail elsewhere (8, 12), but the cannula construction and important methodological points are reiterated here.

Indwelling intrathecal cannulae for rats (Fig. 1) are constructed in four steps in the workers' laboratory. First a sleeve of translucent vinyl tubing (20 mm long, 1.4 mm OD, 0.63 mm ID) is placed over a 23-gauge hypodermic needle (this will form the exteriorized portion of the implanted cannula). Second, the nylon tube (0.63 mm OD, 0.5 mm ID) used for injection is attached to the needle and lightly coated in epoxy resin adhesive. The vinyl sleeve is slowly rotated completely over the nylon tubing, which is then cut to the appropriate length. Finally a collar of epoxy resin (used to secure the implanted cannula) is applied to the junction of the two tubes and allowed to dry (24 hr) before coating with cyanoacrylate adhesive and cutting the needle to form a stilette (Fig. 1).

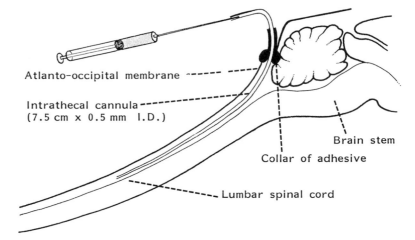

Atlanto-occipital membrane

Intrathecal cannula
(7.5 cm x 0.5 mm I.D.)

Brain stem

Collar of adhesive

Lumbar spinal cord

FIG. 1   Diagrammatic representation of an indwelling intrathecal cannula implanted along the subarachnoid space.

For implantation, the animal is placed in a raised stereotaxic frame, so that the spinal cord can, at least temporarily, be moved into the vertical plane and the cranium tilted forward without interfering with respiration. A small midline rostrocaudal incision is made, and the musculature overlying the occipital bone (from lamboid suture ridge to atlas) is retracted laterally. A sterile hypodermic needle is used to produce a 1.5 mm lateral slit in the atlantooccipital membrane, which is then held open with watchmakers' forceps. Gentle pressure is applied to straighten the cervical arch while the intrathecal cannula, freshly primed with sterile vehicle, is slowly advanced with gentle ±90° rotation. At the sign of any resistance the cannula is withdrawn 1–2 mm, rotated, and advanced at the new angle. Once in position, a small piece of collagen hemostatic pad is applied to prevent further loss of cerebrospinal fluid, and the cannula is permanently secured with dental cement or by sutures to the surrounding muscle. The site of implantation is treated with a plastic dressing and a tribiotic spray to prevent infection.

Following implantation animals should be monitored until they regain consciousness and locomotion, at which time any signs of motor impairment will be readily visible. Animals showing normal motor coordination then require individual housing to prevent destruction of the implant, and they need to be weighed and monitored for locomotor performance daily. With practice at least 75% success should be attained, but any increase in cannula diameter will dramatically affect this criterion. Most workers allow 3 to 5 days of recovery before further experimentation, but 7 days of recovery, used in the current work, ensures return to weight gain. With careful handling (particularly when removing stilettes) such cannulae remain patent for a month, typically allowing eight to ten injections each at 3- to 4-day intervals.

## B. Injection Media

Sterile saline or artificial cerebrospinal fluid (CSF, pH 7.4) is the vehicle medium of first choice, as with other techniques, and many neuropeptides will dissolve in either, especially if the pH is manipulated. Neither saline nor CSF produces any notable behavioral response. The rapid degradation of neuropeptides has lead to the development of a variety of vehicles to facilitate slow release. In particular, neuropeptides can be administered as a complex with 2-hydroxypropyl-$\beta$-cyclodextrin (20–40% CDEX) (13), and hydrophobic neuropeptides can be given in the lipid iophendylate (6).

## C. Injection Protocol

All intrathecal injections should be of sufficient volume to ensure that the drug is expelled from the cannula but should be sufficiently small to prevent any marked increase in intracranial pressure and to minimize the spread of injectate (see Section V,B). The volume of injectate delivered will depend on the solubility of the neuropeptide and the length (dead space) of the cannula. Typically, for a 7.5 cm cannula the neuropeptide solution (5 to 10 $\mu$l) can be washed in with 15 $\mu$l vehicle. In contrast to intracerebroventricular injections, drugs should be rapidly washed through the intrathecal cannula with vehicle to ensure complete delivery to the subarachnoid space. Intrathecal injections are best performed while lightly restraining the rat in a Perspex tube (25 cm long, 5 mm thick with a 5.6 cm internal diameter and a 1.5 cm wide notch cut along the upper edge to allow assess to the cannula, supported on two Perspex strips and occluded at one end by two vertical bars). Rats (up to 350 g) will freely enter restraining tubes after a training visit, and such tubes minimize both the risk of damaging the cannula during injection and the possibility of subsequent behavioral response to the handling procedure. Other restraining devices have been described for other species, including rabbits (6).

For chronic studies of the long-term effects of neuropeptides, or to assess whether behavioral tolerance occurs, either a repeated dosage regimen or continuous intrathecal infusion can be employed. Repeated intrathecal administration of as much as 35 $\mu$l/hr of saline for 2 days fails to produce any notable histopathological change in the spinal cord (14). Alternatively osmotic minipumps allow neuropeptides and the corresponding antibodies to be continuously infused; 1 $\mu$l/hr is the rate of CSF production in the rat, so all smaller rates are suitable. Suitably primed osmotic minipumps (prewarmed to 37°C for 5–8 hr in a water bath) can be implanted subcutaneously over the scapula under anesthesia and connected by a short length of exteriorized polyethylene tubing and a portion of a 23-gauge hypodermic needle to the indwelling intrathecal cannula. Continuous infusion can be interrupted for intrathecal injection of test drugs provided that the dead space of the cannula is immediately reprimed with the contents. The viability of intrathecal delivery and bioactivity of the pump contents are confirmed by placing the minipump in a water bath at 37°C and testing the effluent by radioimmunoassay.

## III. Measurement of Motor Behavior

### A. *Measurement of Locomotor Activity*

An enormous variety of methods has been utilized to assess locomotor activity, but these fall into one of two categories according to whether they require mechanical recording or direct human observation. An objective mechanical record of locomotion can be obtained by (a) photocell cages (counting crossings through an infrared beam), (b) Doppler shift radar (assessing different frequency movements), (c) running wheels (counting revolutions), (d) stabilimeters (monitoring floor movement), (e) open fields (counting floor square crossings), or (f) electromyographic activity [fast Fourier transform analysis of recordings from intramuscular electrodes to assess tremor (15)]; however, these methods give little or no information on the different components of the motor behavior recorded. The respective advantages and limitations of most of the mechanical devices used to measure locomotion are documented in an excellent review by Robbins (16).

Modifications have been made to several of these techniques, such as inclusion of two levels of photocell beams to differentiate rears from horizontal cage crossings, and, in so-called behavioral pattern monitors, nose pokes into holes in the floor and walls of the cage are also automatically monitored and a computer used to determine the spatial pattern of locomotor activity (17). More sophisticated stabilimeters have also been introduced to minimize the problem of momentum causing repetitive artifacts and to ascertain information on stride length and pattern (18). Doppler shift radar actomats (19) also allow separate automatic recording of low-frequency locomotion and higher frequency activity such as grooming and tremor, thus distinguishing between normal exploratory and stereotype behavior. All the mechanical devices offer the advantage of automation, reproducibility, and sensitivity and allow several animals to be examined simultaneously, limited only by the availability of sufficient sets of equipment. However, even the simplest of motor behaviors cannot operate in isolation. Indeed, some behaviors are mutually exclusive (a rat cannot groom when performing wet-dog shakes). Hence, it is particularly important that observation is made along with mechanical records of motor behavior, to ensure that new behaviors do not appear with increasing dose, which would compete with expression of the chosen motor response.

Although mechanical methods are useful for assessing generalized locomotor activity, more complex stereotype behaviors (purposeless and repetitive fragments of normal behavior) are best assessed by direct observation, with each different motor component being independently timed or scored. In the ideal situation direct observation would be used to examine the effect of the neuropeptide on the entire normal behavioral repertoire. In reality, several predominant components of the whole motor behavioral syndrome are measured independently. The behavior monitored needs to be reproducible and have definitive end points to be readily quantifiable. Pilot studies using the particular neuropeptide over a wide dose range, starting with nanogram doses (rather than low microgram doses often required with more "classical" neurotransmitters, probably reflecting respective endogenous synaptic levels), should target components of interest.

As in all fields of research with neuropeptides, the lack of specific antagonists provides an additional problem to the interpretation and pharmacological characterization of results. However, modified peptide antagonists are increasingly becoming available for many neuropeptides including opioids, bradykinin (20), bombesin (21), calcitonin gene-related peptide (CGRP) (22), corticotropin-releasing factor (23), leutinizing hormone-releasing hormone (LHRH) (24), neuropeptide Y (25), neurotensin (26), somatostatin (27), tachykinins (28), vasoactive intestinal peptide (29), and vasopressin (30). Furthermore, metabolically stable nonpeptide antagonists are also available for opioids (naloxone and naltrexone), cholecystokinin (31), angiotensin (32), neuropeptide Y (33), and substance P (34), although the receptor specificity of both peptide and nonpeptide antagonists is often incomplete. Alternatively, a number of studies have successfully utilized either monoclonal or polyclonal antisera to antagonize the behavioral effect of neuropeptides, including the following: angiotensin II, drinking behavior (35); LHRH, lordosis (35); arginine vasopressin, passive avoidance performance (35); γ-endorphin, locomotor activity and passive avoidance behavior (36); and thyrotropin-releasing hormone, hypothermia (37) and reversal of barbiturate anesthesia and respiratory depression (38). It is essential, however, to include either a preimmune or nonimmune antiserum control in such studies to ensure the specificity of any observed effects (see Section IV,A). A more recent development of this approach is to immunize neonates with neuropeptide-directed antisera to alter the development of peptidergic neurons and ascertain the behavioral consequence of this depletion (39).

An alternative and possibly more physiologically germane approach for neuropeptides, which after all may serve as neuromodulators rather than neurotransmitters, regulators rather than initiators of behavior, would be to examine the effect of the neuropeptide on behavior evoked by the more "classical" neurotransmitter with which it is known to coexist. Selective neurotoxins (40) can then be utilized to deplete both the neurotransmitter and the coexistent neuropeptide, such as 6-hydroxydopamine or DSP4 [$N$-(2-chloroethyl)-$N$-ethyl-2-bromobenzylamine] for catecholamines (41), 5,7-dihydroxytryptamine for 5-hydroxytryptamine (5-HT) (40) (see Section IV,A), AF64A (ethylcholine mustard aziridinium ion) or hemicholinium-3 for acetylcholine (42, 43), and capsaicin for primary afferent neurons (13). In comparison, enzyme inhibitors such as $\alpha$-methyl-$p$-tyrosine or $p$-chlorophenylalanine will inhibit catecholamine or 5-HT synthesis, respectively, without affecting the neuropeptides. Peptidergic neuronal cell bodies could be destroyed by discrete microinjection of an excitatory amino acid such as kainate or ibotenic acid (40). Considerable evidence in the literature suggests that at least 70% depletion of many neurotransmitters may be required before a behavioral consequence of neuronal destruction can be demonstrated, probably because of large synaptic reserves. To maintain specificity and obtain such large-scale depletion, it is advisable to use repeated neurotoxin treatment rather than a single large dose.

Whichever approach is utilized, an ordered sequence of steps needs to be taken to establish (a) the specificity of the neuropeptide-induced behavior, (b) the neuronal networks and pharmacological mechanism of action, and (c) the physiological relevance of the observed behavior.

## B. Recording Chambers

With mechanical methods of recording motor behavior, care should be taken that the chamber is clean and consideration given to whether the animal should be habituated. To enable the direct observation of motor behaviors, animals are usually placed in isolation in small Perspex behavioral chambers (25 to 45 cm$^2$, with sides typically 45 cm high) inside a uniformly illuminated soundproof cabinet, with low level background noise (e.g., a ventilatory fan) masking extraneous sounds. Unless the response to a novel environment is being examined, animals are habituated (typically for 30 min) to an individual clean chamber which is placed inside the soundproof cabinet but screened from any other chamber inside the cabinet. Such a protocol also allows intrathecal pretreatment to be performed while behavioral monitoring is continuing on another rat. Drugs should be administered using a restraining device, a blind protocol, and in a random order, immediately prior to returning the animal to the chamber, and behavioral effects should be compared with vehicle controls. The behavior is monitored remotely by camera and recorded on videotape, thus permitting both simultaneous and subsequent analysis.

A rat given intrathecal saline under such conditions will typically spend most of the time asleep or at rest facing the edge of the chamber after an initial period of locomotion and intermittent periods of general grooming, particularly in the first 5 min following injection, probably due to prior handling. Many neuropeptides given intrathecally will enhance the low basal level of grooming (44).

## C. Scoring and Rating Motor Behaviors

Once preliminary studies have identified the predominant features of the motor behavioral response, a method should be devised to quantify the behavior. Each of the separate components could be monitored continuously from a video recording, but the time involved usually precludes this approach. One of two methods is therefore adopted: either a continuous rating scale is used to analyze the whole pattern of behavior, or individual components are measured at fixed time points. For instance, a variety of continuous scales have been used to quantify stereotypy, based on the ordered appearance of various behaviors following increasing doses of amphetamine (45). If such a method is employed it is clearly vital that the scale covers the entire repertoire as a continuum in the appropriate order of ranking. In most instances a better procedure is to assess separately each of the behavioral components for a set time, 20 or 30 sec, over a number of observation periods to produce a cumulative score for each component.

The second approach has been more widely used, for instance, to assess each of the components of the 5-HT behavioral syndrome (46–49). Intermittent 5-HT agonist-induced behaviors [such as lateral head weaving, repetitive sideward head movements, and reciprocal forepaw treading (piano-playing-like movements of the forepaws)] can be scored on a scale of 0–4, where 0 = absent, 1 = occurs once, 2 = occurs several times, 3 = frequent, and 4 = continuous in the observation period. More continuous behavioral components (such as Straub-tail, flat body posture, and hindlimb abduction) can be quantified according to intensity such that 0 = absent, 1 = perceptible, 2 = weak, 3 = medium, and 4 = maximal (50, 51). Another method is to use a simpler scale (0–3) to rate the proportion of each time period spent enacting each individual behavioral component, where 0 = absent, 1 = present less than 50%, 2 = present greater than 50%, and 3 = continuous (52). In general, with such quantitative scales the larger the number of divisions the more subjective the measurement and the less accurate the final score becomes. A check that interindividual concordance in scoring is high (16) should therefore be made.

Grooming is one stereotype behavior frequently produced by neuropeptides and therefore worthy of further comment. Grooming behavior can be dissociated into many different components, such as forepaw and face washing, whole body grooming, caudally directed scratching and biting, anogenital grooming, tail elevation, and tremor, all of which can be timed or subjectively scored on a rating scale (as described above). Methods have been developed to apportion the time spent performing each behavioral component, to allow the effect of the neuropeptide on the composition of the entire behavioral response to be compared with the drug-free condition (44). With such a protocol (which could be applied to other syndromes), if peptidergic neuronal pathways are physiologically involved in mediating the whole behavior, then exogenous neuropeptide would be expected to increase the entire response without altering the overall composition of the syndrome. In contrast, methods based on summation of quantitative scores from different behavioral components (51) can obscure changes in individual constituents and incorrectly assume that each component involves a common mechanism of activation (53), and they are therefore best avoided. For example, the various components of the 5-HT behavioral syndrome require participation of different neuronal pathways and receptor subtypes (47–49).

## IV. Experimental Studies of Neuropeptide-Induced Motor Behavior

To attempt to illustrate some of the methods and potential problems of determining the effect of neuropeptides on drug-induced and spontaneous motor behavior, an account of some recent experiments is given. Several neuropeptides (thyrotropin-releasing hormone, substance P, galanin, enkephalins, and an unknown neuropeptide which cross-reacts with an antiserum raised against proctolin) all coexist to varying extents with 5-HT in bulbospinal raphe neurons which innervate spinal motoneurons (54). The role of these coexistent neuropeptides in the control of spinal motor function remains to be determined, and the following experiments attempt to identify a motor behavioral model of the effect of 5-HT and coexistent neuropeptides on spinal motor function, using the intrathecal injection protocol.

## A. Neuropeptide-Induced Wet-Dog Shakes

One of the most prominent motor behaviors produced by central administration of thyrotropin-releasing hormone (TRH) peptides, enkephalins, or opiate withdrawal, 5-hydroxytryptamine$_2$ (5-HT$_2$) receptor agonist or substance P agonists with affinity for NK3 receptors (55), is paroxysmal rapid sideways shaking of the head (head twitch) or of the head and body which progresses along the rostrocaudal axis (called wet-dog shakes because of the resemblance to the response made by a variety of animals after immersion in water) (46, 56–58). Such wet-dog shakes may represent an augmentation of a physiological response to remove irritants from the pinnae or body surface.

The number of wet-dog shakes can readily be counted using a tally counter, and initial studies established that intrathecal administration of TRH, or a variety of TRH analogs, potently elicited this motor behavior (59). However, the duration of the wet-dog shake response was inversely related to dose with metabolically stable TRH analogs such as CG 3509 (59) or RX 77368, with wet-dog shakes continuing for 30 min with 0.5 $\mu$g but terminating within 20 min with 10 $\mu$g RX 77368, such that there was no significant difference in the total number of shakes observed with any dose (Fig. 2). This is probably in part due to desensitization-induced tachyphylaxis and to the induction of hyperlocomotion and an escape response (competing behaviors) with higher doses (59). The number of wet-dog shakes in the first 6–9 min from intrathecal TRH peptide injection is, however, linearly related to dose, illustrating the importance of examining the time course and assessing several components of any motor response.

Wet-dog shakes must involve multisegmental activation of most lower motoneurons in the brain stem and spinal cord. Yet the response is not maximal for at least 5 min, suggesting that spread of the injectate to a slightly more rostral site may be required to induce this behavior (see Section V,B), in agreement with lesion studies showing that the brain stem is the most sensitive site for initiation of 5-HT- and TRH-induced wet-dog shakes (46, 58).

No TRH antagonists are available to further characterize the pharmacology of TRH-induced wet-dog shakes, so a polyclonal TRH antiserum (which does not cross-react with TRH analogs such as CG 3509 that are modified at the N terminus) was used in attempts to immunologically neutralize TRH and induce postsynaptic receptor supersensitivity. Two weeks of infusion

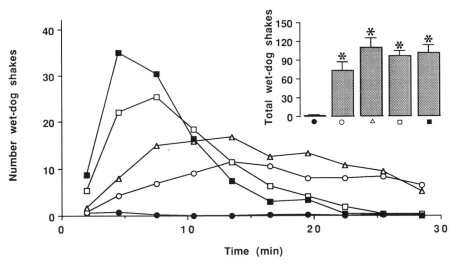

FIG. 2  Time course of the number of wet-dog shakes produced by intrathecal injection of saline (●) or 0.1 (○), 0.5 (△), 2 (□), or 10 $\mu$g (■) of the thyrotropin-releasing hormone analog 3,3'-dimethyl-pGlu-His-Pro-amide (RX 77368), each administered at 4-day intervals ($n = 6$) expressed as a mean for each 3-min interval (without error bars for clarity). The inset shows the total number of wet-dog shakes (mean ± S.E.M.) observed in the same animals over the entire 30 min. Note the progressive decrease in the peak and duration of this motor behavior with increasing dose without any change in the total number of shakes. Asterisks denote $p < 0.01$ with respect to saline.

of a partially purified TRH antiserum significantly increased the number of wet-dog shakes evoked by CG 3509 injection (Fig. 3), indicative of supersensitivity. Although infusion of the control nonimmune serum tended to elevate TRH analog-induced wet-dog shakes, this did not reach significance, confirming the specificity of the effect. Furthermore, apart from a small but significant increase in 5-HT agonist-induced lateral head weaving following TRH serum, none of the other 5-HT-induced motor behaviors was altered by either sera (Fig. 3).

Irrespective of the agent used to elicit wet-dog shakes (TRH, enkephalin, or 5-HT agonist), this motor behavior is attenuated by pretreatment with a 5-HT$_2$ receptor antagonist (52, 57, 60), implying that serotonergic neuronal activation is a common component. TRH coexists with 5-HT in virtually all bulbospinal raphe nerve terminals, which innervate motoneurons in the ventral spinal cord (54), making this a strong candidate for the participating serotonergic pathway (Fig. 4A). Virtually complete destruction of the bulbos-

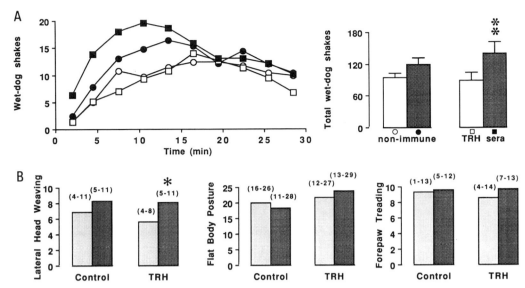

FIG. 3   Effect of chronic infusion (14 days) of partially purified (0.33 $M$ ammonium sulfate precipitation followed by dialysis against phosphate-buffered saline) control (nonimmune, circles) or polyclonal TRH antisera (squares, raised in sheep against a conjugate of TRH to keyhole limpet hemocyanin) on (A) TRH analog- and (B) 5-HT agonist-induced motor behaviors. (A) Time course (mean per 3 min) and total number (mean $\pm$ S.E.M., histograms) of wet-dog shakes evoked by intrathecal injection of the TRH analog (CG 3509, 0.5 $\mu$g). (B) Mean cumulative score (range) for lateral head weaving, flat body posture, and reciprocal forepaw treading (rated on a scale of 0–3 for 20 sec once every 2 min for 20 min, maximum score = 30) recorded before (open symbols and light columns) and again 12 or 14 days after (filled symbols and dark columns) initiating antibody infusion by osmotic minipumps (0.48 $\mu$l/hr, diluted 1:20 in 0.154 $M$ saline). Note the enhancement of CG 3509-induced wet-dog shakes following TRH antisera infusion (double asterisk denotes $p < 0.02$ versus the effect of CG 3509 before TRH antiserum, Student's $t$-test). In contrast, apart from a small, but significant, increase in lateral head weaving following TRH serum (single asterisk denotes $p < 0.05$ versus preserum, Wilcoxon rank test), neither control nor TRH antisera altered any 5-HT-induced behavior.

pinal raphe neurons, causing depletion of both 5-HT and the coexistent neuropeptide TRH, can be achieved by intrathecal injection of the serotonergic neurotoxin 5,7-dihydroxytryptamine (2 $\times$ 150 $\mu$g 5,7-DHT at 48-hr intervals 30 min after 12.5 mg/kg desipramine i.p. to prevent uptake into catecholaminergic neurons). Such 5,7-DHT treatment reduced both ventral horn 5-HT and TRH by approximately 90%, yet it did not affect either 5-HT

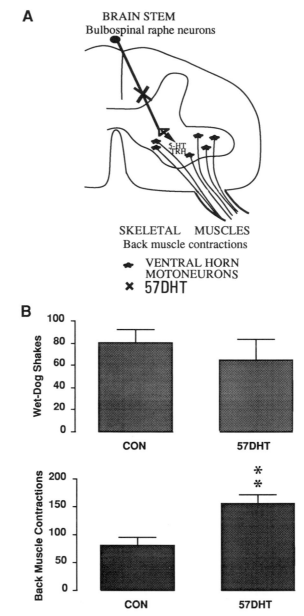

FIG. 4    (A) Diagram illustrating the bulbospinal raphe neurons (containing 5-HT and TRH) which innervate the ventral horn of the spinal cord and which can be destroyed (**X**) by intrathecal injection of a serotonergic neurotoxin. (B) Comparison of the number (mean ± S.E.M. in 30 min) of wet-dog shakes evoked by the TRH analog

agonist- (52) or TRH peptide-induced wet-dog shakes (Fig. 4B), suggesting that this descending serotonergic network was not involved in the expression of the motor behavior.

## B. Neuropeptide Modulation of Back Muscle Contractions

None of the ventral horn neuropeptides (apart from TRH just described) produced any marked motor behavior when given intrathecally in nanogram doses (61); therefore, careful examination of the motor behaviors elicited by intrathecal 5-HT agonist injection was performed. Intrathecal injection of 5-HT (following the uptake inhibitor fluoxetine) or 5-HT agonists such as 5-methoxy-$N,N'$-dimethyltryptamine (5-MeODMT) produced a previously undescribed motor behavior, namely, brief, caudorostral contractions of the superficial muscles of the back without any movement of the underlying vertebrae, which we termed back muscle contractions (60). Unlike wet-dog shakes, back muscle contractions began immediately and reached a peak within 5 min of intrathecal injection, suggesting a spinal site of initiation (compare Figs. 2 and 5), and they were virtually abolished by pretreatment with $5-HT_2$ receptor antagonists (such as the $5-HT_2/5-HT_{1C}$ antagonist ritanserin), implicating this 5-HT receptor subtype (Fig. 5). Furthermore, intrathecal injection of the serotonergic neurotoxin 5,7-DHT (to destroy bulbospinal raphe neurons) markedly potentiated 5-HT agonist-induced back muscle contractions (Fig. 4), as expected if the $5-HT_2$ receptors involved were located postsynaptically in relation to the serotonergic nerve terminals, possibly on interneurons or motoneurons in the ventral horn.

None of the neuropeptides or peptide analogs which coexist with 5-HT in raphe terminals or with acetylcholine in motoneurons (CGRP), the presumed target cell, produced back muscle contractions when given intrathecally. Moreover, only pretreatment (and not concomitant administration) with either proctolin (1 $\mu$g) or, in particular, the TRH analog RX 77368 (10 ng)

---

(CG 3509, 0.5 $\mu$g, $n = 6$) and back muscle contractions evoked by the $5-HT_{2/1C}$ agonist 2,5-dimethoxy-$\alpha$,4-dimethylbenzene ethamine hydrochloride (DOM, 10 $\mu$g, $n = 6$), both given by intrathecal injection 2 days before (CON) and again 10 days after (57DHT) the second of two intrathecal injections of 5,7-DHT ($2 \times 150$ $\mu$g at 48-hr intervals, each given 30 min after desipramine 12.5 mg/kg, i.p.). Note the marked potentiation of 5-HT agonist-induced back muscle contractions but the lack of effect on TRH-evoked wet-dog shakes following 5,7-DHT treatment, which reduced ventral horn 5-HT and TRH levels by 84 and 96%, respectively [see Fone et al. (52) for further details]. Double asterisk denotes $p < 0.01$ from before 5,7-DHT Student's $t$-test.

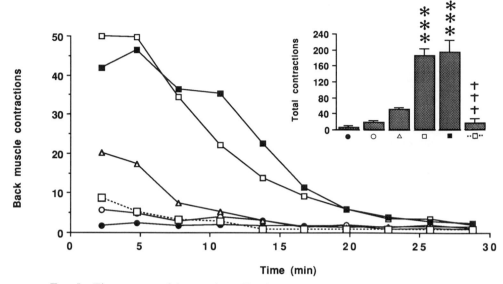

FIG. 5   Time course of the number of back muscle contractions induced by intrathecal injection of saline (●) or 2 (○), 10 (△), 25 (□), or 50 μg (■) of the nonselective 5-HT agonist 5-MeODMT alone and 25 μg 5-MeODMT given 30 min after ritanserin (--□--, 1 mg/kg, i.p., $n = 8$) recorded at 4-day intervals and expressed as a mean for each 3-min interval (without error bars for clarity). The inset shows the total number of back muscle contractions (mean ± S.E.M.) observed in the same animals over the entire 30 min. Triple asterisk denotes $p < 0.001$ versus saline and triple dagger, $p < 0.001$ versus 25 μg 5-MeODMT, Dunnett's $t$-test following analysis of variance (ANOVA) ($F = 36.363$, $p < 0.0001$). Note the dose-related increase in 5-MeODMT-induced back muscle contractions and the marked reduction in this behavior following pretreatment with the 5-HT$_2$ antagonist ritanserin without any alteration in the time course of the response.

potentiated 5-HT-induced back muscle contractions (Fig. 6A), an effect which was further enhanced following intrathecal 5,7-DHT (data not shown). In contrast, when given at the same doses, most of the neuropeptides investigated (most of which are located in dorsal as well as ventral horn nerve teminals) significantly attenuate wet-dog shakes produced by intrathecal TRH peptide injection (Fig. 6B). Taken together, these behavioral results suggest that back muscle contractions may be a useful behavioral model with which to examine the interaction between 5-HT and ventral horn neuropeptides on spinal motor function, whereas TRH peptide-induced wet-dog

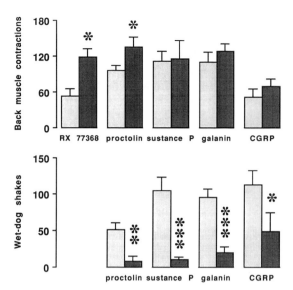

FIG. 6 Effect of intrathecal pretreatment (−30 min, dark columns) with either the TRH analog (RX 77368, 10 ng, $n = 5$), proctolin (1 $\mu$g, $n = 8$), substance P (10 ng, $n = 5$), galanin (100 ng, $n = 6$), or CGRP (1 $\mu$g, $n = 6$) on the total number of back muscle contractions (mean ± S.E.M. evoked by 5-HT) and wet-dog shakes (induced by TRH peptide) evoked in 30 min following intrathecal administration of submaximally effective doses of either 5-HT agonists (5-MeODMT, 15–25 $\mu$g, or, for RX 77368 experiments, the 5-HT$_2$/HT$_{1C}$ agonist (±)-1-(2,5-dimethoxy-4-iodophenyl)-2-aminopropane hydrochloride, DOI, 15 $\mu$g) or the TRH analog (RX 77368, 50 ng), respectively, compared in each case with the effect of administration of the 5-HT agonist or TRH analog alone (left-hand, light columns). Single asterisk denotes $p < 0.05$, double asterisk, $p < 0.02$, and triple asterisk, $p < 0.001$ versus 5-HT agonist or TRH peptide alone, Dunnett's $t$-test following ANOVA. None of the neuropeptides evoked back muscle contractions when given alone, nor did any affect the time course of the behavioral response. Both RX 77368 and proctolin also tended to increase this motor behavior when given concomitantly with the 5-HT agonist, but, this effect failed to reach significance (data not shown).

shakes probably involves activation of supraspinal and/or primary afferent neurons (testing the effect of capsaicin pretreatment on this behavior would be of value). Furthermore, the experiments suggest that TRH (which coexists with 5-HT in bulbospinal raphe neurons) may positively modulate the excitatory influence of 5-HT on motoneurons through receptors located postsynaptic to raphe nerve terminals.

## V. Anatomical Specificity

### A. Theoretical Considerations

Intrathecal administration delivers injectate to the dorsal surface of the spinal cord; therefore, penetration to relatively superficial sensory nerve terminals would be expected to be more immediate than to the deeper, ventral motoneurons. Hence sensory effects might be expected to precede motor effects with this technique. Indeed, at least 16 different neuropeptides have been identified in the dorsal horn (most being in primary afferent nerve terminals in the superficial lamina), many of which are also present in both intermediolateral and ventral horns. Consequently alteration in sensory and autonomic functions may be expected to accompany and indirectly modify motor behavior. For instance, intrathecal TRH and substance P, which elicit sensory (grooming, scratching, and biting) and motor (wet-dog shakes and hyperlocomotion) behaviors also increase sympathetic outflow even when given to anesthetized rats (62, 63). Conversely, behavioral measures used to assess analgesic effects of neuropeptides, such as the tail-flick test and paw withdrawal, rely critically on the assumption that the animal has normal motor coordination and function, yet until recently relatively few studies had attempted to assess this point directly, even though several neuropeptides (e.g., relatively high doses, 10–100 $\mu$g, of somatostatin, several substance P analogs, and opiates) and excitatory amino acids (e.g., $\beta$-$N$-methylamino-L-alanine, $\beta$-NMA) are known to be directly neurotoxic to motoneurons (64). Careful consideration (and appropriate pharmacological intervention) therefore needs to be given to the possible etiology of any observed motor behavior (could, for instance, flat body posture be the consequence of marked systemic hypotension, or could wet-dog shakes result from sensory activation?) before concluding direct activation of spinal motoneurons even with intrathecal administration.

### B. Verification and Spread of Injection

Regardless of the behavioral paradigm used, accomplishment of the objective to limit the spread of intrathecal injectate to the spinal cord is of vital importance to the interpretation of any results. Acute verification of the intrathecal cannula could be performed by the reversible paralysis of hind limbs produced by injection of 2% (w/v) lignocaine, but this is completely unnecessary once the technique has been acquired. The spread of injectate can be assessed postmortem, by the injection of a dye such as methylene blue, but by far the best approach is to determine the distribution of radiolabeled neuropeptide. The spread of radiolabel can be examined either qualitatively, by isolat-

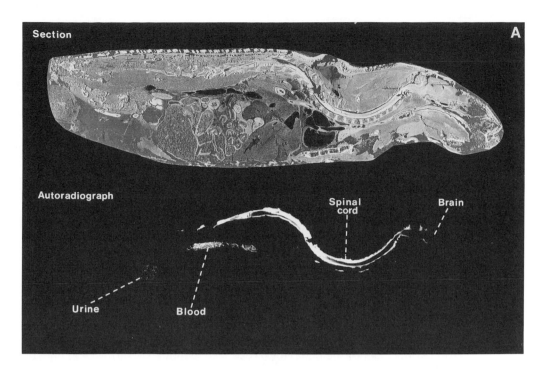

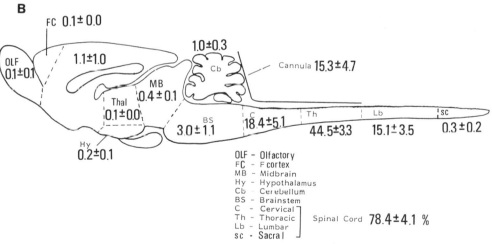

OLF – Olfactory
FC – F cortex
MB – Midbrain
Hy – Hypothalamus
Cb – Cerebellum
BS – Brainstem
C – Cervical
Th – Thoracic
Lb – Lumbar
sc – Sacral

Spinal Cord 78.4±4.1 %

FIG. 7 (A) Midline sagittal section of a rat and corresponding autoradiograph showing the distribution of radioactivity 9 min after the intrathecal injection of a 50:50 mixture of cold and $^{14}$C-labeled RX 77368 [10 $\mu$g (7.59 kBq) in 10 $\mu$l washed in with 20 $\mu$l saline]. (Reproduced with permission from *Neuroscience,* Vol. 229, No. 2, pp. 463–470.) (B) Distribution of radioactivity, expressed as a percentage (mean $\pm$ S.E.M., $n$ = 6) of the total recovered (being 10 $\pm$ 1% of that administered) in the cannula, brain, and spinal cord regions listed, 36 min after intrathecal injection of a 10:1 mixture of cold $CGRP_{1-37}$ and $[^{125}I]$-labeled $Tyr_0$ $CGRP_{28-37}$, 1 $\mu$g (4.18 KBq) in 10 $\mu$l washed in with 15 $\mu$l saline] and 6 min after intrathecal RX 77368.

ing brain and spinal cord regions and using a $\beta$ or $\gamma$ counter as appropriate for the isotope used, or qualitatively, using whole body autoradiography, both of which are illustrated in Fig. 7. Many neuropeptides are commercially available in radiolabeled form; alternatively, providing the neuropeptide contains a tyrosine (or less favorably a histidine), iodine-125 can readily be incorporated in-house into the neuropeptide (as in both examples in Fig. 7) by a single-step reaction using either chloramine-T, chloroamide 1,3,4,6-tetrachlor-3a,6a-diphenylglycoluril (Iodogen), or lactoperoxidase followed by chromatographic purification of the tracer using Sephadex or $C_{18}$ Seppak cartridges (see Refs. 65 and 66 for detailed methodology).

A mixture of unlabeled and labeled neuropeptide (dictated by the specific activity of the label) should be injected, utilizing the behavioral protocol, and the distribution examined at selected times, the peak of the neuropeptide-induced behavior being optimum. Care needs to be taken not to contaminate brain regions during the dissection procedure, and results should be expressed as a percentage of the total radioactivity recovered and compared with total levels injected to estimate the fraction recovered. Rapid anesthetic overdose (e.g., with halothane) followed by freezing in a cardice: isopentane mixture ($-90°C$) is suitable preparation for whole-body autoradiography (67). At the peak of the behavioral response, 9 min after intrathecal injection of the TRH analog ([$^{14}$C]RX 77368), the majority of the radioactivity seen in the whole-body autoradiograph is restricted to the spinal cord and rhombencephalon (Fig. 7A). However, some radioactivity is localized in the inferior vena cava, which could result from direct vascular or lymphatic drainage from the cerebrospinal fluid into the systemic circulation, possibly as a consequence of tissue damage associated with the intrathecal implantation.

In this example, the absolute levels of RX 77368 in the blood were low, amounting to less than 4% of the total administered (estimated by complete tissue combustion in a Packard 306 sample oxidizer and assuming the blood volume to be 7% of body weight). Similarly, even 36 min after intrathecal injection of [$^{125}$I]-labeled $Tyr_0$ $CGRP_{28-37}$ and following a second intrathecal injection ($10\mu$l RX 77368 plus 15 $\mu$l saline 6 min before termination), most of the radioactivity recovered in the central nervous system (CNS) and cannula was confined to the spinal cord ($78 \pm 4\%$) or was in, or adhered to, the cannula ($15 \pm 3\%$), whereas 11.5% of the total radioactivity administered was in the blood (Fig. 7B). Thus, for both of these two very different time courses and neuropeptides (3 compared with 11 amino acids), most of the radioactivity remained relatively restricted to the spinal cord and brain stem, suggesting that any observed behavioral interaction is likely to be the result of an effect on spinal neurons. However, in each case less than 25% of the radioactivity was recovered from the CNS tissue, implying that brain penetration is poor, but circulating levels of radioactivity (not necessarily bioactive neuropeptide) appear to rise relatively slowly in the two cases illustrated.

## VI. Summary

There is thus far little conclusive evidence for the involvement of neuropeptides in normal behavior, but this may be because neuropeptide release modulates rather than initiates behavior. In addition, most behaviors are complex and require activation of diffuse heterogeneous networks of neurons, making neuropharmacological characterization formidable. The paucity of selective neuropeptide antagonists has hindered progress, and the increasing evidence of multiple receptor subtypes for most of the neuropeptides will undoubtedly further complicate developments.

Despite these problems a combination of neuropeptide antibodies, neurotoxins, and discrete methods of administration should stimulate considerable advances in the pursuit of the functional influence of neuropeptides on behavior. This chapter attempts to exemplify some of these approaches to analyze specifically the motor behavioral effects of neuropeptides. Clearly such behavioral data cannot be interpreted in isolation, and complete analysis requires consideration of neuropeptide distribution, ligand binding, and electrophysiological information, and, where possible, associated changes in appropriate neurochemical markers of neurons contributing to observed behavioral effects should be determined (68).

## Acknowledgments

Much of the work from the author's laboratory was supported by Action Research. I also thank Mr. Ian Topham for technical assistance and Phillipa Dix (Nottingham University B.Med.Sci. graduate) for work with the TRH antiserum, Mike Havler (Reckitt and Coleman) for the autoradiography, Professor L. Flohé (Grunenthal GMBH) for CG 3509, and Reckitt and Coleman for RX 77368.

## References

1. K. M. Kendrick, *J. Neurosci. Methods* **34,** 35 (1990).
2. U. Ungerstedt, *in* "Methods in Neurosciences" (P. M. Conn, ed.), Vol. 6, p. 81. Academic Press, New York, 1984.
3. A. W. Duggan, I. A. Hendry, J. L. Green, C. R. Morton, and W. D. Hutchison, *J. Neurosci. Methods* **23,** 241 (1988).
4. M. Herkenham, *Neuroscience (Oxford)* **23,** 1 (1987).

5. A. J. Harmer, *in* "Basic and Clinical Aspects of Neuroscience" (E. Fluckiger, E. E. Muller and M. Thorner, eds.), Vol. 2, p. 17. Springer Sandoz, 1987.

6. L. Langerman, E. Golomb, and S. Benita, *Anesth. Analg. (N.Y.)* **72,** 635 1991).

7. J. L. Vaught and R. Scott, *Life Sci.* **40,** 175 (1986).

8. T. L. Yaksh, *in* "Methods in Neurosciences" (P.M. Conn, ed.), Vol. 6, p. 107.

9. N. Rawal, L. Nuutinen, P. P. Raj, S. L. Lovering, A. H. Gobuty, J. Hargardine, L. Lehmkuhl, R. Herva, and E. Abouleish, *Anaesthesiology* **75,** 1025 (1991).

10. J. L. K. Hylden and A. A. Wilcox, *Brain Res.* **217,** 212 (1981).

11. D. Papir-Kricheli, C. Gilon, M. Chorev, Z. Selinger, and M. Devor, *Neuropharmacology* **29,** 1055 (1990).

12. T. L. Yaksh and T. A. Rudy, *Physiol. Behav.* **17,** 1031 (1976).

13. T. L. Yaksh, J. Jang, Y. Nishiuchi, K. P. Braun, S. Ro, and M. Goodman, *Life Sci.* **48,** 623 (1991).

14. M. Bahar, G. Cole, M. Rosen, and M. D. Vickers, *Eur. J. Anaesthesiol.* **1,** 293 (1984).

15. S. L. Dickinson, P. Slater, and D.A. Longman, *Neuropharmacology* **20,** 757 (1981).

16. T. W. Robbins, *in* "Handbook of Psychopharmacology, Volume 7: Principles of Behavioral Pharmacology" (L. L. Iversen, S. D. Iversen, and S. H. Snyder, eds.), p. 37, Plenum, New York, 1977.

17. M. A. Geyer, P. Russo, and V. L. Masten, *Pharmacol. Biochem. Behav.* **28,** (1986).

18. K. A. Clarke and A. J. Parker, *Med. Sci. Res.* **16,** 901 (1988).

19. C. A. Marsden and B. King, *Pharmacol. Biochem. Behav.* **10,** 631 (1979).

20. F. Lembeck, T. Griesbacher, M. Eckhardt, S. Henke, G. Briepohl, and J. Knolle, *Br. J. Pharmacol.* **102,** 297 (1991).

21. R. T. Jensen, S. W. Jones, K. Folkers, and J. D. Gardner, *Nature (London)* **309,** 61 (1984).

22. M. V. Donoso, A. Fournier, S. St.-Pierre, and J. P. Huidobro-Toro, *Peptides* **11,** 885 (1990).

23. J. Rivier, C. Rivier, and W. Vale, *Science* **224,** 889 (1984).

24. J. Humphries, *J. Med. Chem.* **21,** 120 (1978).

25. A. Balasubramanium and S. Sheriff, *J. Biol. Chem.* **265,** 14724 (1990).

26. R. Quirion, F. Rioux, D. Regoli, and S. St.-Pierre, *Eur. J. Pharmacol.* **61,** 309 (1980).

27. J. L. Fries, *Peptides* **3,** 811 (1982).

28. S. Caranikas, J. Mizrahi, P. D'Orleans-Juste, and D. Regoli, *Eur. J. Pharmacol.* **77,** 205 (1982).

29. I. Gozes, E. Meltzer, S. Rubinrout, D. E. Brenneman, and M. Fridkin, *Endocrinology (Baltimore)* **125,** 2945 (1989).

30. M. Manning, *J. Med. Chem.* **24,** 701 (1981).

31. R. S. L. Chang and V. J. Lotti, *Proc. Natl. Acad. Sci. U.S.A.* **83,** 4923 (1986).

32. A. T. Chui, W. F. Herbin, D.E. MaCall, R. J. Ardecky, D. J. Carini, R. Duncia, L. J. Pease, P. C. Wong, R. R. Wexler, A. I. Johnson, and P. B. M. W. M. Timmermans, *Biochem. Biophys. Res. Commun.* **165,** 196 (1989).

33. M. C. Michel and H. J. Motulsky, *Ann. N.Y. Acad. Sci.* **611,** 392 (1990).
34. R. M. Snider, J. W. Constantine, J. A. Lowe III, K. P. Longo, W. S. Lebel, H. A. Woody, S. E. Drozda, M. C. Desai, F. J. Vinick, R. W. Spencer, and H. J. Hess, *Science* **251,** 435 (1991).
35. R. L. Moss and C. A. Dudley, *in* "Handbook of Psychopharmacology" (L. L. Iversen, S. D. Iversen, and S. Snyder, eds.), Vol. 18, p. 397. Plenum, New York, 1984.
36. J. M. Van Ree, G. Wolterink, M. Fekete, and D. De Weid, *Neuropharmacology* **21,** 1119 (1982).
37. C. Prassad, J. J. Jacobs, and J. F. Wilber, *Brain Res.* **193,** 580 (1980).
38. C. Lighton, G. W. Bennett, and C. A. Marsden, *Brain Res.* **378,** 385 (1986).
39. M. C. De Felipe, M. T. Molinero, and J. Delrio, *Brain Res.* **485,** 301 (1989).
40. G. Jonsson, *Annu. Rev. Neurosci.* **3,** 169 (1980).
41. G. Jonsson, H. Hallman, F. Ponzio, and S. Ross, *Eur. J. Pharmacol.* **72,** 173 (1981).
42. G. Gaal, P. E. Potter, I. Hanin, I. Kakucska, and E. S. Vizi, *Neuroscience (Oxford)* **19,** 1197 (1986).
43. M. Glavinovic, N. Ropert, K. Krnjevic, and B. Collier, *Neuroscience (Oxford)* **9,** 319 (1983).
44. R. L. Isaacson and W. H. Gispen, *in* "Neurobiology of Stereotyped Behaviour" (S. J. Cooper and C. T. Dourish, eds.), p. 117. Oxford Univ. Press (Clarendon), Oxford, 1990.
45. Creese and L. L. Iversen, *Brain Res.* **55,** 369 (1973).
46. P. Bédard and C. J. Pycock, *Neuropharmacology* **16,** 663 (1977).
47. G. Curzon, *in* "Neurobiology of Stereotyped Behavior" (S. J. Cooper and C. T. Dourish, eds.), p. 142. Oxford Univ. Press (Clarendon), Oxford, 1990.
48. R. Green and L. I. Backus, *Ann. N.Y. Acad. Sci.* **600,** 237 (1990).
49. B. L. Jacobs and H. Klemfuss, *Brain Res.* **100,** 450 (1975).
50. C. D. Andrews, J. C. R. Fernando, and G. Curzon, *Neuropharmacology* **21,** 63 (1982).
51. J. F. Stolz and C. A. Marsden, *J. Physiol (Paris)* **77,** 385 (1981).
52. K. C. F. Fone, J. V. Johnson, G. W. Bennett, and C.A. Marsden, *Br. J. Pharmacol.* **96,** 599 (1989).
53. S. L. Dickinson, H. Jackson, and G. Curzon, *Psychopharmacology* **80,** 196 (1983).
54. U. Arvidsson, S. Cullheim, B. Ulfhake, G. W. Bennett, K. C. F. Fone, A. C. Cuello, T. J. Visser, and T. Hökfelt, *Synapse* **6,** 237 (1990).
55. P. J. Elliott and S. D. Iversen, *Brain Res.* **381,** 68 (1986).
56. E. G. Drust and J. D. Connor, *J. Pharmacol. Exp. Ther.* **224,** 148 (1983).
57. S. L. Handley and L. Singh, *Trends Pharmacol. Sci.* **7,** 324 (1986).
58. E. T. Wei, H. H. Loh, and E. L. Way, *Life Sci.* **12,** 489 (1973).
59. K. C. F. Fone, G. W. Bennett, and C. A. Marsden, *Neuropharmacology* **26,** 1147 (1987).
60. K. C. F. Fone, A. J. Robinson, and C. A. Marsden, *Br. J. Pharmacol.* **103,** 1547 (1991).

61. R. A. Cridland and J. L. Henry, *Neuropeptides* **11,** 23 (1988).
62. K. Yasphal, S. G. Gauthier, and J. L. Henry, *Neuroscience* (*Oxford*) **15,** 529 (1985).
63. K. Yasphal, S. G. Gauthier, and J. L. Henry, *Circ. Res.* **65,** 859 (1989).
64. D. M. Gaumann, T. S. Grabow, T. L. Yaksh, S. L. Casey, and M. Rodriguez, *Neuroscience* (*Oxford*) **39,** 761 (1990).
65. A. E. Bolton, "Radioiodination Techniques." Amersham International, Amersham, England, 1985.
66. T. Chard, "Laboratory Techniques in Biochemistry and Molecular Biology" (R. H. Burdon and P. H. van Knippenberg eds.), Elsevier, Amsterdam, 1987.
67. J. V. Johnson, K. C. F. Fone, M. Havler, I. F. Tullock, G.W. Bennett, and C. A. Marsden, *Neuroscience* (*Oxford*) **29,** 463 (1989).
68. K. C. F. Fone and D. M. Dixon, *Brain Res.* **544,** 181 (1991).

# [13] Measurement of Dissociation of Amnesic and Behavioral Effects of Drugs in Mice

Toshitaka Nabeshima and Kenji Ichihara

## Introduction

Many different tasks have been used to evaluate learning and memory in experimental animals. These include the passive (inhibitory) avoidance task, spatial maze tasks such as the Morris water maze and the radial arm maze, and operant matching and nonmatching to sample memory tasks. Many of these models are used to study learning and memory in rats, and simple tasks such as the one-trial passive avoidance task are widely used to study learning and memory in mice. Those tasks that are based on negative or positive reinforcers, for example, electric foot shock or food and water, appear to be sensitive to changes in the motivational impact of the reinforcers during acquisition. The intensity of motivation may be modified by several factors such as pain threshold, rewarding efficacy (e.g., anhedonia), motor activity, and moods. To decrease these risks, some methods have been improved (1), and methods that do not use reinforcers have been developed (2, 3).

This chapter describes two novel learning tasks (4–7) for use with mice in which the animals are not required to perform a special behavior in the nonreinforced preexposure stage (i.e., the training session). We have used these tasks to dissociate the amnesic effects of drugs from other behavioral effects.

## Utility of an Elevated Plus-Maze Task for Evaluating Memory in Mice

The elevated plus-maze was introduced by Pellow et al. (8) and is used to measure anxiety. The maze has two open arms and two enclosed arms, and it is based on the apparent natural aversion of rodents to an open, high, well-lighted space. Thus, rodents spend more time in the enclosed arms than in the open arms, and they will move quickly from the open space to the enclosed space if they have experience with this environment. The amount of time that passes before animals enter the enclosed arm after being placed on the open arm is regarded as an index of memory (4, 5).

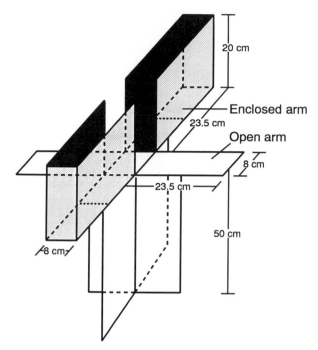

FIG. 1   Design of the elevated plus-maze.

## Materials and Methods

### Animals

Male ddY mice (Nihon SLC, Shizuoka, Japan) weighing 32–37 g are used. Mice are housed in groups of 7 to 8, under standard conditions ($23 \pm 1°C$, $50 \pm 5\%$ humidity, light/dark cycle with the lights on between 800 and 2000 hr), with free access to water and food. The mice are used after they have adapted to the laboratory conditions for at least 1 week. At that time they are still naive to the elevated plus-maze. The experiments are conducted between 1000 and 1600 hr in a "semisoundproof" laboratory.

### Apparatus

The maze is shown in Fig. 1. It is made of plywood and has two open arms ($23.5 \times 8.0$ cm) and two enclosed arms ($23.5 \times 8.0 \times 20$ cm high). The arms extend from a central platform ($8.0 \times 8.0$ cm). The plus-maze is elevated to a height of 50 cm above the floor. The open arms and the central platform are painted white, and the enclosed arms are painted black. A fine white line (about 1 mm) is drawn in the middle of the floor of each enclosed arm. The maze is set on the center of the floor of the laboratory, and the illumination is 700 lux.

### Procedures

On the first day, the first trial (training) is carried out. Mice are placed individually at the end of one open arm facing away from the central platform, and the number of seconds it takes for the mouse to move from the open arm to either of the enclosed arms (the transfer latency) is counted and recorded. Counting begins when the animal is placed in the open arm and ends when its legs cross the white line in the middle of an enclosed arm. If the mouse does not enter an enclosed arm within 90 sec, it is pushed gently on the back into the enclosed arm, and the transfer latency is recorded as 90 sec. The mouse is allowed to move freely in the maze for 10 sec after the transfer latency is measured. The mouse is then gently taken out of the maze and returned to its home cage. Twenty-four hours later, the second trial (retention test) is performed. Mice are again put into the elevated plus-maze and the transfer latency is recorded. If the mouse does not enter the enclosed arm within 90 sec during the retention test, the transfer latency is recorded as 90 sec.

### Statistical Analysis

The results are expressed as the mean ($\pm$S.E.). Statistical significance is determined by a Kruskal-Wallis nonparametric one-way analysis of variance, followed by a nonparametric Dunn-type's multiple range test. Any $p$ values less than 0.05 indicate statistically significant differences.

## Effects of Scopolamine, Diazepam, and Haloperidol on Elevated Plus-Maze Learning

Scopolamine, an anticholinergic agent, and diazepam, a benzodiazepine, are well known to produce anterograde amnesia in several behavioral tests of learning and memory when given to rodents before training. We first examined whether the amnesic effects of these drugs could be detected with the elevated plus-maze task. In addition, the effect of the dopamine receptor antagonist haloperidol, which reduces motor activity, was tested to examine the influence of a change in motor function on the acquisition of memory.

Scopolamine hydrobromide (Sigma, St. Louis, MO), diazepam (injectable solution from ampules, Takeda, Osaka, Japan), and haloperidol (injectable solution from ampules, Dainippon, Osaka, Japan) are dissolved in or diluted with 0.9% saline solution. Scopolamine is given subcutaneously 30 min before training, and diazepam and haloperidol are given intraperitoneally 30 and 60 min before training, respectively, in 10 ml/kg.

The effects of scopolamine, diazepam, and haloperidol on the transfer latencies in the training and test trials are shown in Fig. 2. The transfer latencies of vehicle-treated mice were significantly shorter during the test than during training, which indicates that mice can remember the layout of the elevated plus-maze. In scopolamine-treated mice transfer latencies during both trials were significantly longer than in vehicle-treated mice (Fig. 2A). Diazepam slightly, but not significantly, shortened the transfer latency during training, but the drug significantly prolonged it during the test (Fig. 2B). The transfer latencies of haloperidol-treated mice were significantly longer than those of vehicle-treated mice during training, but not during the test (Fig. 2C). These results indicate that scopolamine-induced and diazepam-induced amnesia can be measured with the elevated plus-maze task.

## Discussion

Because scopolamine decreases the response to electric foot shock and diazepam is an anxiolytic drug, methods that rely on negative reinforcers, such as the passive avoidance task, may be unsuitable for evaluating the effects of these drugs on memory. In the elevated plus-maze test, the amnesic effects of scopolamine and diazepam could be detected, although these drugs induced different behavioral changes during training (i.e., scopolamine prolonged the transfer latency but diazepam did not change it). Haloperidol also prolonged the transfer latency during training. In fact, most mice given haloperidol (0.4 mg/kg) stayed at the start point for 90 sec and had to be pushed to the enclosed arm by the experimenter. Despite the lack of spontaneous motor activity during training, the transfer latencies of haloperidol-treated mice on the test day were quite similar to those of vehicle-treated mice. This indicates that sedation induced by haloperidol does not directly influence memory. Thus, the behavioral aspects during training hardly influenced the transfer latency during the test session. Consequently, the elevated plus-maze may be useful for dissociating amnesic and behavioral effects of drugs in mice. In addition, this method may be a useful screening test for nootropic drugs; the antiamnesic effects of aniracetam, tetrahydroaminoacridine (tacrin), and physostigmine have been evaluated by this method (4).

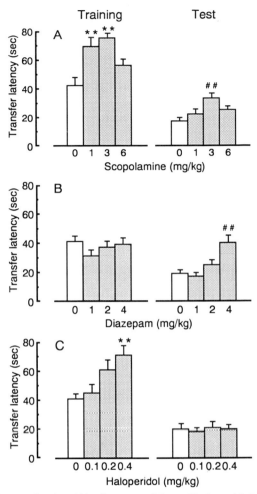

FIG. 2   Effects of scopolamine (A), diazepam (B), and haloperidol (C) on the transfer latency to an enclosed arm of the elevated plus-maze. Scopolamine and diazepam were given subcutaneously and intraperitoneally, respectively, 30 min before, and haloperidol was given intraperitoneally 60 min before training. Each bar represents the mean ± S.E. of 19 to 25 mice. Double asterisks denote $p < 0.01$, compared to vehicle control during training. Double pounds denote $p < 0.01$, compared to vehicle control during the test. From Itoh *et al.* (5).

## Utility of a Water-Finding Task for Evaluating Latent Learning in Mice

The water-finding task has been proposed for evaluating memory in rats (9). We attempted to modify the apparatus for mice and to modify the method to dissociate learning from other behaviors (6, 7). The test apparatus consists of an open field containing a water tube. If animals are allowed to explore the open field when they are normally hydrated and are then deprived of water, they find the water source more quickly than animals that are similarly deprived of water but have no prior experience in the open field. Because the animals are not reinforced positively by water during training, their performance during the retention test is thought to be based on "latent" learning. This concept was proposed in a study of mice in a maze (10).

## Materials and Methods

### Animals

Male ddY mice (Nihon SLC) weighing 30–35 g are used. Mice are housed in groups of 20, under standard conditions (23 ± 1°C, 50 ± 5% humidity, light/dark cycle with the lights on between 800 and 2000 hr), with free access to water and food. The mice are used after they have adapted to the laboratory conditions for at least 3 days. At that time they are naive to the water-finding task apparatus. The experiments are carried out between 900 and 1700 hr in a "semisoundproof" laboratory.

### Apparatus

The test apparatus is shown in Fig. 3. It is made of plywood and consists of an open field (30 × 50 × 15 cm high) with an alcove (10 × 10 × 10 cm) in the middle of one of the long walls of the enclosure. The inside is painted gray, and the floor of the open field is divided into 15 identical squares with fine black lines (about 2 mm) to allow measurement of the animals' motion (ambulation). A metal drinking tube, one of the same type used in the home cages, is inserted into the center of the alcove ceiling with its tip 5 cm (during the training session) or 7 cm (during the test session) above the floor. The open field is set on a table (60 × 90 × 75 cm high) and is illuminated with an incandescent lamp at about 1200 lux.

### Procedures

During the training session, normally hydrated mice are placed individually into one corner of the open field. The time until they start exploring (i.e., start latency) is recorded. The mice are allowed to freely explore for 3 min

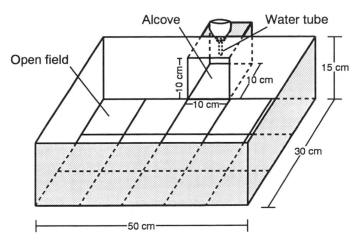

Fig. 3   Apparatus for the water-finding task.

after they begin to do so. During this time, the number of times the animals cross from one square to another in the open field (i.e., ambulation) and the number of times they touch or sniff the water tube in the alcove (i.e., number of approaches) are counted and recorded. Animals that do not begin exploring within 3 min and those that do not find the water tube during the 3 min of exploration are not tested further, but the number of such animals is recorded. Naive mice are prepared for comparison with the trained mice in terms of ability to find the water source in the same environment (latent learning ability). Mice are immediately returned to their home cages after the training procedure and are deprived of water for about 20 hr before the test trial. The test session takes place 48 hr after the end of training.

On the test day, mice are individually placed in the same corner of the test apparatus as during training. The number of seconds until they begin to drink (i.e., drinking latency) after they begin to explore is counted, as is the number of seconds until they enter the alcove (i.e., entering latency) and the time elapsed before they find the water after first entering the alcove (i.e., finding latency). Thus, the drinking latency consists of the sum of the entering and finding latencies.

### Statistical analysis

The results are expressed as the mean (±S.E.). Statistical significance is determined by a one-way analysis of variance followed by Dunnett's multiple range test. Any $p$ values less than 0.05 indicate statistically significant differences.

## Effects of SCH 23390, Pimozide, and Chlordiazepoxide on Water Finding

We examined whether chlordiazepoxide, a benzodiazepine, produces amnesia as tested with this procedure. SCH 23390 and pimozide, dopamine $D_1$ and $D_2$ receptor antagonists, respectively, are used, like haloperidol, to test whether a change in motor activity on the training day directly influences memory.

Pimozide (Fujisawa, Osaka, Japan) is dissolved in a 0.35% (w/v) tartaric acid solution. Chlordiazepoxide (Hoffman-La Roche, Basel, Switzerland) and SCH 23390 (Schering-Plough, Bloomfield, NJ) are dissolved in 0.9% saline. SCH 23390 and pimozide are given intraperitoneally 30 and 120 min before training, respectively, and chlordiazepoxide is given subcutaneously 20 min before training.

In the trained mice drinking latency on the test day was shorter than that in naive mice (vehicle-treated groups versus naive groups in Figs. 4 and 5). This indicates that mice can learn in a weakly reinforced situation (latent learning). The effects of SCH 23390 and pimozide on exploratory behavior are summarized in Table I. The number of animals that did not start to explore within 3 min or could not find the water tube during exploration did not differ significantly from group to group. However, in animals given SCH 23390 or pimozide the start latency was significantly longer, and the number of approaches and the ambulation count were significantly lower than in vehicle-treated mice. Nevertheless, as shown in Fig. 4, pimozide-treated mice had shorter drinking latencies than SCH 23390-treated mice in the test session. The finding latency in pimozide-treated mice was slightly shorter than in vehicle-treated mice. Figure 5 shows the effects of chlordiazepoxide on entering and drinking latencies and on ambulation counts in the training and test trials. Compared to mice given vehicle, those given chlordiazepoxide before training had higher ambulation counts during training and had significantly longer drinking latencies and ambulation counts during the test (Fig. 5). The increase in ambulation suggests that chlordiazepoxide inhibited the habituation of the mice to an environment to which they had previously been exposed.

## Discussion

In the water-finding task, memory is assessed by recording the latencies for entering the alcove and for drinking water. The entering latency appears to be related to the degree of memory of the spatial layout of the test apparatus, because mice tend to like a narrow, dark place rather than a wide, well-

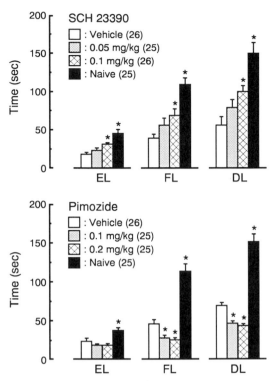

FIG. 4   Effects of SCH 23390 and pimozide on water-finding performance during the test trial. Latencies for entering the alcove (EL), finding the water after first entering the alcove (FL), and drinking the water (DL) are shown. SCH 23390 and pimozide were given intraperitoneally 30 and 60 min, respectively, before training. Each bar represents the mean ± S.E. of the number of animals shown in parentheses. Asterisks denote $p < 0.05$, compared to vehicle control. From Ichihara et al. (6).

lighted place. This is also true of the plus-maze task described above. Also, the finding latency seems to reflect the memory of a specific object (i.e., the water tube) in the apparatus. This parameter indicates learning ability in a nonreinforcement situation (i.e., "latent" learning), because the training trial is conducted without an appetitive stimulus. Although thirst may be totally ruled out as a motivator during training, water deprivation before the test session is indispensable to promote recall of the location of the water tube.

The dopamine $D_1$ receptor antagonist SCH 23390 reduced exploratory behavior on the training day (increased the start latency and decreased ambulation). In addition, SCH 23390-treated mice spent less time around the water tube (fewer approaches to the water tube). On the test day, SCH

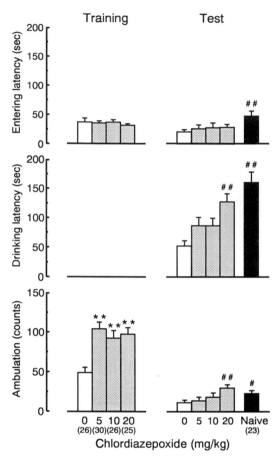

FIG. 5    Effect of chlordiazepoxide on water-finding performance during the training and test trials. Chlordiazepoxide was given subcutaneously 20 min before training. Each bar represents the mean ± S.E. of the number of animals shown in parentheses. Double asterisks denote $p < 0.01$, compared to vehicle control during training. Pound denotes $p < 0.05$ and double pounds $p < 0.01$, compared to vehicle control during the test. From Nabeshima *et al.* (7).

23390-treated mice took longer to find the water than did vehicle-treated mice. Because these mice explored less than control animals during the training session, the environment was perhaps less well known to them during the test session. In contrast, pimozide-treated mice found the water tube rapidly during the test trial, even though their exploratory behavior was as reduced as that of SCH 23390-treated animals during training. These

TABLE I   Effects of SCH 23390 and Pimozide on Exploratory Behavior during
Training for Water-Finding Task[a,b]

| Treatment | Dose (mg/kg) | $N^c$ | Start latency (sec) | Number of approaches (counts) | Ambulation (counts) |
|---|---|---|---|---|---|
| Vehicle | — | 26 (4) | 7.0 ± 1.4 | 5.5 ± 0.2 | 32.7 ± 1.5 |
| SCH 23390 | 0.05 | 25 (5) | 11.6 ± 2.2 | 4.6 ± 0.2$^d$ | 22.0 ± 1.4$^d$ |
|  | 0.1 | 26 (4) | 34.2 ± 6.5$^d$ | 3.7 ± 0.2$^d$ | 14.1 ± 1.1$^d$ |
| Vehicle | — | 26 (4) | 12.0 ± 3.1 | 5.4 ± 0.3 | 31.3 ± 1.8 |
| Pimozide | 0.1 | 25 (5) | 23.9 ± 3.4 | 3.6 ± 0.2$^d$ | 19.3 ± 1.3$^d$ |
|  | 0.2 | 25 (5) | 35.8 ± 5.3$^d$ | 3.0 ± 0.2$^d$ | 14.6 ± 1.1$^d$ |

[a] From Ichihara et al. (6).

[b] SCH 23390 and pimozide were given intraperitoneally 30 and 120 min, respectively, before training. Each value is the mean ± S.E.

[c] Numbers in parentheses show the number of animals that did not start to explore within 3 min or could not find the water tube during exploration.

[d] $p < 0.05$, compared to vehicle control.

findings suggest that finding water on the test day does not always depend on behavior during training. Another dopamine $D_2$ receptor antagonist, haloperidol, does not interfere with latent learning in the maze task with mice (10). Conversely, chlordiazepoxide increased exploratory activity during the acquisition session, perhaps owing to its anxiolytic effect at the doses used. However, these animals had trouble finding the water during the test session, which suggests that chlordiazepoxide disrupts the acquisition stage of latent learning. Moreover, the dopamine releaser methamphetamine had a tendency to interfere with water finding despite an increase in the ambulation count during training (11).

In view of these findings, the water-finding task can be used to dissociate the effects of a drug on memory and on other actions such as anxiety and motor activity. Furthermore, this method can be used to detect independently two different types of memory: memory for space and memory for a specific object.

## General Comments

The elevated plus-maze and the water-finding tasks offer several advantages for evaluating memory: the procedures are simple, single-trial tests that do not require manipulation of appetitive behaviors or aversive stimuli during training. A similar method that takes advantage of the instinctive behavior of animals, the habituation task, has been widely used to estimate learning and memory (2). This task uses the habituation to a novel environment, that is, when mice are placed in a field to which they have been exposed previously, exploratory activity (locomotor activity) is less than it was during the first trial. This decrease in locomotor activity is regarded as a sign of memory. However, exploratory activity varies among mice. For example, some animals are more active on the second day (the test session) than on the first day (the training session) because they often freeze when exposed suddenly to a novel environment. The value obtained for exploratory activity also varies with the method of measurement. For example, ambulation counts measured by an experimenter and motor activity measured mechanically by an activity meter such as a photocell counter are not necessarily the same. In addition, the correlation between the decrease in exploratory activity and a quantitative value for memory is unclear. When motor activity is changed by sedative or stimulant drugs given before training, it is particularly difficult to simply compare the exploratory activity on the test day with that on the training day. The exploratory activity during the test session may be affected by behavior during training. Thus, a decrease in exploratory activity caused by animal habituation may not always be a good indicator of memory.

It is important to consider how best to evaluate learning and memory in mice. Simple tasks like a one-trial learning paradigm that does not require special behaviors may be adequate. The present methods detect memory in an all-or-none manner, so the type of learning may be different from that measured with highly controlled learning paradigms, such as those in which the acquisition process is quantified (e.g., a correct response rate versus the number of trials).

It appears to be hard to detect amnesic effects of drugs completely independently of their behavioral effects, especially when animals are trained or tested under the influence of a drug. Either of the present methods can be used to evaluate memory in mice and to dissociate the effects of drugs on memory from some other behavioral effects of training. However, it is necessary to develop methods to detect clearly the effects of a drug on both acquisition and recall of memory even when the behavior of an animal is markedly altered by the drug.

## References

1. T. Kameyama, T. Nabeshima, and T. Kozawa, *J. Pharmacol. Methods* **16,** 39 (1986).
2. A. Platel and R. D. Porsolt, *Psychopharmacology* (*Berlin*) **78,** 346 (1982).
3. A. J. Hunter and T. K. Murray, *Psychopharmacology* (*Berlin*) **99,** 270 (1989).

4. J. Itoh, T. Nabeshima, and T. Kameyama, *Psychopharmacology* (*Berlin*) **101,** 27 (1990).
5. J. Itoh, T. Nabeshima, and T. Kameyama, *Eur. J. Pharmacol.* **194,** 71 (1991).
6. K. Ichihara, T. Nabeshima, and T. Kameyama, *Eur. J. Pharmacol.* **164,** 189 (1989).
7. T. Nabeshima, K. Tohyama, K. Ichihara, and T. Kameyama, *J. Pharmacol. Exp. Ther.* **255,** 789 (1990).
8. S. Pellow, P. Chopin, S. E. File, and M. Briley, *J. Neurosci. Methods* **14,** 149 (1985).
9. A. Ettenberg, M. Le Moal, G. F. Koob, and F. E. Bloom, *Pharmacol. Biochem. Behav.* **18,** 645 (1983).
10. S. Ahlenius, J. Engel, and M. Zoller, *Physiol. Psychol.* **5,** 290 (1977).
11. K. Ichihara, T. Nabeshima, and T. Kameyama, *J. Pharmacol. Exp. Ther.* **264,** 122 (1993).

# [14] Evaluation of Putative Anxiolytics in the Elevated Plus-Maze Test

## Michel Reibaud and Georg Andrees Böhme

## Introduction

The elevated plus-maze is now widely accepted as an animal model of "anxiety." It features technical simplicity together with a high throughput, thus allowing rapid pharmacological evaluation of drug effects on anxiety. The model is based on the observation of spontaneous activity of rodents placed in an aversive environment produced by height and open spaces.

Exploratory behavior in connection with fear was first studied by Montgomery in the 1950s. Using a Y-maze composed of enclosed and open arms, he observed that rats explore the walled areas significantly more frequently than the open arms of the apparatus. This behavior is believed to result from an approach–avoidance conflict generated by fear and a drive to explore. While both the open and enclosed parts of the maze would evoke the same exploratory drive, Montgomery suggested that the open arms evoke more fear, and this results in less exploration (1).

Several investigators applied these findings to the measurement of anxiolytic and anxiogenic properties of drugs in the mid-1980s. Most teams use a cross-shaped apparatus made of two open arms facing each other and two enclosed arms disposed at right angles to the open arms. This plus-maze is elevated several tens of centimeters above the floor to create an "anxiogenic" environment in the open arms. Rats or mice are placed individually in the center of the apparatus, and their exploratory behavior is observed over several minutes.

The ratio of the number of entries into the open arms to the total number of arm entries and the ratio of the time spent in the open arms to the total time spent in both types of arm are generally used as markers of the natural aversion of rodents for the open arms. It is the variation of these parameters in the absence of modification of the total number of arm entries that is interpreted as an effect of the drug on the level of anxiety. These markers may reflect a level of "anxiety" experienced by the test animals in exploring the maze. Although the degree of similarity between human anxiety and rodent behavior remains difficult to establish (2), these two parameters have been shown by several teams to be increased by clinically effective anxiolytics and to be decreased by anxiogenic drugs (3, 4). The predictive value of

*Methods in Neurosciences, Volume 14*

the plus-maze test has furthermore been validated physiologically by the measurement of plasma levels of stress hormones (3). This chapter reviews the methodology we use in the field. Examples of the "anxiolytic" activity of a cyclopyrrolone γ-aminobutyric acid type A (GABA-A) agonist and a cholecystokinin (CCK) B-type antagonist are presented at the end of the chapter.

## Plus-Maze Apparatus

Dimensions of the plus-maze arms vary little among laboratories. The open arms must be large enough to allow the animals to turn around without falling off, but sufficiently narrow to generate aversion. The apparatus may be made of wood, although plastic offers the advantage of being easily cleaned between each test to avoid odor trails that might interfere with exploratory behavior. We use for the rat an apparatus derived from that described by Pellow *et al.* (3). It is made of black plastic (Fig. 1), which provides a high contrast to albino rats so that they can readily be tracked by the computerized data acquisition system (see below). For mice, we use a scaled-down version of the maze developed by Lister (5) that has transparent walls to ensure a greater degree of light homogeneity between the open and closed areas.

## Experimentation Room

The testing should be performed in a stable, quiet room held at a constant temperature and humidity. Experiments can also be run directly in the animal maintenance rooms of the vivarium.

Illumination seems to play a crucial role in the plus-maze behavior of rats. Therefore, to avoid possible interference from variations in daylight, the rat plus-maze is kept in a windowless room under artificial illumination. Variations in light intensity modify the basal state of "anxiety" of the animals and therefore change the sensitivity of the test to drug effects. Low light intensities (10 lux or less) increase the exploration of the open arms such that open-to-total arm visit ratios may reach 50%, a point at which there is no longer any difference in exploration of the different parts of the apparatus. Low light intensities therefore reduce the sensitivity of the test for anxiolytics but may be useful to evaluate anxiogenics. Conversely, we have found that light intensities above 300 lux produce a complete blockade of open arm exploration. Lighting intensities around

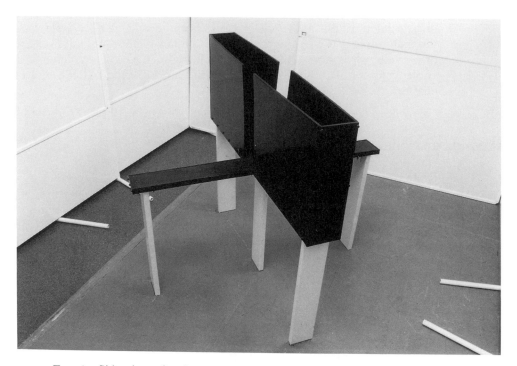

FIG. 1   Side view of a plus-maze apparatus for detecting drug effects on "anxiety" level in rats. The apparatus is made of black plexiglass. Each arm is 50 cm long and 9.5 cm wide. The walls of the two opposite, enclosed arms are 40 cm high and are separated from each other by the central area (9.5 × 9.5 cm). The plus-maze is elevated 50 cm above the floor. A 0.2 cm high strip borders the open arms to prevent the animals from falling off the maze. The apparatus is surrounded by white panels to reduce environmental effects.

200 lux over the center platform of the maze appear to be suitable for the testing of anxiolytics in rats.

The experimenter may be present in the experimentation room, but it is preferable to monitor the behavior from a distance, for example, in an adjacent laboratory via a video camera. Exploration of the open arms appears to be exquisitely sensitive to sound disturbances, which tend to precipitate the animals back to the enclosed arms where they will usually stay for the rest of the observation period.

# Procedure

## Animal Handling

Rats or mice can both be used in the plus-maze procedure, provided that the scale of the apparatus is adapted. The following procedural details refer only to rats. We routinely use male Sprague-Dawley rats weighing 200 to 240 g. After their transportation from the breeding farm, they are kept at least 10 days in the vivarium, prior to experimentation. The night before testing, the rats are delivered in cages of 6 to the experimentation room. As in the vivarium, light and dark periods alternate on a 12-hr cycle (light on at 6:00 a.m.), and the ambient temperature is maintained at 22 ± 2°C. Free access to food and water is also maintained overnight.

As it moves along the apparatus, a rat leaves a number of odor trails which may bias the exploratory behavior of its successors. Several precautions can be taken in order to minimize odor influence. For example, at the beginning of each test session, an additional rat not included in the experimental series can be allowed to explore the apparatus so that the olfactory environment does not affect the first individual of a series differently. Moreover, between each vehicle or drug-treated animal, the apparatus is thoroughly washed with 10% alcohol in water in order to mask the odors (cleaning with water alone appears to be insufficient).

## Drug Treatments and Controls

All systemic routes of administration (intraperitoneal, oral, subcutaneous) are suited for use in the plus-maze paradigm. Some investigators have also successfully developed protocols for intracerebral administration, either in the lateral ventricles (6) or directly into various brain structures (7). We commonly apply drugs in the form of extemporaneously prepared suspensions in distilled water containing 1% (v/v) Tween 80. Administration takes place 30 min and 1 hr before testing for the intraperitoneal and oral routes, respectively.

Although interindividual variations can be seen within a session, groups of around 10 rats appear to be generally sufficient to reliably assess drug effects. The percentage of entries into the open arms varies from one experiment to another. Figure 2 shows a comparison of the performances of vehicle-treated control rats taken from different sessions. These variations may lead to misinterpretation of "anxiolytic" effects and may call for positive controls

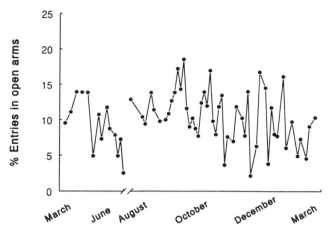

FIG. 2    Interexperimental variations in rat exploratory behavior. Comparison of the percentage of entries into open arms for vehicle-treated control rats taken at different times of the year. Note that variations from less than 5% to greater than 15% can be observed. Values are means for 8 to 14 rats.

to be used as internal standards when investigating new drugs. These variations possibly reflect the presence of individuals with different levels of intrinsic "anxiety," as is commonly observed among humans. Attempts to select the rats on objective behavioral criteria that may reflect their emotional level, such as spontaneous exploratory behavior in an open field, may be a way to reduce interexperimental variations.

## Computerized Data Aquisition

We use an automatic acquisition system which allows animals to be observed at a distance and their performances to be precisely assessed. This system (Videotrack) is based on a computer-driven image processor which analyzes the movements of the rats by detecting the displacement of their image through a video camera (Fig. 3). The detection is based on the contrast between the bright rat image and the background.

The automatic scoring proceeds as follows. The rat is placed at the intersection of the maze arms facing an open arm, so that its head is located approximately in the center of the area. The experimenter then initiates the acquisition and leaves the room. At each computer clock cycle, the system analyzes the image by detecting the parts, the brightness of which is either higher or lower than a given level defined by the user. A masking board restricts this

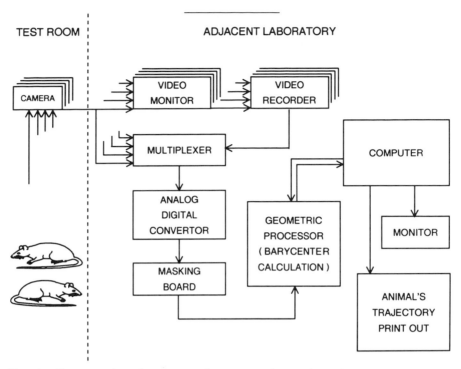

FIG. 3  Computer-based processor for remote observation of plus-maze behavior. Every 40 ms, the 512 × 512 pixel image of the rat through one or more multiplexed video cameras is digitalized on 256 gray levels. The source image is also monitored and recorded for off-line analysis. See text for further details.

image analysis to those areas of the maze of interest for the calculation of exploration parameters. The surface of the rat image is then calculated and the optical center of gravity located. The computer tracks the displacement of this center of gravity in the different parts of the maze and records the number of entries into each arm for the 5-min scoring session. As an example, monitoring of a rat treated with lorazepam is shown in Fig. 4. The rough data are down-loaded on spread sheets and database management software for calculations of mean values and statistical analysis.

## Effect of γ-Aminobutyric Acid Type A Agonists

Anxiolytic benzodiazepines (BZDs) exert their multiple pharmacological actions (including myorelaxant, anticonvulsive, and sedative–hypnotic properties) by binding to specific membrane receptors of the brain which

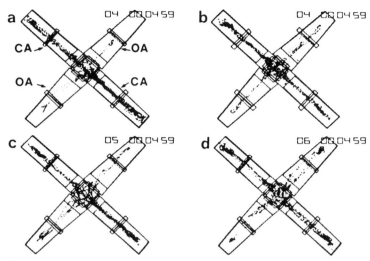

FIG. 4 Computer output of rat barycenter displacement in the plus-maze. (a) vehicle-treated control; (b) lorazepam, 0.15 mg/kg; (c) lorazepam, 0.3 mg/kg; (d) lorazepam, 0.6 mg/kg per os. The plus-maze image is divided into several virtual areas by the software. The shape of the image is such that an entry is scored only when all four limbs of the rat are within a given arm. Note the progressive increase in exploration of the open arms (OA) and the decrease of entries into the closed arms (CA), as shown by the density of dots representing the optical center of gravity of the rat.

are also recognized by compounds of other chemical structures such as the cyclopyrrolones (8). RP 59037 [2-(7-chloro-1,8-naphthiridin-2-yl)-3-(5-methyl-2-oxohexyl)-1-isoindolinone] is a member of this new class of compounds that has been developed as an anxiolytic with reduced incidence of sedation and myorelaxation (9). This compound possesses subnanomolar affinity for the BZD binding site on the GABA-A receptor, and it displays anticonvulsant activity with little or no sedative or myorelaxant effects compared with diazepam. This profile suggests partial agonist activity for RP 59037 [2,3-dihydro-2-(7-chloro-1,8-naphtyridin-2 yl)-3-(5-methyl-2-oxohexyl)-1H-isoindil-1-one] at the GABA-A/BZD/chloride ionophore complex. Figure 5 features the effects of this compound in the elevated plus-maze test. RP 59037 increased both the percentage of entries into and the percentage of time spent in the open arms. This effect appeared to be dose dependent over the dose range tested (0.07 to 1.25 mg/kg per os). Analysis of variance (ANOVA) followed by Dunnett's $t$-test indicated that the first active dose reaching statistical significance (minimal effective

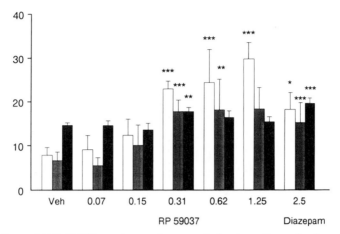

FIG. 5 Effect of RP 59037 on the exploratory behavior of rats in the plus-maze. Rats were scored in the plus-maze apparatus for 5 min. Values are expressed on the same axis as percentages (open bars, entries into open arms; hatched bars, time therein) with the exception that total number of entries is shown as absolute values (solid bars). Diazepam (2.5 mg/kg) was tested in parallel as a positive control (last histogram). Compounds were administered orally 1 hr before testing (doses are given in mg/kg). All values are means ± S.E.M. for 6–12 animals. Asterisks denote significant differences from vehicle-treated controls as follows: single asterisk, $p < 0.05$; double asterisk, $p < 0.01$; triple asterisk, $p < 0.001$. Veh, Vehicle.

dose) was 0.3 mg/kg. Total arm entries were slightly, but significantly, increased at the dose of 0.3 mg/kg, but this effect was not seen at any other higher or lower doses, indicating that a modification in locomotor activity is not likely to account for the observed effects. RP 59037 behaves, therefore, as an anxiolytic in the plus-maze test. This compound appears to be both more active and more potent than the BZD anxiolytic diazepam.

## Effect of Cholecystokinin B-Type Receptor Antagonists

Cholecystokinin octapeptide (CCK-8) is one of the major neuropeptides of the brain. It recognizes there two different type of receptors, referred to as CCK-A and CCK-B receptors, for each of which nonpeptide antagonists have been developed. Recent evidence from both clinical and animal studies indicates that administration of exogenous CCK agonists induces panic attacks in humans and an anxiogenic-like behavior in animals, suggesting a

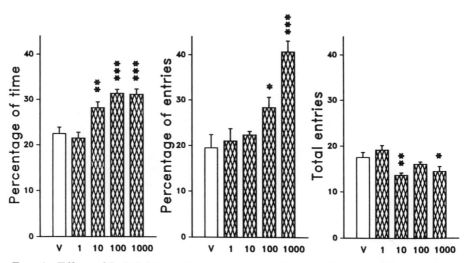

FIG. 6   Effect of L-365,260 on the exploratory behavior of mice in the plus-maze. Performances of control (open bars) and treated (cross-hatched bars) mice (5 min observation) are expressed on three different graphs as the percentage of time spent in open arms, the percentage of entries therein, and the total number of entries. The CCK-B antagonist L-365,260 was given intraperitoneally 30 min before testing (doses in $\mu$g/kg). All values are means $\pm$ S.E.M. for 8 animals. Asterisks denote significant differences from vehicle-treated controls as follows: single asterisk, $p < 0.05$; double asterisk, $p < 0.01$; triple asterisk, $p < 0.001$. [Reprinted by permission from J. Rataud, F. Darche, O. Piot, J. M. Stutzmann, G. A. Böhme, and J. C. Blanchard, *Brain Res.* **548,** 315 (1991); copyright Elsevier.]

possible role for CCK in anxiety (for review, see Ref. 10). To examine this hypothesis, we have tested the effects in the plus-maze of L-365,260, a selective CCK-B antagonist (11). As shown in Fig. 6, L-365,260 produced a dose-dependent increase in both the percentage of time and the percentage of entries into the open arms. This effect reached statistical significance for these parameters at doses of 0.01 and 0.1 mg/kg (i.p.), respectively. The total number of entries was decreased by about 20% for the doses of 0.01 and 1 mg/kg, but this effect did not seem to be dose related. These results indicate that the CCK-B antagonist L-365,260 reverses the natural aversion of mice for the open part of the elevated plus-maze. In the absence of consistent modifications of the total number of entries, this suggests that L-365,260 may have anxiolytic properties. This is an accordance with the recently reported "anxiolytic" activity of CI-988, another selective CCK-B antagonist (6).

## Acknowledgments

We thank Karen Pepper for stylistic revision of the manuscript.

## References

1. M. C. Montgomery, *J. Comp. Physiol. Psychol.* **48,** 254 (1955).
2. R. G. Lister, *Pharmacol. Ther.* **46,** 321 (1990).
3. S. Pellow, P. Chopin, S. E. File, and M. Briley, *J. Neurosci. Methods* **14,** 149 (1985).
4. S. L. Handley and S. Mithani, *Naunyn-Schmiedeberg's Arch. Pharmacol.* **327,** 1 (1984).
5. R. G. Lister, *Psychopharmacology (Berlin)* **92,** 180 (1987).
6. L. Singh, A. S. Lewis, M. J. Field, J. Hughes, and G. N. Woodruff, *Proc. Natl. Acad. Sci. U.S.A.* **88,** 1130 (1991).
7. V. Daugé, P. Steimes, M. Derrien, N. Beau, B. P. Roques, and J. Féger, *Pharmacol. Biochem. Behav.* **34,** 157 (1989).
8. A. Doble, T. Canton, O. Piot, J. L. Zundel, J. M. Stutzmann, C. Cotrel, and J. C. Blanchard, *in* "GABAergic Synaptic Transmission" (G. Biggio, A. Concas, and E. Costa, eds.), p. 407. Raven Press, New York, 1992.
9. O. Piot, M. C. Bardone, J. L. Zundel, J. M. Stutzmann, A. Boireau, and A. Doble, *Br. J. Pharmacol.* **99,** 133P, 1990.
10. S. Ravard and C. T. Dourish, *Trends Pharmacol. Sci.* **11,** 271 (1991).
11. M. G. Bock, R. M. DiPardo, B. E. Evans, K. E. Rittle, W. L. Whitter, D. F. Veber, P. S. Anderson, and R. M. Freidinger, *J. Med. Chem.* **32,** 16 (1989).

# Section IV

## Environmentally Induced Behavior

# [15] Measurement of Behavioral Changes in the Fetus Caused by Vibroacoustic Stimulation

M. J. Parkes

Vibroacoustic stimulation is currently seen as an important new tool for investigation of normal and abnormal function in the central nervous system (CNS) of the fetus *in utero*. The three main advantages of vibroacoustic stimulation are that it is noninvasive, apparently safe, and produces obvious responses in the fetus. This chapter describes how to induce responses in the human and sheep fetus. It also provides a brief introduction to the problems of interpreting the responses.

## Human Fetus

### The Stimulus

Of the physical forces that can be applied against the abdomen and may be used to investigate the fetal CNS, so far only vibroacoustic stimulation has been shown to induce immediate responses in the human fetus. Figure 1A shows the effect of vibroacoustic stimulation on heart rate at 38 weeks of gestation (1). Figure 1B shows that such effects occur significantly more often than expected by chance, that is, they can be said to be caused by the stimulus.

Vibroacoustic stimuli have been generated from a loudspeaker held a short distance from the abdomen, from a modified electric tooth brush motor, or from a model 5C electrolarynx held against the abdomen. The common effect is to generate pressure waves of various frequencies and intenstties that change with time. These waves propagate in all directions through a variety of differentially compressible tissues and incompressible liquids and stimulate the fetus. The distinction between vibration and sound is deliberately not made since the mechanism by which the human fetus detects the stimulus is not clear. Of these various generators, the loudspeaker is potentially least efficient because of the impedance mismatch between air and body tissues. The electrolarynx is preferred since it requires no prior modification, and the model 5C is chosen because of its worldwide availability. (The electrolarynx was developed originally as an means of aiding speech production in patients after laryngectomy.)

*Methods in Neurosciences, Volume 14*

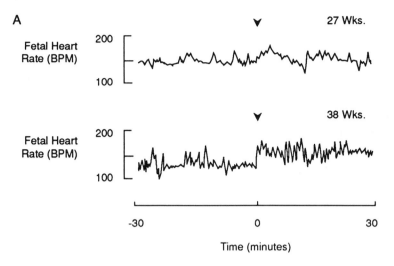

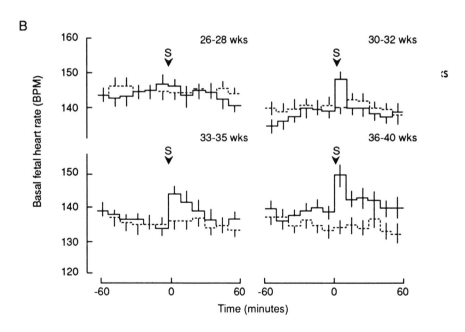

Traditionally the electrolarynx is held by the experimenter against the maternal abdomen and directly over the fetal head (located by ultrasound). The mother is aware of its placement and use but is not in any way disturbed by it. The electrolarynx is switched on for 5 sec. One of the principal sources of variability in transmission of the stimulus is the force with which the electrolarynx is pressed against the abdomen. Too little force enables energy to be dissipated in air, and too much prevents vibration of the piston. An appropriate force is usually judged by experience. Although providing a relatively constant stimulus, the electrolarynx is of limited use for further scientific investigation of stimulus–response relationships since the frequency and intensity of the stimulus cannot be varied. There are no detailed accounts of the effects of varying the duration of the stimulus.

## The Responses

Fetal heart rate is not necessarily the ideal variable for investigation of CNS function, but is chosen for three reasons: (1) it can be measured noninvasively by ultrasound, (2) it is a continuous variable, therefore easily quantified, and (3) in obstetrics clinics computerized data collection and analysis techniques are readily available to analyze fetal heart rate. Even in the unstimulated fetus, heart rate (2) shows continuous short- and long-term variation (Fig. 2), and therefore randomized trial designs are preferred. Furthermore, because it is easiest to detect a change in fetal heart rate when heart rate is at its most stable, vibroacoustic stimulation is usually applied during episodes of comparatively low variability. The response can be quantified simply by numerical analysis of the mean (Fig. 1B), although the response is also obvious using more advanced analysis of heart rate variability (3).

Vibroacoustic stimulation also changes the incidence of other types of fetal activity, including activity which may originate only from the CNS and spinal cord, for example, body movements, breathing movements, and eye blinks (4, 5). There are classification schemes proposed for analyzing fetal behavior in terms of specific and unique combinations of heart rate, body, and

FIG. 1   (A) Examples of an increase in human fetal heart rate after a 5-sec vibroacoustic stimulus with the electronic artifical larynx, in a 27-week fetus and in a 38-week fetus. (B) Mean basal fetal heart rate (±SEM) plotted in 10-min intervals for groups of fetuses at 26–28, 30–32, 33–35, and 36–40 weeks of gestation, before and after either control (dashed line) or vibroacoustic stimulation (solid line). [Reproduced with permission from Gagnon et al. (1).]

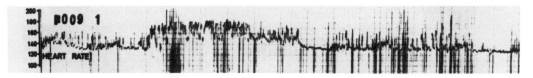

FIG. 2    Polygraph recording (8 hr) showing heart rate in a human fetus. [Reproduced with permission from Junge (2).]

eye movement changes (6), and for analyzing the responses to vibrocoustic stimulation using such combinations (7). The question of whether such combinations are any more meaningful than simply considering each variable alone is discussed elsewhere (6, 8, 9).

## Interpretation of Responses

The ultimate aims of studying these stimulus–response paradigms in the fetus are to understand normal function of the fetal CNS, to detect abnormal function, and to try and diagnose the primary cause of the abnormality so that appropriate treatment might be given. These aims can only be attained if the mechanisms by which the human fetus detects and processes vibroacoustic stimulation are known. It is important, therefore, to understand the limitations in our current knowledge of the mechanisms of detection and processing of vibroacoustic stimuli by the human fetus.

The primary limitation at present is that the mechanism of detection is unknown. It is possible that vibroacoustic pressure waves act directly to compress the fetal heart or brain stem. The body, however, has specialized receptors in the auditory, vestibular, and somatosensory apparatus that are designed to detect a variety of mechanical vibrations. Therefore, it is more likely that vibroacoustic stimuli are detected by these receptors. Subsequent neural traffic from them would then follow the normal central nervous pathways and finally act on heart rate via autonomic efferents.

In the human fetus there are at present only limited techniques available to resolve these issues. Clinically, if a human fetus is compromised with certain defined illnesses, it is acceptable to take a blood sample. This is performed without anesthesia by passing a needle through the maternal abdominal wall, under ultrasound guidance, and entering the umbilical vein. This procedure can be combined with vibroacoustic stimulation (10). Figure 3 shows that vibroacoustic stimulation during such procedures causes no change in fetal plasma catecholamine levels. Thus, the possibility of the

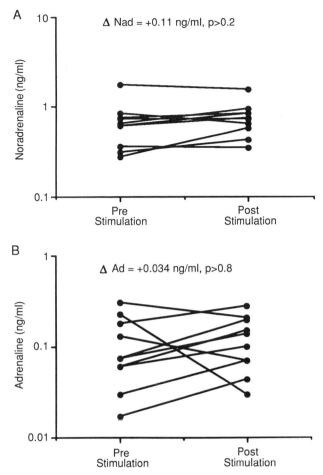

FIG. 3 Concentration of noradrenaline (A) and adrenaline (B) in plasma from the umbilical vein before and 1 min after vibroacoustic stimulation of the human fetus. [Reproduced with permission from Fisk *et al.* (10).]

increase in fetal heart rate being caused by a rise in plasma catecholamine levels, released from the adrenal gland after sympathetic stimulation, can be eliminated.

Modification of this protocol could enable investigation of the role of other parts of the autonomic efferent system, for example, by testing whether vibroacoustic stimulation still stimulates fetal heart rate after pharmacological blockade of transmission from the cardiac or vagal nerves. Furthermore, if blockade of transmission from both types of autonomic efferent could be justified clinically, demonstration that vibroacoustic stimulation still stimulated fetal heart rate would be strong evidence for a direct effect of vibroacoustic waves on the heart itself.

Controlled trials in the unanesthetized human fetus have shown that strong shaking of the uterus from side to side produces no change in fetal heart rate or motor activity (11, 12). Such shaking would be expected to provide a powerful stimulus to the fetal vestibular system. The fact that the fetus does not respond to shaking suggests that vestibular stimulation is not involved in the detection of vibroacoustic stimuli by the human fetus.

It has been suggested that if the fetus responded more to low-frequency vibroacoustic stimulation than to high-frequency stimulation, this would be evidence in favor of "vibratory" rather than "acoustic" stimulation. This extrapolation, however, is unlikely to be valid for two reasons. First, there is some overlap in the low-frequency response range of auditory and classic mechanoreceptors. Second, the tissue and fluid barriers may transform even simple sine waves into more complex waves. An alternative solution might be to identify babies after birth who could be shown to have been deaf through mechanisms which must have also applied *in utero,* and to hope that some conscientious obstetrician had demonstrated failure to respond to vibroacoustic stimulation *in utero*.

A second major limitation is that it is not clear how intense a stimulus to the fetus is provided by the model 5C electrolarynx. Even use of hydrophones *in utero* to measure the intensity of vibroacoustic stimuli relative to background intrauterine noise or maternal (or paternal) speech gives no indication of how sensitive the fetus is to this intensity of stimulation. Thus, the fetal heart rate response to stimulation from the electrolarynx can be viewed either as an extreme response to an intense stimulus (13) or a slight response to a relatively weak stimulus.

## Sheep Fetus

### Chronically Instrumented Sheep Fetus

Much fundamental information on the mechanisms of detection and processing of vibroacoustic stimulation can be gained by study of the responses in a species in which more detailed invasive procedures are possible. Since the 1970s anesthetic, analgesic, surgical, and instrumentation techniques have been available in sheep to enable sterile implantation of catheters into fetal blood vessels, electrodes into fetal muscles and fetal nerve tissue, and a range of other monitoring devices into the amniotic cavity. The basic surgical procedures are described elsewhere. The fetus continues its normal growth and development *in utero,* and its mother is free to move about. An experimental preparation unique to biomedical science is therefore available in which continuous measurements of fetal blood pressure, hormone levels, motor and electroencephalograph (EEG) activity can be made over many weeks in the unanesthetized fetus.

Because the sheep fetus is more accessible than the human fetus, there is more information available on its spontaneous activity. In brief, from about 125 days of gestation until term (147 days) fetal EEG activity is differentiated into states of synchronized (high voltage) and desynchronized (low voltage) activity. The states are of similar duration, and the fetus switches spontaneously between the two states throughout the day and night. There is no obvious entrainment of the switching to maternal activity or other external clues. Much of fetal visceral and somatic motor activity appears to coincide with the desynchronized EEG state (e.g., heart rate variability, eye movements, breathing movements; see Fig. 4). The precise relationship of these EEG states to the postnatal states of sleep and wakefulness, however, is unclear (8, 9).

Fetal EEG, muscle, and eye activity is measured simply using Teflon-coated stainless steel electrodes (Cooner Wire Company, Chatsworth, CA, wire number AS 633). Paired electrodes (plus a ground) are threaded through a protective PVC sleeve and cut sufficiently long (1.1 m) to pass through the maternal flank to reach preamplifiers anchored to the back of the ewe. With differential amplification, screened cables are unnecessary. The incidence of fetal eye movements is measured from bare electrodes sewn subcutaneously above and below each orbit. These detect the large dc voltage shifts ($\sim 1$V) as the eyeball moves. Muscle electromyograph (EMG) activity ($\sim 100~\mu$V) is recorded from bare electrodes sewn about 5 mm apart into the body of each muscle, with the signal passed through a high-pass filter of 100–300 Hz and a leaky integrator ($T = 30$–100 msec).

There is much dispute about how best to measure EEG activity in fetal sheep. This is because although the electrode location and electronic signal filtering determine the precise frequencies at which most power is found, it is not clear from where in the CNS the EEG activity originates, or what is the physiological significance of power at any one frequency. Since much fetal visual and somatic motor activity, however, coincides with the desynchronized EEG state, it is simplest to choose electrode position and analysis combinations that produce the greatest difference in EEG state when this motor activity is present or absent. Such a combination is described by Clewlow *et al.* (14) with electrodes passed through the parietal bone, 5 mm either side of the midline, and glued in position with cyanoacrylate glue. After amplification, waves of approximately 50–300 $\mu$V are obvious without further signal processing, although the record looks best if only frequencies between 3 and 12 Hz are selected.

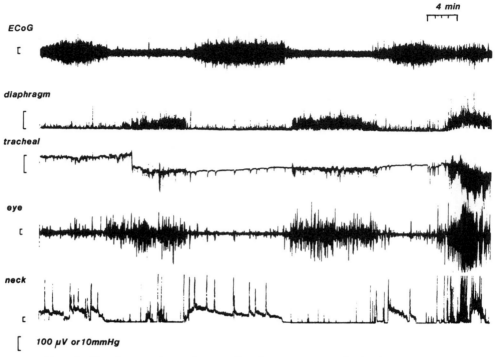

FIG. 4   One-hour polygraph recording from an unanesthetized sheep fetus *in utero* at 139 days of gestation showing, electrooculogram (eye), electroencephalogram (ECoG), fetal breathing movements (diaphragm electromyogram and tracheal pressure), and neck muscle electromyogram. [Reproduced with permission from Parkes (9).]

## The Stimulus

Because of its use in humans, vibroacoustic stimulation was initially applied to the sheep fetus across the maternal abdomen. This is easiest with the ewe standing against an opaque screen, through which the experimenter passes the stimulator. The vibroacoustic stimulator is held against the ventral surface of the abdomen, on the midline and 5 cm rostral to the umbilical scar. Daily practice of these maneuvers during the postoperative recovery period results in the ewe ignoring the stimulus during experimentation. Controlled trials, using 45 min of vibroacoustic stimulation from a model 5C electrolarynx (power consumption 0.36 W) combined with infusions of radioactively labeled glucose, have shown that vibroacoustic stimulation causes an increase

TABLE I    Increase in Glucose Uptake in Auditory Pathway Following Vibroacoustic Stimulation[a]

| | Glucose uptake ($\mu$mol/100 g/min) | | |
| Structures | Control | Electronic artificial larynx | Control versus artificial larynx: $p$ value |
| --- | --- | --- | --- |
| Auditory − sensorimotor cortex | 6.1 | 15.4 | <0.05 |
| Medial − lateral geniculate body | 9.4 | 18.8 | <0.50 |
| Inferior − superior colliculus | 70.6 | 168.2 | <0.05 |
| Lateral lemniscus − Periaqueductal gray | 37.1 | 84.6 | <0.05 |
| Superior − inferior olive | 6.9 | 67.0 | <0.01 |
| Cochlear − vestibular nuclei | −8.6 | 53.8 | <0.01 |

[a] Mean differences in local cerebral glucose utilization 100 g/min) between central auditory structures and nonauditory structures were determined in unanesthetized fetal sheep during cortical stimulation (5 fetuses) or 45 min of stimulation with a model 5C electrolarynx (5 fetuses). Reproduced with permission from R. A. Abrams, *Dev. Brain Res.* **48,** 1 (1989).

in glucose uptake along the auditory pathway from the periphery to the cerebral cortex as well as in other regions of the CNS (15) (Table I).

Two independent studies have failed, however, to detect any effect of vibroacoustic stimuli from the model 5C electrolarynx on fetal behavior (15, 16). We therefore developed a more powerful stimulator using a mechanical oscillator (Fig. 5). This is coupled to the abdomen with ultrasound jelly and driven from a wave form generator via a standard 40 W audio amplifier (16). The stimulus was initially designed to give as wide a frequency spectrum as possible. Therefore, a square wave was used whose frequency was continuously varied manually between 0.4 and 10 kHz, and the oscillator was driven at maximum power (2 W at 400 Hz increasing to 20 W at 10 kHz). Since pilot experiments indicated that the most obvious effect was to desynchronize fetal EEG activity (see later) the duration of the stimulus was chosen so that, at a recorder speed producing clear differentiation between states of synchronized and desynchronized EEG activity, the stimulus produced a visible change in fetal EEG activity. A protocol was used in which, once the fetus had been in either state for 2 min, a random number generator dictated whether vibroacoustic stimulation or no vibroacoustic stimulation was applied. As with the electrolarynx, the transduction of this stimulus also depends on the force used to hold it against the abdomen. Stimulus–response curves for different frequencies, intensities, and durations have yet to be determined.

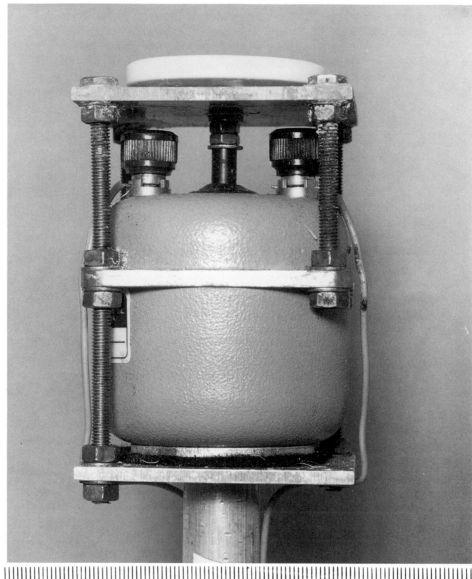

FIG. 5   Mechanical oscillator (Model 101, Ling Dynamic Systems, Royston, Herts, England) and piston assembly for vibroacoustic stimulation in fetal sheep [as used by Parkes *et al.* (16)].

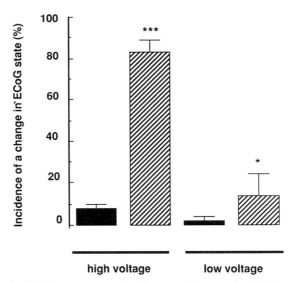

FIG. 6 Mean (±SEM) percentage incidence of change in EEG state during trials with (hatched bars) or without stimulation (solid bars) for trials beginning in the high-voltage EEG state (five fetuses) or the low-voltage EEG state (five fetuses) ($n = 252$ trials; single asterisk denotes $p < 0.05$ and three asterisks, $p < 0.001$, for stimulation versus no stimulation in high-voltage EEG or low-voltage EEG states). ECoG, Electroencephalograph activity. [Reproduced with permission from Parkes *et al.* (16).]

## The Responses

Figure 6 shows that the most obvious response of fetal sheep to vibroacoustic stimulation is to change from the synchronized to the desynchronized EEG state, and the altered state is accompanied by most of its appropriate motor activity (16). Because stimulation and recording can be performed easily in either EEG state, the response can also been shown to be largely unidirectional, that is, a desynchronization but not a resynchronization. An obvious difference with the response to vibroacoustic stimulation of the human fetus is that heart rate in fetal sheep barely changes; if anything, it falls rather than rises during vibroacoustic stimulation (16). Since both species can therefore at least detect vibroacoustic stimuli, this species difference would appear to be at a more central processing stage.

## Interpretation of Responses

Because of greater accessibility, important advances in understanding the mechanisms of detection and processing of vibroacoustic stimuli are possible

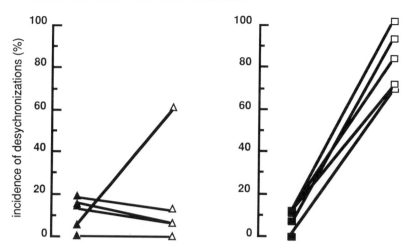

FIG. 7   Incidence of desynchronization in stimulation (open symbols) and no stimulation trials (filled symbols) begun in the high-voltage EEG state in five cochlear-ablated
fetuses ($n = 120$ trials) and five intact fetuses ($n = 155$ trials, replotted from Fig. 6).
[Reproduced with permission from Parkes, *et al.* (16).]

in the sheep fetus. In particular there is extensive experience in applying
lesion and ablation techniques to the fetal CNS.

Bilateral ablation of the fetal cochleae provides the simplest means of
inactivating the peripheral auditory apparatus of the fetus *in utero*. This is
performed during the operation for catheterization. It simply involves exposure of the bulla on each side, drilling through the bone, and ablation of the
cochlea by blunt dissection and aspiration under a low-power dissecting
microscope (16). The bone is sealed with a flap of skin. The success of the
technique can be assessed at postmortem by examination of serial sections
of the turns of the cochlea under a light microscope. Success could also be
assessed *in utero* by demonstrating failure to evoke potentials over the auditory cortex.

In 4 of 5 fetal sheep with cochlear ablation, we found complete failure to
respond to vibroacoustic stimulation (Fig. 7), and in the fifth fetus the responses were abnormally reduced. These results therefore provide strong
evidence that in fetal sheep the auditory system is necessary in detecting
vibroacoustic stimulation. Techniques for chronically recording activity from
nerves or auditory cells in the fetus are under development. It still remains
to be shown that nerve activity in the auditory pathways increases during
vibroacoustic stimulation sufficient to desynchronize fetal EEG activity.
Furthermore, in fetal sheep it should be possible to address the crucial

question of how intense the vibroacoustic stimuli are to the fetus. This could be done by recording the cochlear microphonic potential *in utero* and deriving stimulus–response curves during vibroacoustic stimulation at various frequencies, intensities, and durations.

## Summary and Conclusions

Techniques are described using vibroacoustic stimuli to stimulate the human and sheep fetus and to record the various behavioral responses. A brief introduction is given to the problems of interpreting these responses. In humans, the intensity of the stimulus as detected by the fetus and the mechanism by which the fetus detects the stimulus are not known. In the sheep fetus techniques are described which reveal that detection of vibroacoustic stimuli involves the peripheral auditory apparatus. Techniques are also described which could indicate how the intensity of the stimulus, as detected by the fetus, might be measured.

## Acknowledgments

The experimental work of the author described here was supported by a Wellcome Trust Lectureship. The author is currently supported by the British Heart Foundation.

## References

1. R. Gagnon, C. Hunse, L. Carmichael, F. Fellows, and J. Patrick, *Am. J. Obstet. Gynecol.* **157,** 1375 (1987).
2. H. D. Junge, *J. Perinat. Med.* **7,** 85 (1979).
3. L. D. Devoe, P. Gardner, P. Arnold, and N. Searle, *Am. J. Obstet. Gynecol.* **160,** 1086 (1989).
4. R. Gagnon, C. Hunse, L. Carmichael, F. Fellows, and J. Patrick, *Am. J. Obstet. Gynecol.* **155,** 1227 (1986).
5. J. C. Birnholtz and B. R. Benacerraf, *Science* **222,** 516 (1983).
6. J. G. Nijhuis, H. F. R. Prechtl, C. B. Martin, and R. S. G. M. Bots, *Early Hum. Dev.* **6,** 177 (1982).
7. R. Gagnon, J. Foreman, C. Hunse, and J. Patrick, *Am. J. Obstet. Gynecol.* **161,** 1479 (1989).
8. M. J. Parkes, in "The Fetal and Neonatal Brainstem" (M. A. Hanson, ed.), p. 230. Cambridge Univ. Press, Cambridge, 1991.
9. M. J. Parkes, *Q. J. Exp. Psychol.* **44,** 231 (1992).

10. N. M. Fisk, P. K. Nicolaidis, S. Arulkumaran, M. W. Weg, Y. Tannirandon, U. Nicoli, M. J. Parkes, and C. H. Rodeck, *Early Hum. Dev.* **25,** 11 (1991).

11. G. H. A. Visser, H. J. Zeelenberg, J. P. de Vries, and G. S. Dawes, *Am. J. Obstet. Gynecol.* **145,** 579 (1983).

12. B. Richardson, K. Campbell, L. Carmichael, and J. Patrick, *Am. J. Obstet. Gynecol.* **139,** 344 (1981).

13. G. H. A. Visser, H. N. Mulders, H. P. Wit, E. J. H. Mulders, and H. F. R. Prechtl, *Early Hum. Dev.* **19,** 285 (1989).

14. F. Clewlow, G. S. D. Dawes, B. M. Johnston, and D. W. Walker, *J. Physiol. (London)* **341,** 463 (1983).

15. R. A. Abrams, *Dev. Brain Res.* **48,** 1 (1989).

16. M. J. Parkes, P. J. Moore, D. R. Moore, N. M. Fisk, and M. A. Hanson, *Am. J. Obstet. Gynecol.* **164,** 1336 (1991).

# [16] Behavioral Analysis of Rats Exposed to High Pressure

J. C. Rostain, J. H. Abraini, and C. Tomei

## Introduction

A diver is exposed to an increase in pressure that is dependent on the depth of seawater. At 10 m the absolute pressure equals 0.2 MPa [0.1 MPa of atmospheric pressure plus 0.1 MPa for the 10 m of seawater (msw)], and at a depth of 500 msw the absolute pressure is 5.1 MPa.

The relationship between pressure and volume requires that the diver must breathe a gaseous mixture at the ambient pressure. Both the pressure itself and the partial pressure of each gas in the breathing mixture affect the organism. Increasing the pressure of ambient air induces nitrogen narcosis at pressures above 0.6 MPa. For deeper dives synthetic mixtures with a less narcotic gas must be used, such as a helium–oxygen mixture where helium replaces nitrogen and oxygen is maintained at a partial pressure not higher than 0.04 MPa (1, 2). However, when humans or other mammals are exposed to high-pressure helium–oxygen mixtures, disturbances that differ from those produced by nitrogen narcosis occur above pressures of 1.5 to 2 MPa. These disturbances have been called the high-pressure nervous syndrome (HPNS) (3). The principal symptoms include electroencephalographic (EEG) changes and motor and behavioral disturbances such as muscle tremor, locomotor hyperactivity, and myoclonus, followed, in animals exposed to pressures higher than 8 MPa, by convulsions and epileptic seizures (3–8). In humans, HPNS includes other problems such as motor incoordination, disorientation, nausea, and loss of attention (1, 2).

Hyperbaric tremor is first observed at the extremities of the limbs during activity. It begins at pressures of 1.5 MPa, increases with pressure, and progressively spreads to all the limbs and the trunk. The tremor frequency in animals is between 10 and 16 Hz (9). Hyperbaric locomotor hyperactivity (LHA) consists of ambulatory activity that occurs around 3 MPa. (10). Behavioral disturbances have also been reported in rats, such as hoarding of sawdust and pellets, undifferentiated eating of pellets and dried excrement, tail holding with forepaw and mouth, and body flattening against the home cage (11). Myoclonus consists of muscular jerks that first occur at the extremities (limbs, head) and then spread throughout the body with increasing pressure.

Epileptic seizures are of the grand mal type and consist of tonic–clonic crises.

In most cases, behavioral motor disturbances in restrained or free-moving animals have been visually estimated only by their pressure onset (6–8). This method requires that one individual make all the observations; also it is difficult to quantify the intensity of motor disturbances by observation alone when arbitrary scale scores have been used (12). There are some reports of quantitative studies of one or more behavioral symptoms of HPNS by using noninvasive techniques. These have included the use of a mechanical transducer (13), a magnetic induction device (14), and a small strain gauge (9). To quantify objectively the behavioral motor disturbances of HPNS, we have developed a computerized device which enables continuous long-term analysis (15) of free-moving rats exposed to high pressure.

## Method

Behavioral analysis of rats exposed to high pressure is made by recording motor activities and disturbances using three piezoelectric sensors (No. P 188, Quartz et Silice, Paris, France), connected in series, which are fixed under the floor of an Altuglass cylinder designed for one rat. The signals from the sensors are amplified before being sent to an analog-to-digital converter connected to a PC-AT compatible computer. The program (written in Quick Basic) samples the signal at 120 Hz in 4-sec epochs, which are displayed in real time on a graphic monitor and operated on two channels (i.e., analysis is performed simultaneously in two freely moving rats, each placed in one cylinder). The signal is analyzed both in terms of its amplitude and in frequency bands between 0 and 60 Hz. Myoclonic jerks are detected as signals of unusually high amplitude using a trigger with a detection threshold that is adjusted for each rat after visual observation of the intensity of myoclonus, which can vary among animals. These signals are subtracted from the total signal by the computer and integrated to obtain the score for myoclonus. The remaining signal (i.e., the whole signal minus myoclonus) is analyzed using the fast Fourier transform (FFT) to obtain the power spectra. The power spectra obtained are split into three frequency bands, 0–10, 10–16, and 16–60 Hz. The power obtained in the 0–10 and 16–60 Hz bands is considered as being LHA and the power in the 10–16 Hz band as being tremor.

The different values for LHA, tremor, and myoclonia are calculated on-line and displayed every 4 sec on a graphic monitor, which shows the signal obtained from the sensors, the trigger level, the power spectra, and the values in each of the frequency bands. These data are averaged every 10 min, displayed in bar graph histograms, and stored in files on floppy disk for further calculations such as group means, percentage changes from reference values, curves, and statistical tests. The data are expressed in arbitrary units.

## Procedure

Free-moving rats are placed in individual Altuglass cylinders in a 50-liter pressure chamber (maximum pressure 20 MPa). Behavior is recorded for 24 hr to determine circadian rhythms at atmospheric pressure. The rats are then compressed up to 8 MPa relative pressure (800 msw) with helium at a compression rate of 0.1 MPa/min (duration 80 min). Oxygen is maintained at 0.04 MPa, which is the partial pressure generally used in human dives. Carbon dioxide is controlled to be less than 0.0003% using soda lime absorbers. Humidity is maintained between 50 and 80% with Cecagel. The ambient temperature is progressively increased from 25 to 33°C by circulating warm water inside the chamber. This temperature increase has been shown to maintain the body temperature of rats in a helium–oxygen atmosphere within the thermal range. Two fans situated at each end of the pressure chamber provide adequate gas mixing.

The stay at maximum pressure (8 MPa) lasts 4 hr. The decompression is carried out first at a rate of 0.006 MPa/min between 8 and 1.2 MPa, then 0.004 MPa/min from 1.2 MPa to atmospheric pressure. The partial pressure of oxygen is kept constant at 0.05 MPa during the decompression. Food and water are available throughout the experiments.

During pressure experiments the analysis of behavioral motor disturbances is carried out as described previously. The data obtained from 4-sec sampling epochs are averaged every 10 min (i.e., every 1 MPa) during the compression. Further calculations are performed on the data to obtain a mean for every hour during the 4-hr stay at 8 MPa and a mean for every 1 MPa during the decompression.

Median values, the 25th–75th percentiles, and the Wilcoxon signed-ranks test for paired samples are used to compare data obtained under pressure with those obtained during control measurements at atmospheric pressure with the same rats. The $U$ test is used to compare data obtained at pressure and those obtained at atmospheric pressure with different samples (16).

In the animal house and during experiments rats are maintained under a 12-hr light/dark cycle (lights on from 0700 to 1900 hr). At atmospheric pressure, the circadian activity of the rats is scored from 1000 hr in the morning to 1000 hr the next day. Compression is always started at the same time, namely, 1000 hr in the morning.

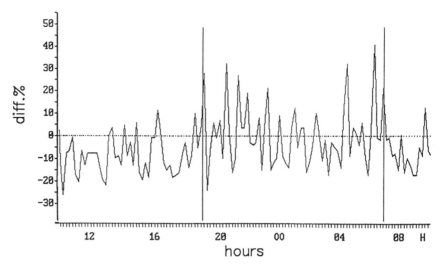

FIG. 1    Motor activity of free-moving rats at atmospheric pressure during a 24-hr recording. The dark period between 1900 and 700 hr is represented by the straight lines. Motor activity is plotted as the median value calculated from four rats and expressed as the percentage difference from the mean calculated for the 24-hr recording (y axis). The x axis shows the time from 1000 hr in the morning to 1000 hr the next day.

During the high-pressure experiments, the animals are observed through the portholes of the chamber by the same experimentor. The estimated onset pressures of the behavioral disturbances are compared to the data obtained with the sensors and subsequent computer analysis.

## Results

The behavioral recording carried out during 24 hr at atmospheric pressure (Fig. 1) showed that the activity of the rats increased during the dark period (1900–700 hr) and decreased during the light period (700–1900 hr). When we consider the mean of the motor activity during the 24-hr recording, the range of activity expressed as the percentage difference from the mean was of −25% during the light period to +35% during the dark period.

Rats compressed to 80 bars showed all the behavioral motor disturbances of HPNS except epileptic seizures. Locomotor hyperactivity (Fig. 2) occurred at 3.8 MPa (percentile range 2.9–4.7 MPa), and the maximum increase was recorded at the end of compression [631 arbitrary units (U);

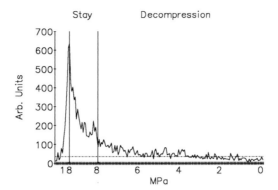

FIG. 2  Quantitative analysis of locomotor hyperactivity (LHA) in free-moving rats ($n = 6$) exposed to a high-pressure helium–oxygen mixture (8 MPa). The progression from (left) predive control (1 hr) and compression up to 8 MPa (1 hr 20 min) to (middle) a 4-hr stay at 8 MPa and finally (right) decompression from 8 MPa to atmospheric pressure (duration 24 hr) is shown. The dark period is between 1900 and 700 hr (i.e., between 6.8 and 2.8 MPa). The $y$ axis shows median LHA value calculated from six rats expressed in arbitrary units, and the $x$ axis shows pressure expressed in megapascals. The area from the $x$ axis to the dotted line represents the range of the motor activity at atmospheric pressure. Increases in LHA are recorded during compression. The increase is significant ($p < 0.05$) compared to the motor activity recorded during 24 hr at atmospheric pressure ($U$ test).

percentile range 460–1088 U; Wilcoxon paired test: $p < 0.05$]. The pressure onset of LHA was visually estimated at a mean pressure of 4.3 MPa. Hyperbaric tremor (Fig. 3) occurred at 5.1 MPa (percentile range 4.7–5.4 MPa). The maximum increase was recorded at the end of compression (170 U; percentile range 70–290 U; Wilcoxon paired test: $p < 0.05$). The onset pressure of tremor was visually estimated to occur around 5 MPa. Myoclonia (Fig. 4) were recorded at 7.3 MPa (percentile range 6.5–7.8 MPa). The maximal value was 40 U (percentile range 33–50 U; Wilcoxon paired test: $p < 0.05$) on arrival at 8 MPa, with a subsequent increase to 50 U during the first hour of the stay. The pressure onset visually estimated was approximately 7 MPa.

During the 4-hr stay at 8 MPa, LHA and tremor decreased but did not disappear. Myoclonus began to decrease later, between 1 and 2 hr after the beginning of the stay, but did not decrease below 10–20 U. During decompression, tremor and LHA disappeared rapidly (7.6–7.5 MPa), but myoclonus did not disappear until 6.5–6.4 MPa.

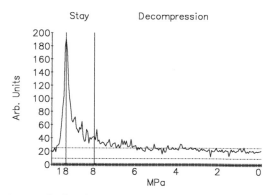

FIG. 3   Quantitative analysis of tremor in free-moving rats ($n$ = 6) exposed to a high-pressure helium–oxygen mixture (8 MPa). The progression from (left) predive record (1 hr) and compression to 8 MPa at 1000 hr in the morning (duration 1 hr 20 min) to (middle) a 4-hr stay at 8 MPa and finally (right) decompression from 8 MPa to atmospheric pressure (duration 24 hr) is shown. The dark period is between 1900 and 700 hr (i.e., between 6.8 and 2.8 MPa). The $y$ axis shows median tremor value calculated from six rats expressed in arbitrary units, and the $x$ axis shows pressure expressed in megapascals. Dotted lines represent the range of activity recorded in the 10–16 Hz frequency band at atmospheric pressure. Increases in tremor are recorded during compression. The increase is significant ($p < 0.05$) compared to the data recorded during 24 hr at atmospheric pressure ($U$ test).

## Discussion

This chapter describes a detailed method to quantify the behavior of free-moving rats exposed to high pressure. The clinical symptoms of HPNS have generally been estimated visually (7, 8); however, that technique makes it difficult to obtain an objective assessment of the severity of the symptoms of HPNS, or to correlate these with the hyperbaric parameters. Our method is based of the use of three piezoelectric sensors placed symmetrically under the floor of the cage. The behavioral disturbances are then quantified in real time using a computer and appropriate software (15).

Under normal conditions, the motor activity of the rats increased as expected during the dark period of the nyctohemeral. At high pressure, the quantitative analysis of behavior is of great interest to measure the intensity of the clinical symptoms of HPNS. The behavioral disturbances recorded with this system were found to occur in the range of onset pressures similar to those reported previously in the rat (6–8).

In all the animals, LHA, tremor, and myoclonus occurred and increased during the compression up to 8 MPa, decreased during the stay at 8 MPa,

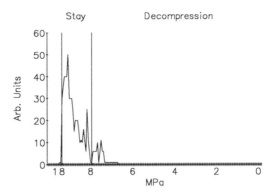

FIG. 4 Quantitative analysis of pressure-induced myoclonus in free-moving rats ($n = 6$) exposed to a high-pressure helium–oxygen mixture (8 MPa). The progression from (left) predive record (1 hr) and compression to 8 MPa at 1000 hr in the morning (duration 1 hr 20 min) to (middle) a 4-hr stay at 8 MPa and finally (right) decompression from 8 MPa to atmospheric pressure (duration 24 hr) is shown. The dark period is between 1900 and 700 hr (i.e., between 6.8 and 2.8 MPa). The $y$ axis shows median myoclonus value calculated from six rats expressed in arbitrary units, and the $x$ axis shows pressure expressed in megapascals. Increases in myoclonus are recorded during compression from 7 MPa. The increase is significant ($p < 0.05$) compared to the data recorded during 24 hr at atmospheric pressure ($U$ test).

and disappeared during the decompression. Results obtained with this system at higher pressures indicated a greater intensity of these disturbances (17). When the compression was stopped at 8 MPa, LHA and tremor decreased rapidly in the next 10 min. In contrast, myoclonus did not decrease until 2 hr of the stay had elapsed. These data comfirm that compression enhances the physiological disturbances, as suggested earlier by several workers (7, 18–20). The results obtained with this system are similar to those of Walker *et al.* (13), who partially quantified the disturbances using a mechanical sensor, but they did not attempt subsequent numerical and computer analysis.

The method we have used has indicated that tremor increases considerably during compression and that the onset pressure is always the same for the same species with the same rate of compression. As suggested by Brauer *et al.* (21), this symptom is a useful variable with which to monitor HPNS because of its reproductible onset pressure. Moreover, this method clearly demonstrated that myoclonus both occurs and decreases at around the same range of pressure (6.5–7.5 MPa). Abraini *et al.* (17) suggested that the development of myoclonus would be sensitive to constant high pressure, whereas LHA and tremor would be more sensitive to increasing pressure. Moreover, as previously suggested by Rostain *et al.* (22), the results obtained with this method confirm that there are probably two types of myoclonia in HPNS (epileptic and nonepileptic).

In conclusion, this study has demonstrated that the simple piezoelectrical sensor device that we have developed can (1) distinguish the various clinical symptoms of HPNS; (2) evaluate the gradual changes in behavioral symptoms; and (3) quantify the amplitude and correlate the symptoms and the severity of the symptoms with pressure. This system is a useful tool for monitoring behavioral disturbances in rats exposed to high pressure and could be of great interest for further studies in the field of behavioral neurochemistry and pharmacology. This method eliminates the need for a single observer and can operate without interruption.

## Acknowledgments

Research was supported by Grants Direction des Recherches, Etudes et Techniques 87/168 and 90/176.

## References

1. C. Lemaire and J. C. Rostain, "The High Pressure Nervous Syndrome and Performance." Octares Publ., Marseille, 1988.
2. P. B. Bennett and J. C. Rostain, *in* "The Physiology and Medicine of Diving" (P. B. Bennett and D. H. Elliott, eds.), Baillere Tindall, New York, 1993 (in press).
3. R. W. Brauer, S. Dimov, X. Fructus, A. Gosset, and R. Naquet, *Rev. Neurol.* **121**, 264 (1969).
4. P. B. Bennett and E. J. Towse, *Electroencephalogr. Clin. Neurophysiol.* **31**, 383 (1971).
5. J. C. Rostain and R. Naquet, *Rev. EEG Neurophysiol. Clin. (Rev. Electroencephalogr. Neurophysiol. Clin.)* **4**, 107 (1974).
6. R. W. Brauer, *in* "The Physiology and Medicine of Diving and Compressed Air Work" (P. B. Bennett and D. H. Elliott, eds.), p. 231. Baillere Tindall, London, 1975.
7. M. J. Halsey, *Physiol. Rev.* **62**, 1341 (1982).
8. J. C. Rostain, *in* "Current Perspectives in High Pressure Biology" (H. W. Jannasch, R. E. Marquis, and A. M. Zimmerman, eds.), p. 137. Academic Press, London, 1987.
9. J. A. Baker, M. J. Halsey, B. Wardley-Smith, and R. T. Wloch, *in* "Underwater Physiology VII" (A. J. Bachrach and M. M. Matzen, eds.), p. 415. Undersea Medical Society, Bethesda, Maryland, 1981.
10. J. H. Abraini, C. Tomei, and J. C. Rostain, *Pharmacol. Biochem. Behav.* **39**, 773 (1991).

11. J. H. Abraini and J. C. Rostain, *J. Appl. Physiol.* **71**, 638 (1991).
12. B. Wardley-Smith, S. Hudson, C. J. Doré, A. Charlett, A. Fletcher, N. T. Brammer, M. C. W. Minchin, and K. T. Wann, *Undersea Biomed. Res.* **17**, 275 (1990).
13. E. Walker, C. Mackie, and A. G. Macdonald, *Undersea Biomed. Res.* **4**, 1 (1977).
14. M. J. Ackerman and S. P. Gruenau, *Undersea Biomed. Res.* **5**, 301 (1978).
15. C. Tomei, J. H. Abraini, and J. C. Rostain, *Physiol. Behav.* **49**, 393 (1991).
16. S. A. Glantz, "Primer of Biostatistics." McGraw-Hill, New York, 1981.
17. J. H. Abraini, C. Tomei, and J. C. Rostain, *Ann. Physiol. Anthrop.* **10**, 183 (1991).
18. R. W. Brauer, R. W. Beaver, R. M. Mansfield, F. O'Connor, and L. W. White, *J. Appl. Physiol.* **38**, 220 (1975).
19. J. C. Rostain, J. C. Dumas, B. Gardette, J. P. Imbert, and C. Lemaire, *J. Appl. Physiol.* **122**, 482 (1984).
20. R. W. Brauer, R. W. Beaver, S. Lasher, R. D. Mccall, and R. Venters, *J. Appl. Physiol.* **46**, 128 (1979).
21. R. W. Brauer, R. W. Beaver, C. D. Hogue, B. Ford, S. M. Goldman, and R. T. Venters, *J. Appl. Physiol.* **37**, 844 (1974).
22. J. C. Rostain, B. Gardette, M. C. Gardette-Chauffour, and C. Forni, *J. Appl. Physiol.* **57**, 341 (1984).

# [17]   Measurement of Behavioral Thermoregulation

Christopher J. Gordon and Roberto Refinetti

## Introduction

The behavioral regulation of body temperature is possibly the best example of the use of behavioral effectors in homeostatic processes. The behavioral sensing of changes in ambient temperature can be exquisitely sensitive. Moreover, behavioral motor outputs play a major role in controlling heat transfer between an animal and its environment. Thus, the monitoring of behavioral thermoregulatory reflexes can provide measures of both the acuity of temperature sensation as well as the status of thermoregulatory control, such as whether a drug-induced change in body temperature is mediated via a change in the set point (1, 2).

Behavioral thermoregulation can be divided into two major categories: natural and instrumental. Natural behavior generally refers to those responses which are inherent. Although these behaviors can be complex, they can generally be observed without specialized apparatus. Natural behavioral responses would include huddling and nest building to reduce heat loss in the cold and grooming of saliva to increase heat loss during heat stress (3). Instrumental behaviors, which are the focus of this chapter, involve those responses that are observed with the use of specialized laboratory instruments: the temperature gradient and operant chamber.

## Temperature Gradients

Temperature gradients have been in use since the 1930s to measure the thermoregulatory behavior of rodents (for review, see Ref. 4). The temperature gradient has certain advantages and drawbacks compared to the operant systems (see below) for measuring thermoregulatory behavior (3, 5). First, the environment of a gradient is most conducive for the display of inherent thermoregulatory behavior, whereas extensive training is usually required for an animal to use an operant system. Temperature gradients are ideal for measuring thermoregulatory behavior of young animals that lack the muscular coordination needed to operate a lever for heat or cold reinforcements. In a temperature gradient, the animal simply moves to an area of optimal

*Methods in Neurosciences, Volume 14*

thermal comfort without expending additional effort, whereas in an operant system a motor task must continually be performed in order to regulate environmental temperature. Because of these limitations, operant systems are generally run over relatively short time periods. However, in a temperature gradient, an animal can sleep, eat, drink, and display other behaviors without having to perform a task to regulate temperature. In a gradient, an animal could be left in the system for days or weeks if needed.

Second, temperature gradients provide a gross measure of temperature selection, whereas operant systems give a precise measure of a given thermoregulatory behavior (e.g., number of thermal reinforcements). Third, animals in an operant system must maintain a steady rate of behavioral thermoregulation in order to prevent marked heat or cold stress. In a temperature gradient an animal could display a variety of behaviors that are not necessarily related to thermoregulation. Thus, an operant chamber is perhaps the system of choice when an experimenter wants behavioral thermoregulatory responses operative continuously. Finally, temperature gradients are generally cumbersome and do not allow easy access to the test subject, whereas operant systems permit good access to the animal and are often used where animals must be tethered with cables and/or catheters.

## Apparatus

The operational concept of temperature gradients (or thermoclines) is relatively simple. Essentially, an animal is provided with a continuum of ambient temperatures from which it can select. It is assumed that the location occupied by an animal for an adequate time period corresponds to its preferred thermal environment.

To explain the basic operation of a temperature gradient, it is necessary to understand the basic mechanisms of heat exchange (2–4). Heat transfer between an animal and its environment occurs by one or more of the following routes: conduction $(K)$, convection $(C)$, radiation $(R)$, and evaporation $(E)$. When the core temperature is stable, it can be assumed that metabolic heat production $(M)$ is equal to the total heat loss $(H_t)$, which is equal to the sums of individual routes of heat exchange [Eq. (1)]:

$$H_t = M = K + C + R + E \tag{1}$$

The temperature gradient essentially provides a continuum of environments whereby movement by the animal results in major shifts in the value of $K$, $C$, or $R$ which, subsequently, leads to changes in metabolic rate. If the selected environment lowers heat loss too much, then $M > H_t$ and the animal will become hyperthermic. If the selected environment results in relatively large heat loss, then the animal is forced to increase metabolic heat production above basal levels to maintain normal body temperature. Clearly, the overall "strategy" of behavioral thermoregulation is the selection of an environment that places minimal metabolic expenditure on the animal without causing heat stress.

Radiant heat gradients are rarely used in mammalian studies because it is difficult to measure the intensity of radiant energy preferred by an animal. It is possible to use a radiant energy gradient in a cold room to assess a degree of behavioral thermoregulation (6). On the other hand, radiant energy gradients are often used in the study of ectothermic species such as reptiles and amphibians. These species normally shuttle in and out of the sunlight to regulate their core body temperature above ambient levels (7). In such studies, the preferred core or skin temperature as a function of the animal's behavior is monitored.

In some systems the gradient is conductive. The gradient is generated on the floor where all direct contact will be made by the animal with its environment. Such gradients are made from highly conductive materials such as copper or aluminum. Thus, the animal can move about in the gradient and select a floor temperature which is presumed to be optimal for conductive heat transfer. Of course, it is difficult to keep the air temperature isothermal in a conductive gradient, but special provisions such as increased air circulation can be employed so that one looks specifically at conduction.

Conductive gradients are useful for studying the thermoregulatory behavior of neonatal rodents (8–10). Newborns of precocial rodents, including mouse, hamster, and rat, have little insulation and are poikilothermic from the age of 0 to about 14 days. This means that their body temperature is closely related to ambient temperature when they are not protected by the dam. When placed on a conductive surface, their body temperature will rapidly equilibrate with that of the surface temperature. Thus, conductive heat transfer is clearly a motivating factor in a newborn rodent moving in a gradient. Newborn rats as young as 1 day of age show thermotactic behavior in a conductive gradient (8). It is important to note that preweaned rodents are weak and have limited mobility in a gradient. Because of this, one must be careful in not placing the animal in an area of the gradient that is too warm or too cold. For example, a 1-day-old newborn rat placed at a surface temperature of 20°C will succumb to hypothermia before it can move to a warmer surface temperature. Likewise, when placed at a surface temperature of 45°C, it may succumb to hyperthermia before it can escape (8).

Although conductive gradients are useful for short-term measures of thermoregulatory behavior in neonates, studying the responses of mature animals requires a more elaborate system, preferably, a convective type of gradient. Several problems can arise in using conductive gradients with adult species.

Animal wastes on the solid floor can wet the fur or cause the animal to avoid certain parts of the gradient. The furless surface of the mature rat that comes in contact with the floor of a gradient accounts for less than 5% of the total surface area. During sleep the feet are curled into the body, and there is very little conductive heat transfer; thus, the floor temperature selected by the animal may not necessarily be related to its thermoregulatory behavior. For adult rodents with fur, it is best to measure behavioral thermoregulatory responses in terms of selected air temperature (i.e., selected $T_a$) rather than floor temperature.

The laboratory of one of the authors (C.J.G.) has spent considerable effort developing temperature gradients to measure thermoregulatory behavior in the mouse, hamster, rat, and guinea pig (11–15). Through these experiences we have found several design problems that should be noted in the development of a gradient for adult rodents: (1) the ideal gradient should be large enough to allow the animal to turn freely; (2) the gradient should be designed to restrict the animal from coming in contact with the temperature- and position-sensing equipment; (3) the gradient should be made of material that optimizes convective heat transfer and does not accumulate animal wastes; and (4) the gradient should provide access to food and water.

With these requirements in mind, we built a convective gradient that permits the continuous monitoring of selected $T_a$ of rodents for long periods of time (Fig. 1) (16). Basically, the system is made from a large copper tube in which a convective temperature gradient is generated by heating one end and cooling the other with circulating water baths. The animal is placed in a runway made of wire screen which is then inserted inside the copper tube. The wire screen material is ideal for minimizing conductive heat transfer and allows fecal material and urine to drop to the bottom of the copper tube. The runway also prevents the animal from interfering with the position- and temperature-measuring circuitry. Position sensors activated by an infrared light source are placed at 10-cm intervals along the length of the gradient. Immediately above each sensor is a thermocouple which continuously measures the air temperature in the immediate vicinity of the area occupied by the animal. Computer software is designed to locate the position of the animal in the runway and then measure the air temperature at that location. Food and water can be placed in the gradient chamber, allowing for the measurement of thermoregulatory behavior for several days. Motor activity of the animal in the gradient can also be determined by measuring the change in position at constant time intervals.

## Analytical Procedures

Temperature gradient data are often analyzed by determining arithmetic means of selected $T_a$ values over a given interval of time in the gradient.

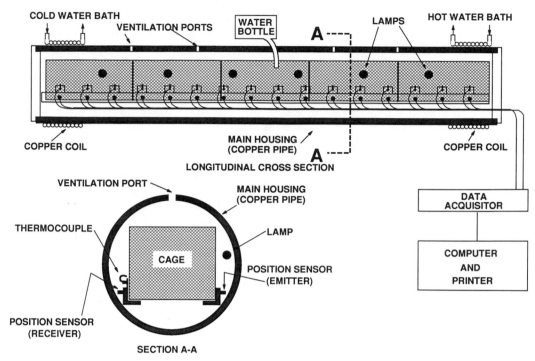

FIG. 1   Diagram of convective temperature gradient chamber. See text for details. [Modified from Gordon *et al.* (16).]

This is especially appropriate with adult species, which can be quite active in the gradient. In such cases, the selected $T_a$ is sampled at regular intervals. (~1–5 min), and the data collected over a period of 30 min or more are averaged. Rats are peculiar in that they do not reach a steady-state selected $T_a$ for at least 2 to 6 hr after placement in a temperature gradient (16). The pigmented Long-Evans rat is especially active and requires the most time to adapt to a gradient. Thus, the measuring period for some rodents may have to be extended for several hours if one wishes to measure a steady-state response. Indeed, we have recently found that rodents in a temperature gradient exhibit a daily oscillation in selected $T_a$, but such a response is not apparent until the rat is left in the gradient for at least 24 hr (Fig. 2) (17). When preweaned animals are tested, their activity in a gradient will not be as marked as that of an adult. It is often the case that the selected $T_a$ of very young animals is assessed when they occupy a given part of the gradient for more than a 5-min period. Because of their dependence on the dam, neonates are generally tested for shorter periods compared to adults.

## Physiological Significance

Data from animals in temperature gradients can provide a wealth of information. Selected ambient temperature has been found to be affected by a variety of endogenous and exogenous factors, including age, temperature acclimation, genetic background, time of day, drugs and toxic chemicals, hypoxia, and other factors. Other than the rat, rodents in temperature gradients select $T_a$ values associated with a minimal metabolic rate (i.e., thermoneutrality) (Table I) (18–21).

One of the most useful aspects of the temperature gradient is to establish if a given treatment affects the set point for thermoregulation (1, 2). When the body temperature of an endotherm changes as a result of exposure to a chemical agent, the change can be forced (i.e., no change in set point) or regulated. Just measuring autonomic thermoregulatory variables may not allow one to decide whether there is a change in set point. If temperature increases because of an elevation in set point, then one should see an increase in selected $T_a$ along with an elevation in body temperature. Such relationships have been noted in guinea pigs treated with exogenous pyrogen and placed in a temperature gradient (21). On the other hand, during a forced elevation in core temperature, the selected $T_a$ should decrease as the animal tries to dissipate the excess heat load.

## Operant Conditioning

Because behavioral selection of thermal microhabitats is a natural thermoregulatory response of most animals, the use of temperature gradients in the laboratory allows the study of a natural response in a controlled environment. Although this is often an advantage, on many occasions the investigator would rather have an artificial response that is not affected by natural variations in the state of the organism. The artificial response of lever pressing by rats was introduced by experimental psychologists in the 1930s (22) and adopted by thermal physiologists in the 1950s (23, 24). In its simplest form, thermal operant conditioning consists of placing an animal in a cold environment and allowing it to obtain a short pulse of heat every time it presses a lever. If the cold stress is of significant magnitude and the heat reward provides a transient relief, the animal will soon learn to press the lever and prevent a fall in its body temperature.

## Apparatus

Operant systems to measure behavioral thermoregulation are relatively simple and typically consist of three major components: operant chamber, heat

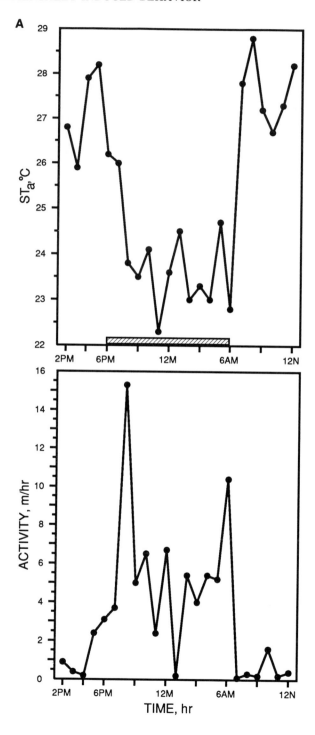

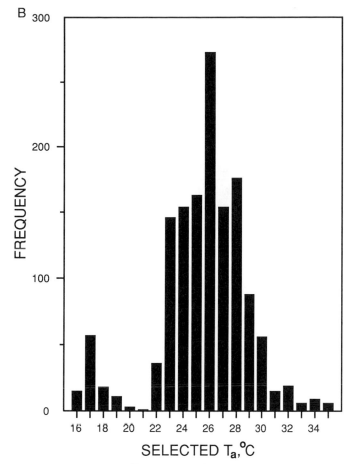

FIG. 2 (*continued*)

lamps, and control equipment (Fig. 3). A microswitch attached to the lever provides the connection between the animal chamber and the control equipment. In the past, electromechanical devices were used to record lever presses, control the contingencies of reinforcement, and deliver the rewards. Today these tasks can be more conveniently performed by personal comput-

FIG. 2   (A) Time course of selected $T_a$ and motor activity of a rat of the Fischer 344 strain tested in a convective temperature gradient chamber (cf. Fig. 1). The rat had been adapted to the gradient for 48 hr. Data were collected at 60-sec intervals and averaged into 60-min bins. (B) Frequency distribution of selected $T_a$ data when collected at 60-sec intervals, illustrating the dynamic nature of behavioral thermoregulation. [Data from Gordon (17).]

TABLE I  Ambient Air Temperatures
Selected by Laboratory Rodents

| Species | Selected $T_a$ (°C) | Ref. |
|---|---|---|
| Mouse | 31.8 | 18 |
| Mouse | 29.5–31.9 | 13 |
| Gerbil | 32.3 | 19 |
| Hamster | 28.2 | 11 |
| Rat | 27.2 | 20 |
| Rat | 19.8–24.9 | 15 |
| Guinea pig | 29–30 | 21 |
| Guinea pig | 30.6 | 14 |

[a] Animals were tested for relatively brief times (<2 hr) in a temperature gradient.

ers fitted with commercially available hardware for data acquisition and control (e.g., Alpha Products, Darien, CT; Intelligent Instrumentation, Tucson, AZ; Keithley Instruments, Cleveland, OH). In its simplest configuration, the system requires no computer or control devices, and it may be limited to a heat lamp and an event counter connected to a power supply through the microswitch. In this case, a response will be counted and the heat lamp will be turned on every time the animal depresses the lever. To complete this simple setup, all that is needed is a climatic chamber (or an ordinary freezer) in which to house the animal chamber.

The dimensions of the animal chamber (referred to as the "operant chamber" or "Skinner box") are necessarily dependent on the species chosen by the experimenter. Although thermal operant behavior has been studied in a variety of species, including molluscs (25), fish (26), birds (27), pigs (28), and primates (29), small rodents are the most commonly used experimental subjects (30). For a rat, a cylindrical wire mesh cage 26 cm in diameter and 20 cm high provides enough freedom of movement and air circulation. A small Plexiglas lever (3 × 4 cm) attached to the wall 5 cm above the floor provides a convenient operandum.

Although a single heat lamp positioned above the chamber can serve as the source of rewards, better results are obtained if two red-bulb infrared lamps (250 W each) are positioned laterally at 45° to the lever (Fig. 3). The intensity of the stimulus can be varied with a voltage controller and monitored with a radiometer. An irradiance of 170 mW/cm$^2$, as measured in the center of the cage in front of the lever, is appropriate.

For non-cold-adapted rats, an ambient temperature of 0°C constitutes an appropriate level of cold stress. Colder temperatures produce higher levels of motivation of lever press but may lead to frostbite of the tail if the rate

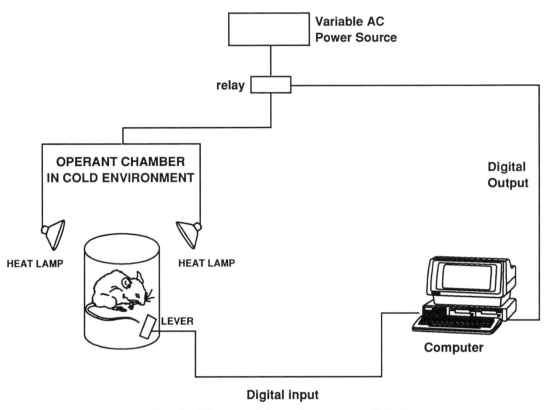

FIG. 3   Diagram of thermal operant conditioning apparatus

of reinforcement is too low. Typically, much colder temperatures are required during the first learning session than during later sessions. If rats are placed in the chamber and allowed to learn to bar press by themselves, they will do so in about 2 hr at −10°C. Shaving the body fur before the session is often necessary (31).

To maintain precise control of the rewarding stimulus, many investigators avoid the procedure of allowing the animal to obtain heat for as long as the lever is depressed (the so-called *ad libitum* or water fountain schedule of reinforcement). Usually, a constant-duration pulse of heat of a few seconds is delivered when the lever is depressed, and the animal must release the lever and press it again to obtain another reward. If a reward is delivered for each response, the schedule is called "continuous reinforcement" (CRF). If a fixed number of responses is required before a reward is delivered, the schedule is called "fixed ratio" (e.g., FR3 or FR10).

Placing the animal in a cold chamber and using a heat lamp as the source of heat rewards is very convenient, as both the environmental temperature and the parameters of the rewarding stimulus can be easily and accurately controlled. However, other setups are sometimes necessary. Some authors have used a constant flow of cold air as the cold stress and pulses of hot air as the heat reward (32, 33). To study heat escape behavior, investigators have used a heat lamp as the source of heat stress and pulses of cool air as the reward (34) or a hot room as the heat stress and refreshing water showers as the reward (35).

Finally, although the rate of responding is usually the variable of interest in studies of thermoregulatory operant behavior, the efficacy of the behavioral response cannot be assessed if the body temperature of the animal is not monitored. Deep body temperature can be measured at the beginning and end of the session with a thermometer (or electronic thermal probe) inserted rectally 6 cm past the anal sphincter for rats. If a continuous record of body temperature is desired, intraabdominal temperature can be easily recorded by telemetry using commercially available implantable transmitters (e.g., Data Sciences Inc., St. Paul, MN; Mini-Mitter Co., Sunriver, OR).

## Analytical Procedures

The basic dependent variable in studies of thermoregulatory operant behavior is the number of lever presses. In most instances, the total number of responses in a session is highly correlated with the total amount of heat obtained, so that number of responses is a variable that can be directly analyzed. If necessary, the amount of heat obtained can be computed given a few basic assumptions (36). The amount of heat absorbed by a shaved animal during a session can be calculated as shown in Eq. (2):

$$E = kndIM^{0.6} \tag{2}$$

where $E$ is the absorbed heat (kJ), $k$ is a species-dependent constant (equal to $3.64 \times 10^{-6}$ for rats), $n$ is the number of rewards obtained, $d$ is reward duration (sec), $I$ is the irradiance provided by the heat lamps (mW/cm$^2$), and $M$ is the body mass of the animal (g). When an *ad libitum* schedule is used, total time of lamps on (sec) is substituted for the terms $n$ and $d$.

Although changes in response rate during a session may be of interest in a few situations, the mean rate for the session (or simply the total number of responses) is what researchers usually look at. Alternatively, they look at the rate or total number of rewards obtained. Data analysis is quite simple and involves mostly comparisons of means of different animals tested under

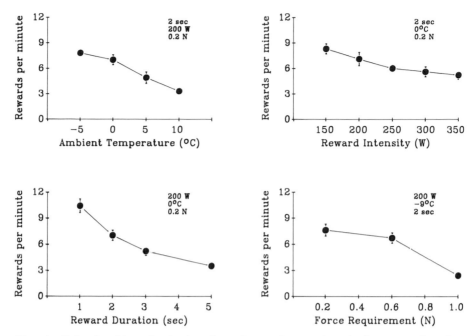

FIG. 4   Behavioral heat intake (number of rewards per minute) of rats as a function of ambient temperature, reward intensity, reward duration, and force required to depress the lever (data from Refs. 37 and 38).

different conditions. The data shown in Fig. 4 (37, 38) illustrate how the number of rewards obtained is affected by changes in ambient temperature, reward intensity, reward duration, and the force required to depress the lever. Clearly, rats obtain more heat rewards when the ambient temperature is lower, when the intensity of the reward is lower, when the duration of the reward is shorter, and when the effort to depress the lever is lower.

## Physiological Significance

Data such as those shown in Fig. 4 strongly suggest that rats use the operant response as a homeostatic mechanism to regulate their body temperature. Further support for this interpretation comes from studies that demonstrate a trade-off between behavioral heat intake and metabolic heat production (39). As illustrated in Fig. 5, rats maintained in a mildly cold environment (3°C) maintain a constant body temperature by increasing their heat intake

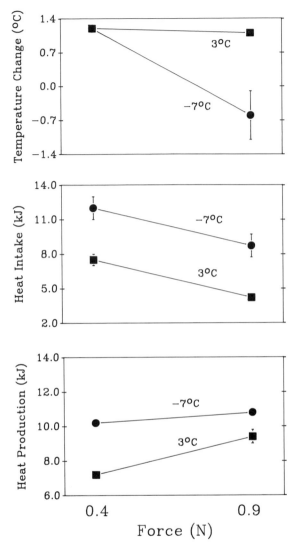

FIG. 5   Heat production, heat intake, and change in body temperature during 30 min for two levels of force requirement at two ambient temperatures. The same group of rats was tested in the four conditions (data from Ref. 39).

and reducing their heat production when the force required to press the lever is reduced from 0.9 to 0.4 N. At the force requirement level of 0.4 N, a reduction in ambient temperature to −7°C does not affect body temperature, as both heat intake and heat production are increased. That heat production alone cannot preserve normothermia at this ambient temperature is indicated by the fact that body temperature falls when a reduction in heat intake is caused by increasing force requirement to 0.9 N.

## Applications in Neuroscience Research

The measurement of thermoregulatory behavior by the techniques of thermal gradient and operant conditioning allows the study of many parameters of the behavioral control of body temperature in particular species as well as the comparative study of thermoregulatory capabilities of different species. In neuroscience research, these techniques provide a reliable dependent variable for the investigation of the action of specific neurotransmitters and the specificity of particular brain loci involved in the regulation of body temperature. Much of the current knowledge of the neural control of thermoregulatory behavior has been obtained by using thermal gradients and operant conditioning in conjunction with systemic and intracerebral injections of neurotransmitters, local thermal stimulation of brain regions, and selective destruction of particular brain sites by electrolytic lesions (2, 40).

## Acknowledgments

This chapter has been reviewed by the Health Effects Research Laboratory, U.S. Environmental Protection Agency. Mention of trade names or commercial products does not constitute endorsement or recommendation for use.

## References

1. C. J. Gordon, *Life Sci.* **32,** 1285 (1983).
2. E. Satinoff, *in* "Body Temperature: Regulation, Drug Effects, and Therapeutic Implications" (P. Lomax and E. Schonbaum, eds.), p. 151. Dekker, New York, 1979.
3. C. J. Gordon, *Physiol. Behav.* **47,** 963 (1990).
4. J. S. Hart, *in* "Comparative Physiology of Thermoregulation" (G. C. Whittow, ed.), Vol. 2, p. 1. Academic Press, New York, 1971.
5. J. S. Laughter, Jr., and C. M. Blatteis, *Physiol. Behav.* **35,** 993 (1985).
6. S. Yehuda, and R. J. Wurtman, *J. Pharmacol. Exp. Ther.* **190,** 118 (1974).
7. M. J. Kluger, R. S. Tarr, and J. E. Heath, *Physiol. Zool.* **46,** 79 (1973).
8. N. Kleitman and E. Satinoff, *Physiol. Behav.* **29,** 537 (1982).
9. I. B. Johanson, *Physiol. Behav.* 23, 871 (1979).
10. C. A. Goodrich and C. Wilk, *Physiol. Behav.* **26,** 1041 (1981).

11. C. J. Gordon, K. S. Fehlner, and M. D. Long, *Am. J. Physiol.* **251,** R320 (1986).
12. C. J. Gordon, *Am. J. Physiol.* **254,** R229 (1988).
13. C. J. Gordon, *Physiol. Behav.* **34,** 687 (1985).
14. C. J. Gordon, *Physiol. Behav.* **38,** 827 (1986).
15. C. J. Gordon, *Am. J. Physiol.* **252,** R1130 (1987).
16. C. J. Gordon, K. A. Lee, T. A. Chen, P. Killough, and J. S. Ali, *Am. J. Physiol.* **261,** R705 (1991).
17. C. J. Gordon, *Physiol. Behav.* **53,** 257 (1993).
18. D. M. Ogilvie and R. H. Stinson, *Can. J. Zool.* **44,** 511 (1966).
19. C. Akins and D. D. Thiessen, *Percept. Mot. Skills* **71,** 1177 (1990).
20. P. R. Marques, R. L. Spencer, T. F. Burks, and J. N. McDougal, *Behav. Neurosci.* **98,** 858 (1984).
21. C. M. Blatteis and K. A. Smith, *Experientia* **36,** 1086 (1980).
22. B. F. Skinner, "The Behavior of Organisms." Appleton-Century-Crofts, New York, 1938.
23. B. Weiss, *J. Comp. Physiol. Psychol.* **50,** 481 (1957).
24. P. L. Carlton and R. A. Marks, *Science* **128,** 1344 (1958).
25. P. Downey and B. Jahan-Parwar, *Am. Zool.* **12,** 507 (1972).
26. P. Rozin and J. Mayer, *Science* **134,** 942 (1961).
27. I. Schmidt and W. Rautenberg, *J. Comp. Physiol.* **101,** 225 (1975).
28. B. A. Baldwin and D. L. Ingram, *Physiol. Behav.* **2,** 15 (1967).
29. E. R. Adair, J. U. Casby, and J. A. J. Stolwijk, *J. Comp. Physiol. Psychol.* **72,** 17 (1970).
30. M. Cabanac, *in* "Microwaves and Thermoregulation" (E. R. Adair, ed.), p. 307. Academic Press, New York, 1983.
31. H. J. Carlisle, *Physiol. Behav.* **3,** 827 (1968).
32. R. Szymusiak, A. DeMory, E. M. W. Kittrell, and E. Satinoff, *Am. J. Physiol.* **249,** R219 (1985).
33. R. Refinetti and H. J. Carlisle, *Physiol. Behav.* **36,** 1099 (1986).
34. J. M. Lipton, *Physiol. Behav.* **3,** 165 (1968).
35. A. N. Epstein and R. Milestone, *Science* **160,** 895 (1968).
36. R. Refinetti and H. J. Carlisle, *J. Therm. Biol.* **12,** 263 (1987).
37. H. J. Carlisle, *J. Comp. Physiol. Psychol.* **61,** 388 (1966).
38. R. Refinetti and H. J. Carlisle, *Anim. Learn. Behav.* **15,** 228 (1987).
39. R. Refinetti and H. J. Carlisle, *Behav. Neural. Biol.* **46,** 64 (1986).
40. M. Cabanac, *J. Physiol. (Paris)* **75,** 115 (1979).

# [18]  Measurement of Stress-Induced Analgesia

Richard J. Bodnar

## Introduction

Four discoveries in the 1970s were critical for our understanding of endogenous pain-inhibitory systems and the role of opiates in their mediation. First, morphine and other opiate drugs were found to produce analgesic responses following direct intracerebral microinjections into the brain, particularly into midline mesencephalic, pontine, and medullary structures (for review, see Ref. 1). Second, electrical stimulation of many of the same structures also elicited analgesic responses (2). These analgesic responses revealed an opioid-mediated pain-inhibitory system which originated in the mesencephalic periaqueductal gray and projected to the medullary nucleus raphe magnus and nucleus reticularis gigantocellularis, which in turn projected to the dorsal horn of the spinal cord (3, 4), where local intrathecal application of opiates produced analgesia as well (5). Third, following isolation of the opiate receptor (6–8), the receptor was found in anatomical sites implicated in pain-inhibitory responses (9). Finally, the enkephalin pentapeptides (10) and subsequently the endorphin and dynorphin opioid families were discovered and either were intrinsic or projected to those anatomical sites implicated in pain inhibition (9). Such discoveries led to speculation as to the normal functional significance of opioid systems in general and with respect to modulation of pain inhibition. One answer was framed by the findings that stressful stimuli such as inescapable foot shock produce profound release of endogenous opioids (11) and an analgesic response (12). Although there are certain classes of stressful stimuli which fail to elicit significant analgesia (13), the elicitation of higher nociceptive thresholds following selective environmental stimuli has been classified as "stress-induced analgesia" (SIA; 14).

SIA has been studied across animal species, including humans, and its pharmacological, neuroendocrine, and behavioral substrates have been reviewed elsewhere (see, e.g., Refs. 14–20). The purpose of this chapter is to examine the measurement of SIA, and the following issues are addressed. First, it is important to review whether stress is a necessary and sufficient condition to produce analgesia. Second, because one of the initial assumptions for the study of SIA was that stress activated endogenous opioid systems which in turn activated analgesic responses, it is important to review the methodology used to delineate opioid-mediated SIA and nonopioid-mediated

SIA. Third, independent of the pharmacological substrates mediating SIA, it is important to ascertain whether other physiological systems are activated by stress and whether they participate in, or even mediate, SIA. Finallly, the consistency of SIA is evaluated across nociceptive tests to determine whether different pain-inhibitory systems or different levels of pain inhibition are influenced in SIA.

## Is Stress a Necessary and Sufficient Condition to Produce Analgesia?

Analgesia is elicited following acute exposure to different physical environmental stimuli, including inescapable foot shock, centrifugal rotation, hypertonic saline injections, cold-water swims, food deprivation, 2-deoxy-D-glucose (2DG) and insulin glucoprivation, restraint, and body pinch (12–17). In contrast, neither ether inhalation nor horizontal oscillation alter nociceptive thresholds, although both are considered stress responses in terms of increased corticosterone release (13). A wide range of other psychological stress stimuli elicit analgesia as well, including vaginocervical probing, social defeat, and various forms of fear conditioning (12, 15, 18–24). In contrast, other physical and psychological stress stimuli such as light restraint, novelty, and tail pinch produce hyperalgesic responses (25, 26). Indeed, tail-pinch hyperalgesia attenuates the analgesic responses following morphine and cold-water swims (26). It is apparent that the typical hormonal hallmarks of stress, such as increased levels of adrenocorticotropic hormone, corticosterone, and adrenaline as well as increased sympathomedullary and adrenocortical activity, are poor predictors of which stressors will either produce analgesia or elicit more profound analgesia.

The lack of correlation between hormonal markers of stress and the occurrence of stress-induced analgesia extends to the study of adaptation. Repeated exposure to such stressors as inescapable foot shock, 2DG glucoprivation, cold-water swims, and food deprivation result in either an elimination or significant attenuation in the magnitude of the analgesic response, indicating that it is the stressful consequences of the stimuli and not the physical characteristics of the stimuli themselves that are eliciting analgesia (14–17). Indeed, "cross-adaptation" or "cross-tolerance" has been demonstrated for analgesia between pairs of stressors, including cold-water swims and 2DG glucoprivation (27), cold-water swims and cervical probing (28), and two forms of inescapable foot shock (29). Repeated exposure to the same stressors, however, may still elicit intact hormonal responses.

This is not to say that the pituitary–adrenal and sympathomedullary systems are not involved in the maintenance or initiation of stress-induced analgesia. Hypophysectomy significantly reduces analgesia following cold-water swims, some parameters of inescapable foot shock, food deprivation, and restraint (30–33), and these forms of analgesia have been characterized as neurohormonally mediated. However, other parameters of inescapable foot shock produce analgesia that is unaffected by alterations of the pituitary–adrenal axis, and they have been characterized as neurally mediated (15, 17). For the neurohormonal forms of stress-induced analgesia, some, like cold-water swim analgesia, utilize the pituitary–adrenal axis for their expression (34–36), whereas others, like inescapable foot shock, use the sympathomedullary axis for their expression (37, 38).

## Opioid and Nonopioid Determinants of Stress-Induced Analgesia

One original and still compelling reason for the use and measurement of stress-induced analgesia is the determination as to whether a particular stressor activates the endogenous opioid system to mediate the analgesic response. The two most common pharmacological techniques utilized to evaluate whether particular stressors elicited opioid-mediated analgesia were the ability of naloxone to block or reverse the analgesic response, and the ability of the stressor to display two-way analgesic cross-tolerance with morphine. Indeed, the initial determinations that neither continuous cold-water swim (CCWS) analgesia nor analgesia following classical conditioning were cross-tolerant with morphine or altered by naloxone (39–42) indicated that these manipulations were nonopioid.

The use of naloxone reversibility and morphine cross-tolerance was then applied to the seminal studies demonstrating that alterations in the parameters of a stressor could alternatively elicit either opioid or nonopioid responses. Lewis and co-workers (43, 44) found that whereas analgesia elicited by prolonged intermittent foot shock was cross-tolerant with morphine analgesia and significantly reduced by naloxone, brief continuous foot shock analgesia was not. Watkins and Mayer (15) found that whereas analgesia elicited by shock delivered to the forepaws was cross-tolerant with morphine analgesia and significantly reduced by naloxone, hind paw shock analgesia was not. Maier and co-workers (45) found that whereas 80 tail shocks elicited a naloxone-reversible analgesia, analgesia induced by 20 tail shocks did not. Indeed, parametric variation of swim analgesia displayed opioid and nonopioid forms of analgesia such that CCWS (2°C, 3.5 min) analgesia was neither cross-tolerant with morphine analgesia nor blocked by naloxone (39, 40), whereas intermittent cold-water swim (ICWS, 2°C, 18 10-sec swims, 10-sec recovery) analgesia was cross-tolerant with morphine analgesia and blocked by naloxone (46, 47). Further, analgesia induced by either food deprivation (48) or restraint (49) was significantly reduced by naloxone pretreatment.

Important issues in the interpretation of these data include (1) the magnitude of the opioid effect on SIA, (2) the dose of the opioid antagonist employed, and (3) the use of alternative methods to confirm opioid involvement. When one assesses the ability of opioid antagonists to block opiate analgesia, general antagonists are capable of *eliminating* opiate analgesia under both pretreatment (blocking) and posttreatment (reversal) conditions. In assessing the above-cited and other studies investigating opioid involvement in SIA using opioid antagonists, it is clear that while opioid-mediated stressors display significant reductions in analgesia following antagonist treatment, the antagonists fail to eliminate the analgesic response.

One notable exception is the ability of intrathecal administration of naloxone to eliminate forepaw analgesia (15). Indeed, in the above foot shock analgesia paradigms (15, 43, 45), short-term analgesia (1–2 min) is unaffected by naloxone, but longer term effects (3–20 min) are significantly reduced. This indicates that although opioids participate in these analgesic responses, they often do not account for the entire effect.

Second, the dose of antagonists is critical. General opioid antagonists block morphine analgesia at antagonist doses of 0.01–1 mg/kg. In contrast, systemic doses between 7 and 14 mg/kg of naloxone or naltrexone have often been used to reduce SIA, implicating the opioid system in these effects (19, 43, 45, 47–49). Further, opioid mediation of SIA effects also results from an interaction between opioid antagonist dose and parameter of the stressor. For instance, our laboratory (50) found that altering the CCWS temperature affects the subsequent analgesic sensitivity to naloxone, with a 2°C CCWS displaying nonopioid analgesia and a 15°C CCWS displaying opioid analgesia. These data were verified subsequently by specifically studying stress severity in mice (51) and rats (52). The selectivity of opioid receptor antagonists varies monotonically with the dose employed. At higher doses, the selectivity of these antagonists can be compromised both in terms of whether only opiate receptors are being blocked (9, 53, 54) and, if so, which opioid receptor subtypes ($\mu$, $\mu_1$, $\delta$, $\kappa$) are blocked (55–57). Naloxone and naltrexone bind to all opioid receptors but do so somewhat preferentially: $\mu > \delta \gg \kappa$.

Third, the use of alternative methods to confirm opioid involvement is necessary. The demonstrations that CCWS analgesia is nonopioid based on lack of cross-tolerance with morphine analgesia (39) and the failure of naloxone to significantly reduce its magnitude (40, 50) have been supported by other corroborative data. Whereas an inhibitor of endopeptidase 24.11 that prevents the degradation of enkephalin pentapeptides potentiates morphine analgesia (58) and opioid-mediated SIA (59, 60), it significantly reduces nonopioid CCWS analgesia (61). In contrast, whereas naloxazone, a selective and irreversible $\mu_1$ antagonist significantly reduces morphine analgesia (62), it significantly potentiates nonopioid CCWS analgesia (63). Moreover, whereas chronic opioid antagonism with naltrexone pellets eliminates morphine analgesia, it significantly increases nonopioid CCWS analgesia (64).

From these data, our laboratory hypothesized a model of collateral inhibition between opioid and nonopioid analgesic systems (17, 20, 63) in which activation of the opioid system produces opioid analgesia but inhibits the nonopioid system, and in which activation of the nonopioid system produces nonopioid analgesia but inhibits the opioid system. To test this hypothesis, we (65) compared the analgesic effects of CCWS and morphine alone, and, following simultaneous treatment, we found that the analgesic response following CCWS and morphine together was significantly reduced relative to the analgesic treatments alone. This is in striking contrast to the additive and sometimes synergistic actions following simultaneous treatment of morphine and an opioid-mediated form of SIA (66–68).

There have been other instances of independent confirmation of opioid involvement in SIA. The opioid involvement in ICWS analgesia suggested by cross-tolerance and naloxone reversal studies (46, 47) has been confirmed by the increase in ICWS analgesia following central pretreatment with an inhibitor of endopeptidase 24.15 which prevents the degradation of longer chained opioid peptides (69). Whereas opioid involvement in analgesia induced by the antimetabolic glucose analog 2DG has been supported by its cross-tolerance and synergy with morphine analgesia, this is questioned by the inability of naloxone to block 2DG analgesia and by its full and reciprocal cross-tolerance with CCWS analgesia (67, 70).

Although the existence of opioid receptor subtypes has been demonstrated for some time (55–57), and selective opioid receptor subtype antagonists have been developed, very little research has addressed the question as to which opioid receptor subtype is involved in the opioid mediation of SIA. An opioid form of foot shock analgesia has been blocked by relatively selective $\kappa$ antagonists (71, 72). As cited above, the $\mu_1$ receptor antagonist naloxazone potentiated CCWS analgesia (63). Indeed, when the laboratory of Watkins and Maier intrathecally administered a range of selective $\mu$, $\kappa$, and $\delta$ receptor antagonists alone and in combination with each other (73), they found that naltrexone-insensitive forms of analgesia can be blocked by intrathecal administration of combinations of selective $\mu$, $\delta$, and $\kappa$ antagonists, $\mu$ and $\delta$ antagonists, and $\mu$ and $\kappa$ antagonists, implying that all forms of SIA are "opioid-mediated," but differ in the level of the neuraxis at which this mediation takes place.

Thus, an important issue in the measurement of SIA, namely, whether the endogenous opioid system is involved in its mediation, needs to be addressed using a series of pharmacological and behavioral techniques rather than the typical use of high doses of general opioid antagonists.

## Roles of Other Physiological Systems in Stress-Induced Analgesia

A critical assumption in the study of SIA is that changes in the response output (dependent nociceptive measure) by a given manipulation (independent stressful situation) are necessarily the exclusive result of the activation by that manipulation of the output system (i.e., pain-inhibitory system; 74). Typically, the foot shock analgesia paradigms described previously (14–20) employ a single nociceptive measure, the spinally mediated tail-flick test (75), and then examine whether particular physiological or pharmacological manipulations alter the analgesic changes following foot shock. Any latency changes induced by pairing foot shock with the manipulation are typically described as intrinsic alterations in a pain-inhibitory system. It is also likely that the manipulation could be altering the coping or response strategies to the foot shock, which in turn could account for subsequent changes in latencies. The latter scenario attributes changes to stress responsiveness and not to analgesic output systems. Using a single nociceptive measure without any other physiological response following stress fails to address this possibility and can result in a false assumption that a given manipulation directly affects intrinsic pain inhibition rather than the possibility that changes in pain inhibition are epiphenomena of some other effect of the manipulation. Our laboratory has addressed this possibility by examining concomitant physiological measures following stress in addition to nociceptive tests. In studying CCWS analgesia, we evaluated another major response following this stressor: hypothermia. In studying 2DG analgesia, we evaluated another major response following glucoprivation: hyperphagia.

Manipulations altering CCWS analgesia and CCWS hypothermia have been demonstrated to dissociate these two measures in some instances and to show associations between the measures in other instances. CCWS analgesia, but not CCWS hypothermia, is reduced following repeated exposure to CCWS (39), following hypophysectomy (30), and following D-phenylalanine treatment (61). CCWS analgesia is reduced and CCWS hypothermia is potentiated following muscarinic receptor antagonism (76), during aging (77), and by gender differences (78). CCWS analgesia, but not CCWS hypothermia, is potentiated following pretreatment with the noradrenergic receptor uptake blocker desipramine (79), the $\alpha_2$-noradrenergic receptor antagonist yohimbine (80), and thyrotropin-releasing hormone (81). In contrast, associations between CCWS analgesia and CCWS hypothermia have been observed with corresponding potentiations in both responses following pretreatment with either the $\alpha$-noradrenergic receptor agonist clonidine (82) or the $H_2$ receptor antagonist cimetidine (74). These data suggest either that the potentiated CCWS analgesia induced by these agents was the epiphenomenological result of increased hypothermia or that both effects were produced by an altered coding of the stressful consequences of the CCWS stimulus. Correspondingly, medial hypothalamic damage produced by neonatal administration of monosodium glutamate significantly reduced the magnitude of both CCWS analgesia and CCWS hypothermia (83), suggesting again that the impaired analgesic response might be due to an impaired hypothermic response and/or an impaired coding of the stressful consequences of the stimulus.

In addition to its analgesic effects (84), the antimetabolic glucose analog 2DG induces peripheral sympathomedullary discharge, hyperglycemia, and hyperphagia (85, 86). 2DG analgesia and 2DG hyperphagia have been shown to be dissociable in that the former is potentiated and the latter reduced following damage to the hypothalamohypophyseal axis (83, 87, 88), following dopamine depletion or inactivation (89, 90), following muscarinic receptor antagonism (76), and following exposure to inescapable foot shock (91). 2DG analgesia, but not 2DG hyperphagia, is reduced following chronic exposure to either 2DG (92) or morphine (70, 91). 2DG analgesia, but not 2DG hyperphagia, is potentiated following treatment with the $H_2$ antagonist zolantidine (93). In contrast, 2DG hyperphagia, but not 2DG analgesia, is reduced following lesions placed in the lateral hypothalamus or zona incerta (91, 94, 95). Thus, across a wide range of physiological and pharmacological manipulations, two major responses to 2DG glucoprivation, analgesia and hyperphagia, can be dissociated, strongly suggesting that the changes induced in the analgesic responses by particular manipulations appear to alter the mechanisms by which 2DG acts to activate intrinsic pain-inhibitory systems.

Another intriguing suggestion that attempts to explain some forms of SIA is the hypothesis that vagal afferents serve as ''an important natural conduit for the modulation of pain'' (96). In this model, Randich and Maixner argue that the sympathetic responses to stress, including cardiopulmonary changes, activate this vagal modulatory action on central pain-inhibitory systems. In further studies, Gebhart and Randich demonstrated that central and spinal opioid antinociception are also modulated by vagal inputs (97). This would suggest that, in SIA research, it would also be important to ascertain whether any manipulation affecting SIA does so through this vagal mechanism rather than directly on central pain-inhibitory processes.

## Stress-Induced Analgesia and Nociceptive Tests

An assumption indicated in the previous section suggests that changes in the dependent nociceptive measure by an independent stressful situation is the exclusive result of activation by that manipulation of a pain-inhibitory system (74). Hence any foot shock analgesia observed on the tail-flick test might be either reduced or potentiated by a given manipulation, and such

TABLE I   Alterations in Magnitude of Analgesia Induced by Stressors as Measured by the Tail-Flick and Jump Tests

| Manipulation | Tail flick | Jump | Ref. |
|---|---|---|---|
| Continuous cold-water swim | | | |
|   Cimetidine | 78% ↑ | 118% ↑ | 74 |
|   Monosodium glutamate (adults) | 127% ↑ | 55% ↑ | 100 |
|   Yohimbine | 70% ↑ | 70% ↑ | 80 |
|   Methysergide | 50% ↓ | 58% ↓ | 101 |
|   Females < males | 33% ↓ | 40% ↓ | 102 |
|   Alloxan | 11% ↓ | 31% ↑ | 103 |
|   Pirenpirone | 57% ↓ | 26% ↓ | 101 |
| Intermittent cold-water swim | | | |
|   Endopeptidase 24.15 inhibition | 135% ↑ | 123% ↑ | 69 |
|   Cimetidine | 63% ↑ | 163% ↑ | 74 |
|   Methysergide | 21% ↓ | 32% ↓ | 101 |
|   Pirenpirone | 34% ↓ | 50% ↑ | 101 |
|   Females < males | 0% | 33% ↓ | 102 |

changes are typically described as changes in pain inhibition. The reliance on a single nociceptive measure in the assessment of analgesia could be problemmatic in terms of the nociceptive system activated (e.g., thermal, mechanical, or chemical) and the level of the neuraxis at which the nociceptive response acts (e.g., spinal and supraspinal). In the evaluation of central opioid analgesia, Jensen and Yaksh (98) found that microinjections of both $\mu$-selective and $\delta$-selective opiates and opioids into the periaqueductal gray and ventromedial medulla produce similar degrees of inhibition on the spinally mediated tail-flick test, the supraspinally mediated jump test, and the chemosensitive writhing test.

Such data are quite important in allowing for the generalizability of analgesic actions, and our laboratory has adopted this strategy for the study and measurement of SIA. To this end, our laboratory examined whether a given stressor produces analgesia on two measures, the spinally mediated tail-flick test (75) and the supraspinally mediated jump test (99), to determine if particular manipulations that alter a given form of SIA do so in a consistent manner across the two tests, or if the two tests dissociate in terms of effects.

Table I (100–103) summarizes the significant alterations in either CCWS or ICWS analgesia on the tail-flick or jump tests following various manipulations. Consistent potentiations in CCWS analgesia are observed on both tests following the $H_2$ antagonist cimetidine (74), following monosodium glutamate treatment in adult animals (100), and following the $\alpha_2$-noradrenergic antagonist yohimbine (80). Consistent potentiations in ICWS analgesia are observed on both tests following cimetidine (74) and following endopeptidase 24.15 inhibition (69). Consistent reductions in CCWS analgesia are observed following the general 5-hydroxytryptamine (5-HT) antagonist methysergide (101) and in the responses of female rats relative to male rats (102). Consistent reductions in ICWS analgesia are also observed following methysergide (101).

In the above four scenarios the two nociceptive tests are in congruence in relating the effects of the physiological and pharmacological manipulations on SIA. In contrast, CCWS analgesia is decreased on the tail-flick test and increased on the jump test in a diabetic model using alloxan (103). Further, the reduction in CCWS analgesia following the 5-HT$_2$ antagonist pirenpirone is significantly more effective on the tail-flick test than on the jump test (101). Moreover, ICWS analgesia is decreased on the tail-flick test and increased on the jump test by pirenpirone (101). Finally, female rats display significantly less ICWS analgesia on the jump test, but similar analgesia on the tail-flick test (102). In these paradigms, the two nociceptive tests fail to display congruence, suggesting that the particular manipulation is acting at one level of the neuraxis but not at the other level.

Alternatively, nonspecific effects could compromise the interpretation of actions on a particular nociceptive test. For instance, to avoid tissue damage, a cutoff latency is usually imposed on the tail-flick test. If the analgesic effect approaches this ceiling, a given manipulation will fail to display potentiating actions, and it could fail to display reductions because of this ceiling. Further, many nociceptive responses can be influenced by motor dysfunction, but the level of dyskinesia or hypoactivity may manifest itself differentially across tests as a function of the manipulation. The ability to perform multiple independent assessments of nociception may illustrate these problems by dissociable effects but, more importantly, may attest to the generalizability of effects across nociceptive measures when the results are congruent.

## Conclusions

This review examines some of the major issues relating to the measurement of stress-induced analgesia and the degree to which these issues may limit or expand our interpretation of data. First, it is important to recognize that stress is neither a necessary nor sufficient condition to produce analgesia in all circumstances, and that independent measurement of adrenocortical or sympathomedullary hormones are not reliable predictors of either the occurrence or magnitude of stress-induced analgesia. However, it should be recognized that these axes can be important in the maintenance of some forms of SIA. Second, the issue of opioid and nonopioid forms of SIA is addressed. An important point is the degree to which manipulations of the opioid system can affect a particular analgesic response. It is indicated that very few "opioid" forms of SIA resulted in either the complete blockade or reversal of the analgesic effect. The importance of other confirmatory methods in addition to the use of morphine tolerance and naloxone reversibility is discussed, especially in terms of identifying receptor subtype roles and loci of action. Third, the effects of manipulations affecting SIA are evaluated in terms of either their direct actions on pain-inhibitory systems, their indirect actions on other physiological responses, or their alterations of the coding of the consequences of the stressor. Finally, the importance of using multiple measures of nociception is addressed in terms of identifying which systems may be affected and whether such effects occur at spinal or supraspinal levels. These factors, together with organismic variables altering pain inhibition (104), play important roles in our ability to understand a model which attempts to evaluate analgesic systems activated by environmental rather than pharmacological or physiological stimuli.

# References

1. T. L. Yaksh and T. A. Rudy, *Pain* **4,** 299 (1978).
2. D. J. Mayer and D. D. Price, *Pain* **2,** 379 (1976).
3. H. L. Fields and A. I. Basbaum, *Anny. Rev. Physiol.* **40,** 193 (1978).
4. A. I. Basbaum and H. L. Fields, *Annu. Rev. Neurosci.* **7,** 309 (1984).
5. T. L. Yaksh, *Pain* **11,** 293 (1981).
6. C. Pert and S. H. Snyder, *Science* **179,** 1011 (1973).
7. E. J. Simon, J. M. Hiller, and I. Edelman, *Proc. Natl. Acad. Sci. U.S.A.* **70,** 1947 (1973).
8. L. Terenius, *Acta Pharmacol. Toxicol.* **32,** 317 (1973).
9. H. Akil, S. J. Watson, E. Young, M. E. Lewis, H. Khachaturian, and J. M. Walker, *Annu. Rev. Neurosci.* **7,** 223 (1984).
10. J. Hughes, T. W. Smith, H. W. Kosterlitz, L. A. Fothergill, B. A. Morgan, and H. R. Morris, *Nature (London)* **258,** 577 (1975).
11. R. Guillemin, T. M. Vargo, J. Rossier, S. Minick, N. Ling, C. Rivier, W. Vale, and F. E. Bloom, *Science* **197,** 1367 (1977).
12. J. N. Madden, H. Akil, R. L. Patrick, and J. D. Barchas, *Nature (London)* **265,** 358 (1977).
13. R. L. Hayes, G. J. Bennett, P. G. Newlon, and D. J. Mayer, *Brain Res.* **155,** 69 (1978).
14. R. J. Bodnar, D. D. Kelly, M. Brutus, and M. Glusman, *Neurosci. Biobehav. Rev.* **4,** 87 (1980).
15. L. R. Watkins and D. J. Mayer, *Science* **216,** 1185 (1982).
16. G. W. Terman, Y. Shavit, J. W. Lewis, J. T. Cannon, and J. C. Liebeskind, *Science* **226,** 1270 (1984).
17. R. J. Bodnar, *Ann. N.Y. Acad. Sci.* **467,** 345 (1986).

18. M. S. Fanselow, *Ann. N.Y. Acad. Sci.* **467,** 40 (1986).

19. S. F. Maier, *Ann. N.Y. Acad. Sci.* **467,** 55 (1986).

20. R. J. Bodnar, *Crit. Rev. Neurobiol.* **6,** 39 (1990).

21. B. R. Komisaruk and J. Wallman, *Brain Res.* **137,** 85, (1977).

22. B. R. Komisaruk and B. Whipple, *Ann. N.Y. Acad. Sci.* **467,** 30 (1986).

23. K. A. Miczek, M. L. Thompson, and L. Shuster, *Science* **215,** 1520 (1982).

24. K. A. Miczek, M. L. Thompson, and L. Shuster, *Ann. N.Y. Acad. Sci.* **467,** 14 (1986).

25. C. Vidal and J. Jacob, *Life Sci.* **31,** 1099 (1982).

26. D. A. Simone and R. J. Bodnar, *Life Sci.* **30,** 719 (1982).

27. A. Spiaggia, R. J. Bodnar, D. D. Kelly, and M. Glusman, *Pharmacol. Biochem. Behav.* **10,** 761 (1979).

28. R. J. Bodnar and B. R. Komisaruk, *Physiol. Behav.* **32,** 653 (1984).

29. G. W. Terman, J. W. Lewis, and J. C. Liebeskind, *Brain Res.* **368,** 101 (1986).

30. R. J. Bodnar, M. Glusman, M. Brutus, A. Spiaggia, and D. D. Kelly, *Physiol. Behav.* **23,** 53 (1979).

31. A. J. MacLennan, R. C. Drugan, H. L. Hyson, S. F. Maier, J. Madden, and J. D. Barchas, *Science* **215,** 1530 (1982).

32. R. J. Hamm, J. S. Knisely, A. Watson, B. G. Lyeth, and D. F. B. Bossut, *Physiol. Behav.* **35,** 879 (1985).

33. S. Amir and Z. Amit, *Life Sci.* **24,** 439 (1979).

34. R. J. Bodnar, N. S. Sharpless, J. H. Kordower, M. Potegal, and G. A. Barr, *Physiol. Behav.* **29,** 1105 (1982).

35. P. Marek, I. Ponecka, and G. Hartmann, *Pharmacol. Biochem. Behav.* **16,** 403 (1982).

36. S. Mousa, C. H. Miller, and D. Couri, *Psychopharmacology (Berlin)* **79,** 199 (1983).

37. M. J. Millan, R. Przewlocki, and A. Herz, *Pain* **8,** 343 (1980).

38. J. W. Lewis, M. G. Tordoff, J. E. Sherman, and J. C. Liebeskind, *Science* **217,** 557 (1982).

39. R. J. Bodnar, D. D. Kelly, S. S. Steiner, and M. Glusman, *Pharmacol. Biochem. Behav.* **8,** 661 (1978).

40. R. J. Bodnar, D. D. Kelly, A. Spiaggia, C. Ehrenberg, and M. Glusman, *Pharmacol. Biochem. Behav.* **8,** 667 (1978).

41. W. T. Chance and J. A. Rosecrans, *Pharmacol. Biochem. Behav.* **11,** 639 (1979).

42. W. T. Chance and J. A. Rosecrans, *Pharmacol. Biochem. Behav.* **11,** 643 (1979).

43. J. W. Lewis, J. T. Cannon, and J. C. Liebeskind, *Science* **208,** 623 (1980).

44. J. W. Lewis, J. E. Sherman, and J. C. Liebeskind, *J. Neurosci.* **1,** 358 (1981).

45. J. W. Grau, R. L. Hyson, S. F. Maier, J. Madden, and J. D. Barchas, *Science* **203,** 1409 (1981).

46. M. N. Girardot and F. A. Holloway, *Pharmacol. Biochem. Behav.* **20,** 631 (1984).

47. M. N. Girardot and F. A. Holloway, *Physiol. Behav.* **32,** 547 (1984).

48. R. McGivern, C. Berka, G. G. Berntson, J. M. Walker, and C. A. Sandman, *Life Sci.* **25,** 885 (1979).

49. S. Amir and Z. Amit, *Life Sci.* **23,** 1143 (1978).
50. R. J. Bodnar and V. Sikorszky, *Learn. Motiv.* **14,** 223 (1983).
51. P. O'Connor and R. E. Chipkin, *Life Sci.* **35,** 631 (1984).
52. G. W. Terman, M. J. Morgan, and J. C. Liebeskind, *Brain Res.* **372,** 167 (1986).
53. J. Sawynok, C. Pinsky, and F. S. LaBella, *Life Sci.* **25,** 1621 (1979).
54. R. S. Zukin and S. R. Zukin, *Life Sci.* **29,** 2681 (1981).
55. W. R. Martin, C. G. Eades, J. A. Thompson, R. E. Huppler, and P. E. Gilbert, *J. Pharmacol. Exp. Ther.* **197,** 517 (1976).
56. J. H. Lord, A. A. Waterfield, J. Hughes, and H. W. Kosterlitz, *Nature (London)* **267,** 495 (1977).
57. G. W. Pasternak and P. L. Wood, *Life Sci.* **38,** 1889 (1986).
58. B. P. Roques, M. C. Fournie-Zaluski, E. Soroca, and J. M. LeComte, *Nature (London)* **288,** 286 (1980).
59. R. E. Chipkin, M. B. Latranyi, L. C. Iorio, and A. Barnett, *Life Sci.* **31,** 1184 (1983).
60. R. Greenberg and E. H. O'Keefe, *Life Sci.* **31,** 1185 (1982).
61. R. J. Bodnar, M. Lattner, and M. M. Wallace, *Pharmacol. Biochem. Behav.* **13,** 829 (1980).
62. G. W. Pasternak, S. R. Childers, and S. H. Snyder, *J. Pharmacol. Exp. Ther.* **214,** 455 (1980).
63. A. L. Kirchgessner, R. J. Bodnar, and G. W. Pasternak, *Pharmacol. Biochem. Behav.* **17,** 1175 (1982).
64. B. C. Yoburn, L. S. Truesdell, B. Kest, C. R. Inturrisi, and R. J. Bodnar, *Pharmacol. Biochem. Behav.* **27,** 525 (1987).
65. J. L. Steinman, P. L. Faris, P. E. Mann, J. W. Olney, B. R. Komisaruk, W. D. Willis, and R. J. Bodnar, *Neurosci. Biobehav. Rev.* **14,** 1 (1990).
66. B. D. Appelbaum and S. G. Holtzman, *J. Pharmacol. Exp. Ther.* **231,** 555 (1984).
67. R. J. Bodnar, D. D. Kelly, and M. Glusman, *Pharmacol. Biochem. Behav.* **11,** 297 (1979).
68. J. E. Sherman, H. Strub, and J. W. Lewis, *Behav. Neurosci.* **98,** 293 (1984).
69. B. Kest, M. Orlowski, and R. J. Bodnar, *Physiol. Behav.* **50,** 843 (1991).
70. A. Spiaggia, R. J. Bodnar, D. D. Kelly, and M. Glusman, *Pharmacol. Biochem. Behav.* **10,** 761 (1979).
71. A. E. Panerai, A. Martini, P. Sacerdote, and P. Mantegazza, *Brain Res.* **304,** 153 (1984).
72. M. J. Millan, *Pain* **27,** 303 (1986).
73. L. R. Watkins, E. P. Wirtelak, J. E. Grisel, L. H. Silbert, and S. F. Maier, *Brain Res.* **594,** 99 (1992).
74. J. A. Robertson, L. B. Hough, and R. J. Bodnar, *Pharmacol. Biochem. Behav.* **31,** 107 (1988).
75. F. E. D'Amour and D. L. Smith, *J. Pharmacol. Exp. Ther.* **72,** 74 (1941).
76. E. S. Sperber, E. Kramer, and R. J. Bodnar, *Pharmacol. Biochem. Behav.* **25,** 171 (1986).
77. E. Kramer and R. J. Bodnar, *Physiol. Behav.* **36,** 875 (1986).

78. M.-T. Romero, K. L. Kepler, M. L. Cooper, B. R. Komisaruk, and R. J. Bodnar, *Physiol. Behav.* **40,** 39 (1987).
79. R. J. Bodnar, P. E. Mann, and E. A. Stone, *Pharmacol. Biochem. Behav.* **23,** 749 (1985).
80. K. L. Kepler and R. J. Bodnar, *Pharmacol. Biochem. Behav.* **29,** 83 (1988).
81. P. D. Butler and R. J. Bodnar, *Peptides* **8,** 299 (1987).
82. R. J. Bodnar, K. P. Merrigan, and E. S. Sperber, *Pharmacol. Biochem. Behav.* **19,** 447 (1983).
83. D. Badillo-Martinez, N. Nicotera, P. D. Butler, A. L. Kirchgessner, and R. J. Bodnar, *Neuroendocrinology* **38,** 438 (1984).
84. R. J. Bodnar, D. D. Kelly, M. Brutus, A. Mansour, and M. Glusman, *Pharmacol. Biochem. Behav.* **9,** 543 (1978).
85. R. L. Himsworth, *J. Physiol.* (*London*) **198,** 451 (1970).
86. G. P. Smith and A. N. Epstein, *Am. J. Physiol.* **217,** 1083 (1969).
87. R. J. Bodnar, D. D. Kelly, A. Mansour, and M. Glusman, *Pharmacol. Biochem. Behav.* **11,** 303 (1979).
88. M. T. Lowy and G. K. W. Yim, *Life Sci.* **26,** 2553 (1980).
89. R. J. Bodnar and N. Nicotera, *Pharmacol. Biochem. Behav.* **16,** 411 (1982).
90. E. M. Stricker and M. J. Zigmond, *J. Comp. Physiol. Psychol.* **86,** 973 (1974).
91. R. J. Bodnar, E. Kramer, D. A. Simone, A. L. Kirchgessner, and J. Scalisi, *Int. J. Neurosci.* **21,** 225 (1983).
92. R. J. Bodnar, D. D. Kelly, M. Brutus, and M. Glusman, *Pharmacol. Biochem. Behav.* **9,** 763 (1978).
93. J. E. Koch, L. B. Hough, and R. J. Bodnar, *Pharmacol. Biochem. Behav.* **41,** 371 (1992).
94. S. P. Grossman and L. Grossman, *J. Comp. Physiol. Psychol.* **85,** 515 (1973).
95. L. L. Walsh and S. P. Grossman, *Physiol. Behav.* **15,** 481 (1975).
96. A. Randich and W. Maixner, *Neurosci. Biobehav. Rev.* **8,** 343 (1984).
97. G. F. Gebhart and A. Randich, *APS J.* **1,** 26 (1992).
98. T. S. Jensen and T. L. Yaksh, *Brain Res.* **372,** 301 (1986).
99. W. O. Evans, *Psychopharmacology* (*Berlin*) **2,** 318 (1961).
100. N. Sathaye and R. J. Bodnar, *Physiol. Behav.* **46,** 217 (1989).
101. J. M. Kiefel, D. Paul, and R. J. Bodnar, *Brain Res.* **500,** 231 (1989).
102. M.-T. Romero and R. J. Bodnar, *Physiol. Behav.* **37,** 893 (1986).
103. E. Lubin and R. J. Bodnar, *Pharmacol. Biochem. Behav.* **34,** 511 (1989).
104. R. J. Bodnar, M.-T. Romero, and E. Kramer, *Brain Res. Bull.* **21,** 947 (1988).

# Section V

## Feeding Behavior

# [19]   Assessment of Feeding Behavior

Theodore J. Kalogeris and Thomas W. Castonguay

## Introduction

Food intake research has grown into a truly multidisciplinary field. Accordingly, the number and variety of paradigms for studying ingestive behavior have grown to such an extent that they cannot all be covered in sufficient detail within the limits of a single chapter. Such fundamental techniques as standardization of experimental diets, correction for diet spillage when quantifying amounts ingested, and expression of intake measurements have been described elsewhere (1). Here the focus is on two important methods: (1) assessment of whether an experimental treatment which reduces food intake might be acting on a physiological satiety mechanism or, rather, might be producing a deleterious effect which could also explain the food intake suppression; and (2) the setup of computer-automated systems for monitoring feeding behavior and for measurement and analysis of feeding patterns. The former method is an important and useful adjunct to studies of potential satiety mechanisms. The latter approach is increasingly becoming the method of choice for studying feeding behavior, since it takes advantage of the widespread availability of relatively inexpensive computer technology to achieve automatic monitoring of experimental subjects with minimal disturbance under "normal," *ad libitum* feeding conditions.

## Conditioned Taste Aversion Testing

One difficulty in studying feeding behavior is that a reduction in food intake in response to an experimental treatment does not necessarily imply the existence of a physiological satiety mechanism. Instead, that treatment could inhibit ingestion by eliciting an avoidance response to the food as a result of some deleterious posttreatment consequence such as malaise. This possibility can be tested by determining whether the treatment can serve as an unconditioned stimulus in the acquisition of an avoidance response (or conditioned aversion) in the presence of a particular stimulus—usually a flavor, an odor, or other sensory property of a food item. We consider the use of flavor as the conditioned stimulus since it has been used most extensively in work with the rat. Failure of a compound

*Methods in Neurosciences, Volume 14*

to act as an unconditioned stimulus in the formation of an avoidance response is generally considered to be evidence in favor of its being a physiologically relevant intake suppressant.

Several approaches to conditioned taste aversion (CTA) testing in rats, with varying sensitivities, have been described (2–6). The general procedure is conducted in two phases: the conditioning phase, then the testing phase. For conditioning, the general approach is to pair the consumption of a specific flavor with the experimental treatment versus the pairing of either the same or a different flavor with a control treatment. For testing, the intake by subjects of the treatment-paired flavor versus the control-paired flavor is compared. More than one pairing of a given treatment may be required for the animal to associate a flavor with a particular treatment.

It is important to maximize sensitivity of the experimental model, so that relatively mild aversions will not be missed. First, the conditioned stimulus (e.g., a flavor mixed in water) must not be so preferred by the subject that it is consumed regardless of any postingestive aftereffects. An example is the use of flavors for which rats have neutral preference, such as banana versus almond flavors (6), rather than highly palatable sucrose solutions (4, 5). Second, the appropriate control treatment is critical. In some studies (4, 5), the effects of administration of a putative satiety agent (i.e., injection of a gastrointestinal peptide) were compared with those of apomorphine or lithium chloride, both of which are potent aversive agents. The most that can be logically concluded from negative results in such studies (i.e., no CTA demonstrated in response to peptide injection as determined by comparing intake of peptide-associated flavor with the intake of apomorphine- or LiCl-paired flavor) is that the peptide caused less of an aversion than apomorphine or LiCl. A more appropriate control would be injection of an innocuous substance such as physiological saline. Finally, since most CTA testing is performed on rats that have been deprived of water for varying lengths of time (e.g., 12–24 hr), it is important that thirst not be a factor in the results. If a treatment causes only mild deleterious effects, allowing a rat access to only one of either the treatment-paired flavor or the control-paired flavor during testing (the so-called one-bottle test) may result in the animal's thirst overcoming any aversive effects. A more sensitive assay is the two-bottle choice test (2, 3, 6), in which the subject has access to both flavors during testing.

The procedure described here, an example of a two-bottle choice test, is that of Deutsch and Hardy (6), who used it to determine whether bolus intraperitoneal injections of large doses of cholecystokinin octapeptide (CCK-8) administered to rats depressed feeding by causing deleterious effects.

## Procedure

A group of rats are deprived of water for 22 hr; half the rats are offered almond-flavored water, and the other half are offered banana-flavored water (0.5% Schilling, readily available in grocery stores). Both groups are then allowed 24 hr *ad libitum* access to water. They are again water-deprived for 22 hr, then each is given the alternate flavored water to drink (i.e., rats offered banana-flavored water the first day are given almond-flavored water on the second day, and vice versa). Immediately after drinking, half the rats in each group are given a control treatment, and the other half are given the experimental treatment. After 24 hr of *ad libitum* access to water, rats are again fluid-deprived for 22 hr, then given a 30-min, simultaneous choice between the two flavored solutions. In the absence of a conditioned taste aversion, animals should demonstrate no preference between the flavors. Significant preference of the control treatment-paired flavor compared to the flavor paired with the experimental treatment suggests some form of sickness or discomfort produced by the experimental treatment, which rats associate with its paired flavor. Two or more conditioning sessions (i.e., repeated pairings of treatments with specific flavors) may be necessary for relatively mild aversions to become manifest.

## Meal Pattern Analysis

The methodological objective of meal pattern analysis is to obtain a moment-by-moment measurement of *ad libitum* intake. Most investigators have used such data to quantify eating episodes in terms of frequency, size, and temporal distribution across the day/night cycle. The practical use of this technique is to provide the researcher with measures of meal frequency, meal size, rate of meal intake, and the interval separating successive meals (intermeal interval). These parameters can then serve as a sensitive set of baselines against which the effects of environmental or physiological manipulations can be assessed.

Historically, meal pattern analysis has grown out of a research tradition that had as its objective the discovery of the mechanism(s) that controls daily food intake. Although Richter and colleagues (7) were among the first to describe the periodic or episodic nature of eating behavior of laboratory rats, it was LeMagnen and colleagues (8) who first introduced the idea that meal eating could be under physiological controls. They reasoned that these measurements permit an evaluation of the relationships between the amount of food intake consumed at any one bout (meal) and the interval before the meal (the hunger/meal size relationship) as well as the relationship between the amount of food intake consumed at a meal and the interval after a meal (the meal size/satiety relationship).

Although these ideas have developed into one of the central themes of the question of whether food intake is physiologically regulated, almost 30 years has elapsed without a conclusive answer to the question. Most published meal pattern analysis papers in recent years have elected to make use of the techniques first introduced by LeMagnen (8) and Collier *et al.* (9) without entering into the regulation versus control controversy. Rather, most contemporary investigators have made use of a system that allows the investigator to examine several physiological or environmental manipulations on meal taking without having to disturb the animal being studied.

In its most rudimentary form, meal pattern analysis has been used to quantify food intake in terms of amount of food consumed in a fixed interval. Investigators have then compared within the same animal the amount of food that is habitually eaten during a particular interval with the amount of food eaten after experimental treatment. As daily food intake patterns have been found to be extremely stable within an animal over time, small changes in intake can be related to experimental treatment with impressive accuracy. In an attempt to examine this property of meal patterns, we have recently applied time series analysis, a mathematical technique used principally by economists, to the meal patterns of laboratory rats given *ad libitum* access to standard maintenance diet (Purina Chow). Results revealed that the single best predictor of meal size was the size of the meal initiated 24 hr prior to the time in question (T. W. Castonguay, unpublished observations).

## Intake versus "Meals"

As noted above, meal pattern analysis allows the investigator to quantify intake not only in terms of amount eaten during a given interval, but also in terms of the frequency, size, and duration of eating episodes during that interval. Unlike measuring intake in terms of total amount consumed during an interval, quantifying "meals" within that interval provides several measures that, although related to one another, can vary independently. For example, rats fed diets that are missing an essential amino acid typically eat much less than they do when fed a balanced diet, yet they are not completely aphagic. By examining the patterns of eating, Gietzen *et al.* (10) noted that rats fed an amino acid-imbalanced diet initiated many very small meals; that is, meal frequency increased and meal size was dramatically reduced. In another example, we (11) have reported that obese Zucker rats, unlike their lean littermates, eat both day and night and initiate three distinct types of meals: (1) meals within the normal range of meal sizes (i.e., similar to meal size in lean rats), (2) meals significantly larger than those initiated by lean rats, and (3) a significant number of very small meals. Thus, obese rats eat more meals and bigger meals than do lean rats. However, a large part of the increased frequency can be attributed to a large number of very small meals that contribute only marginally to the overall daily caloric intake of the rat. Thus, it is reasonable to conclude that obese rats initiate larger than average meals but eat as often as do lean controls.

## Instrumentation

### Overview

The essential requirements for meal pattern analysis include a means of either individually housing the animals to be monitored or some means of identifying one animal from another as they feed from a monitored food source. Food containers are typically monitored by weight directly, using either load cells or electronic balances. Others have placed a photocell directly in front of a food cup, so that a feeding animal will break a photobeam as it takes food from the cup. Output from these devices is directed to strip chart recorders, computers, or printout counters.

### Hardware

Figure 1 shows a scheme of a system that is currently in use in several laboratories. It consists of two separate computer-based systems: one for data acquisition and the other for subsequent analysis. Input from individual stations (e.g., electronic balances) is collected and written to an output device with minimal processing. These archival data files are then accepted as input by a second computer that sorts the data, combines individual episodes into meals, and generates a report either on disk or on paper. Although it is possible to use a single, multiuser computer to perform both acquisition and analysis, several issues preclude the use of a single machine.

The first disadvantage is economic. Larger computers, such as minicomputers or multiuser supermicrocomputers, could be used as a single acquisition and analytical system; however, the costs associated with purchasing not only the computer, but also the additional input/output lines make this option unattractive. Such computers range in cost from $10,000 to $50,000. These hardware costs do not compare favorably with the cost of personal computers, especially the PC clones that are readily available for a fraction of the cost. Further, software packages specific for the larger machines

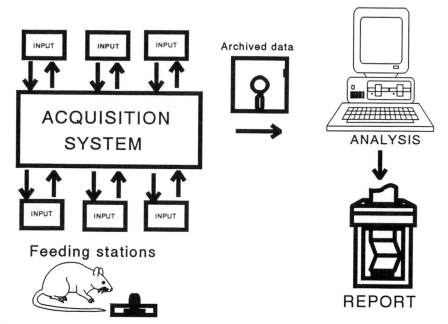

FIG. 1   System for meal pattern data acquisition and analysis. Input from feeding stations (e.g., food cups placed on electronic balances) is collected by an acquisition computer and written to an archival data file. The data archive then serves as input to a second computer, which performs data analysis.

are also much more expensive than the software available for the personal computer. License fees for specific languages such as Fortran or Pascal can exceed $1000 each for some applications. Similarly, databases that work on larger machines are both expensive and difficult to install and maintain.

A second reason to develop a dual-computer based meal pattern system is related to the development of customized software that can interfere with the other functions of computer. It has been our experience that software development is a continuously evolving process. As investigators become familiar with their data, different questions from those initially planned need to be examined, usually with the development of additional software routines. Frequently a computer will "hang up" while performing a task using commercially available, "debugged" software. Usually such problems are remedied by stopping the computer and then restarting it. Developing customized software increases the likelihood of similar interruptions in computer operation. Thus, data collection via the acquisition function could be interrupted for reasons of software development. Because most facilities will be used by more than one investigator, unplanned interruptions in data acquisition can be problematic. A dual-computer based system allows for independent acquisition and analysis functions.

Finally, given the initial investment of both time and money that occurs with developing a meal pattern analysis system, it may be desirable to share developments and capabilities with colleagues. For example, colleagues from different departments of the same institution may want to adapt a rodent meal pattern system for use with ponies, cattle, chickens, or other livestock. Such requests can be easily and relatively inexpensively met if only the acquisition system needs to be modified, usually with differently designed feeders or balances with different ranges (kilograms versus milligrams). Similarly, data analysis programs can be easily modified to use different units of measurement.

## Input Devices

Several different electronic balances have been evaluated for use with this application. Although most work with little trouble, there are a few considerations that make some balances more suitable. Typically, less expensive electronic balances "drift" or fail to maintain a constant output over time. Others are overly sensitive to the environment so that animal hair, airborne particulate matter, and changes in room air currents can interfere with stability and trouble-free operation of the balance. Balances that can be dampened will minimize these environmental effects. Electronic balances with capacities of 300–600 g (accurate to 0.1 g) manufactured or supplied by Ohaus, Mettler, and Sartorius have proved to be most reliable and are recommended on the basis of size, capacity, price, reliability, and maintenance.

## Interface Hardware

Our laboratory makes use of a multiple serial port board, manufactured by Stargate Technologies Inc. (Solon, OH). The card is expandable, allowing for the serial monitoring of 64 individual electronic devices. It interfaces with DOS-equipped, IBM-compatible computers. Software generated for use with the serial port card is typically written in either Pascal or BASIC. An alternative to this board is a code-activated electronic switch, such as the CAS 161 code-activated switch marketed by Western Telematic Inc. (Irvine, CA). Input from standard RS-232 serial interfaces, such as those found in most electronic balances, is accepted by the switch, which in turn is controlled by the acquisition computer. Each line is polled consecutively; data are forwarded from balance to computer, evaluated for change since last polling, and subsequently processed.

## Software

### Data Acquisition Schemes

Two schemes have been successfully used in collecting data: the continuous stream sampling approach and the poke/peek approach. Input devices such as electronic balances can often be set for continuous output, so that a computer polling a particular device need only capture a stream of data, evaluate it, and then switch to another device. The advantage of this scheme is that both input and output operations are not required of the balance, a process that takes time and a sophisticated input/output board on the balance. The disadvantage of this scheme is that the software needed to distinguish between the end of one cycle of output and the start of a second can be difficult and is not always error-free (negative numbers, "bounce in readings", load cell drift, etc., all present interesting challenges to software developers).

A second option for capturing data from input devices is to set the balance to accept a print command; the acquisition machine issues the command and then immediately captures the output from the balance. This poke/peek, output/input routine requires a bidirectional interface, since the computer would send information (the "print" command) to the balance as well as receive information from that source. The main advantage associated with this option is that the process is not dependent on a timing or data editing routine. Among the disadvantages of the scheme is that the two operations require more time to complete than the sampling of a continuous stream of data. If the user needs to monitor output more frequently than every few seconds (such as in the case of monitoring licking from a water spout), the poke/peek scheme may be inappropriate. On the other hand, if timing of bout duration is not so critical (10–15 sec inaccuracy), then the poke/peek scheme is often preferred.

Duty cycles, defined as the time required to complete a polling of all input devices, vary in duration as a function of the number of stations to be polled, the activity at each station (whether the last reading is the same or different from the current reading), and the stability of the input device. These three factors significantly contribute to the error in timing individual bouts. Because several input devices will be monitored sequentially, it may be desirable to include a set of commands in the acquisition software identifying stations that have provided changes in readings (such as during a meal) since the last polling. Rather than to start polling the balances at the beginning once completing a cycle, the computer could be instructed to cycle through only those balances that were active during the next cycle, and to return to a full scanning of all stations after completing the active-only list. In that way, more precise definition of the end of a bout could be recorded. Many balances will not output data following a print command unless the value has been stable for a fixed period of time. During feeding the balances will not respond as quickly to the print command as they would during intermeal intervals since animals will be eating from food cups placed on the balance. By directing the computer to poll only previously active stations one or two additional times before initiating a complete cycle, some precision is restored.

*On the Fly Analysis*

The most basic approach to data collection is to write out to a file the value of each balance each time it was polled. Although simplicity favors this approach, the resulting files are so large and filled with meaningless data that some minimal amount of data editing during the acquisition stage is recommended. One method of data editing is to have the software store the first reading of each balance and the date and time into arrays within memory. After completion of a duty cycle, the system will initiate its second set of readings of each balance. The software can then compare the second reading with the stored value. If the difference between the first and second values is not more than some predetermined value (usually 1 or 2 cg at most), then the second value and its date and time are substituted for the first in the arrays, and the acqustion software moves on to sample the next input device. However, if the second value is different from the first, the computer makes note of the activity (it flags that cell in the array as active). As noted above, once the software has polled all input lines the second time, a list of active stations is generated. Each of these active stations is then polled, and differences in values are once again noted. Eventually, each of the active lines will once again return to inactivity (i.e., polling results in the same value observed in the previous polling). Once this criterion has been met, the difference in values between the reading at the start of the bout and the last reading is calculated. Start times and ending times, the device number, and the difference in readings (usually in mg) are written out to a floppy disk as one data line (see Fig. 2).

Using this method, a chronologically sorted file is created. However, since these records are written "on the fly," they will not be sorted by device. Note that no additional spaces or notes are written on the data file. This compacted file is recommended out of consideration for the size of the data files that are typically generated. Each line of the output data file represents an individual "bout." It remains to be determined if that bout will meet the criteria established to be scored as a "meal," or if it will be only one of several bouts that will ultimately be combined to form a "meal." Because the user may wish to examine intake data using a variety of definitions, constructing a data file of moment-to-moment records is recommended. In that way, bout data can be evaluated under several combinations of meal

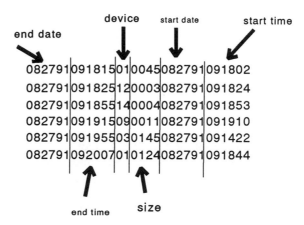

FIG. 2   Segment of a chronologically sorted acquisition file as saved "on the fly" to a data archive (see text). Each line represents a meal "event" at a particular feeding station (i.e., for one rat). Included in each data line are the start and ending times of the feeding bout, the station, rat, or device number, and the meal size or amount of food eaten (in mg). Acquisition software further organizes data from this log into a secondary output file (organized by station and time but leaving the original data archive intact) prior to analysis.

pattern definitions (different minimum meal size definitions and different end-of-the-meal definitions).

### Data Acquisition

Elements that must be recorded by acquisition software include episode or "bout" size (with the highest resolution possible, usually to the nearest 0.1 g, but data accurate to 1 mg have often been evaluated), the input device generating the record, the start date and time, and the stop date and time. Data collected are typically much more resolved than is required to classify eating episodes into meals. At first consideration, the provision of date within each record might seem unnecessary. However, because data are acquired over the course of several days, and because these records will be written to disk chronologically, without sorting by station, having full date and time in each record can be helpful. An example of a typical archival data file is presented in Fig. 2.

### Analysis of Bout Data

Postacquisition "analysis" software performs several tasks. First, it should sort through the logged output generated by the acquisition program and organize it by station, date, and time. It is recommended that a second output file, namely, one sorted by station and time, be generated as a first step. This sorted file can then be used and reused, all the while maintaining the integrity of the original archive file. Although it is possible to edit master archive files prior to submission to the analysis program, it is often helpful to build into the analysis program several options such as start date and time and end date and time. In that way the large sorted secondary file generated by the acquisition software can be used directly.

Once the data file is sorted by station and time, it becomes feasible to combine bouts into meals based on a few simple rules or definitions. The definitions of intermeal intervals and minimum meal size can play an important role in evaluating meal pattern data. How much food must be consumed to distinguish the eating episode as a meal rather than an investigatory behavior emitted by a freely feeding caged animal that has little else to do than eat, drink, and sleep? More importantly, how much time must elapse between episodes to determine when a meal has stopped and when a second has begun? These issues have been examined in a series of computational reports that have concluded that the most appropriate between-meal interval definition is often close to 10 min in the rat under most conditions. Minimum meal size definitions of as low as 0.05 g and as high as 0.25 g can be used without loss of accuracy or sensitivity. The interested reader should refer to the reports by Castonguay *et al.* (11, 12) for supporting data.

*Other Computational Details*

The correlation between intervals and meal size should be performed on individual animals, as pooling interval and size data from many animals can have the effect of erroneously finding a relationship between intervals and meal size (8). Some animals eat larger meals than others. Those that eat larger meals tend to eat less frequently, and thus have longer intermeal intervals than do rats that eat smaller but more frequent meals. By pooling intermeal data from these extremes, it is possible to calculate a correlation coefficient that is anchored in both the upper left and lower right quadrants of an $x$–$y$ plot. Distributions with extremes promote high correlation coefficients, but they do not describe the relationship between interval and size within individual animals, only in pooled data generated by many individuals. The correct method of calculating the correlation is to consider data generated by each animal individually, usually collected over many days. In that way, meal size and interval data can be examined free from statistical artifact directly attributable to unrelated factors such as differences in body weight or idiosyncratic feeding history.

# References

1. T. W. Castonguay, *in* "Feeding and Drinking" (F. M. Toates and N. E. Rowland, eds.), p. 429. Elsevier, Amsterdam, 1987.
2. W. Dragoin, G. E. McCleary, and P. McCleary, *Behav. Res. Methods Instrum.* **3,** 309 (1971).
3. F. W. Grote and R. T. Brown, *Behav. Res. Methods Instrum.* **3,** 311 (1971).
4. J. Gibbs, R. C. Young, and G. P. Smith, *Nature (London)* **245,** 323 (1973).
5. J. Holt, J. Antin, J. Gibbs, R. C. Young, and G. P. Smith, *Physiol. Behav.* **12,** 497 (1974).
6. J. A. Deutsch and W. T. Hardy, *Nature (London)* **266,** 196 (1977).
7. C. P. Richter, *Q. Rev. Biol.* **2,** 307 (1927).
8. J. LeMagnen and S. Tallon, *J. Physiol. (Paris)* **58,** 323 (1966).
9. G. H. Collier, E. Hirsch, and P. Hamlin, *Physiol. Behav.* **9,** 399 (1972).
10. D. W. Gietzen, P. M. B. Leung, T. W. Castonguay, W. J. Hartman, and Q. R. Rogers, *in* "Interaction of the Chemical Senses with Nutrition" (J. G. Brand and M. R. Kare, eds.), p. 415. Academic Press, New York, 1986.
11. T. W. Castonguay, D. E. Upton, P. M. B. Leung, and J. S. Stern, *Physiol. Behav.* **28,** 911 (1982).
12. T. W. Castonguay, L. Kaiser, and J. S. Stern, *Brain Res. Bull.* **17,** 439 (1986).

# [20] Continuous Analysis of Feeding Patterns in Rats Receiving Total Parenteral Nutrition as Measured by the Automated Computerized Rat Eater Meter

Michael M. Meguid, Zhong-Jin Yang, J. Lee Beverly, and John R. Gleason

## Introduction

To fully understand the physiological and learning mechanisms that regulate feeding, the individual components which constitute food intake need to be isolated and studied. Simplistically, food intake is the amount of food ingested over a given period, and it is the product of meal number and meal size. The amount of food ingested is also the result of integrating different variables reflecting the states of the internal and the external environments. Thus, defining intra- and intermeal components provides clues to the relative contributions of the individual variables in regulating food intake. These variables may have an immediate impact influencing either a single meal or even meal size or meal number; alternatively, the variables may have a more long-term impact by modifying several meals during a period. Such information would be relevant to the ultimate goal of gaining insight into the mechanisms that alter food intake and that may lead to the related energy imbalances occurring in human pathological conditions such as surgical stress, cancer, diabetes, obesity, and eating disorders.

The characteristics of a meal have been useful in establishing similarities or differences in food intake regulation. Thus, the reduction in meal size in response to changes in diet composition (1) or the introduction of putative satiety agents (2) may reflect similar neurochemical responses such as an increase in dopamine activity at postsynaptic $D_2$ receptors, which has been associated with smaller meal size in free-feeding rats (3). Furthermore, differences in the mechanism of action between two factors having similar effects on food intake may become apparent on analysis of the meal patterns. For example, cholecystokinin (CCK) and bombesin (BBS) are putative "satiety" hormones that are increased after a meal. Both hormones reduce food intake when administered to rats, with the effect of CCK, but not BBS, being to reduce meal duration and increase the intermeal interval (4). Another example of the usefulness of meal pattern analysis is the finding that genotype differences (5) and the effect of different conditioning paradigms (6) also became apparent when meal patterns were analyzed.

In this chapter we illustrate how the analysis of meal patterns, employing the Automated Computerized Rat Eater Meter (ACREM), is being used to identify the possible cause(s) of anorexia associated with a clinically relevant phenomenon, the delayed return to normal food intake following extended periods of intravenous nutrition in the form of total parenteral nutrition (TPN).

## Development of the Automated Computerized Rat Eater Meter

Various automated methods, falling into two broad groups, have been used to examine feeding behavior in the rat. Meal size has indirectly been measured by determining the cumulative time spent eating. In these "contact" types of systems (7), the amount consumed was measured independently. From the frequency of contact and the total amount eaten, the average meal size was estimated. This method is limited because it measures only contact time, not the amount eaten per contact; it also fails to eliminate non-feeding-related activity. Direct measurement of meal size has been accomplished with bar-press operant systems (8) and a Pellet Eatometer (9). These methods record the time and number of pellets dispensed or removed. Because pellets are of known weight, food consumption is easily estimated. However, calculated food intake is based on two tenuous assumptions that (a) the rat ate all of each pellet and (b) the rat did not hoard the pellets. Furthermore, this method does not account for nonconsumption of the pellet due to spillage. Finally, the training and conditioning required for operant systems may well induce altered spontaneous feeding characteristics, thus affecting feeding behavior and pattern.

Because of the limitations inherent in the available devices, we designed an Automated Computerized Rat Eater Meter (10). The ACREM device enables us to measure each discrete access to food and to distinguish between an access which results in food consumption and one which does not. Through computerized calculations, the ACREM permits continuous recording of meal number, meal size, meal duration, and other meal-related events including intermeal intervals for either 24 hr or for each 12-hr light/dark cycle in the spontaneously free-feeding rat without the need for pretraining or reflex conditioning. It is equally applicable to normal and to pathophysiological rat models.

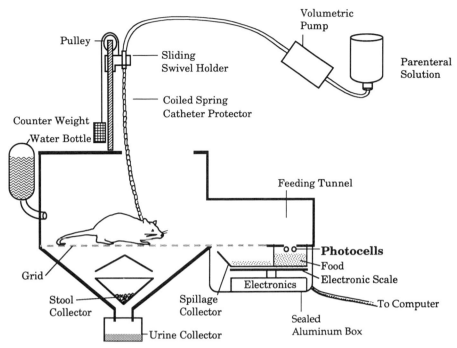

FIG. 1   Schematic illustration of the Automated Computerized Rat Eater Meter consisting of a commercially available metabolic cage with a feeding tunnel modified by an electronic food scale and photoelectric cells.

## Description of the Device

The Automated Computerized Rat Eater Meter is based on a commercially available metabolic cage (Lab Products, Inc., Maywood, NJ; Model LC-176) in which the supplied feeding cup has been replaced by an electronic scale balance and two photocells (Fig. 1). To obtain continuous data on spontaneous feeding characteristics, (a) two photoelectric sensors are centered just above the food dish to detect an access, (b) a sensitive electronic scale is used to measure the amount of food consumed, and (c) a real-time remote computerized data collection device is used. The latter integrates the information on feeding characteristics, as detected from the photoelectric cells and the electronic scale, with real time.

## Modified Feeding Cup

The feeding cup supplied with the commercially available metabolic cages is replaced with a unit containing an electronic scale balance and a combination food dish and spillage-collector tray (Fig. 1). Infrared photoelectric sensors are used to sense the presence of the rat's head anywhere within the 3 cm diameter opening in the tunnel floor centered over the food dish. The balance is constructed of a precision aluminum bending beam to which a full Wheatstone bridge strain gauge has been glued; any imbalance in the strain gauge signal is linearly related to the applied weight. Food consumption thereby results in a decrease in the food tray weight. The entire unit is housed in a $6 \times 10 \times 16$ cm sealed aluminum box that slides under the feeding tunnel on the rails provided for the commercially supplied food cup.

## Signal Processing

The analog signal from the strain gauge is routed through a differential amplifier, a multiplexor (Burr-Brown MPC801, Tucson, AZ), an analog-to-digital converter (Burr-Brown ADC80, Tucson, AZ), and then stored in computer memory (Fig. 2). The bending beam has an inherent resonant frequency of 20 Hz, which is electronically filtered. A circuit is included to maintain scale calibration throughout normal ambient temperature. Data transmission to the computer (IBM PC-XT) occurs at 100-msec intervals. The 100-msec interrupt causes the computer to record the time-of-day counter and to collect data. The first 10 msec of the 100-msec period is used to convert the analog signal to a digital signal for each bending beam strain gauge. The converted data and infrared sensor information are then transferred into a 16-bit register. The remaining 90 msec is used to read the register and process the data.

## Software Program

The signals generated by the ACREM are processed by the computer and recorded on magnetic disk. Specially written data collection programs provide for system interface initialization, scale calibration, data capture, and data storage. These programs take advantage of an interrupt technique simultaneously to track time as well as to read the data each 100 msec. With the interrupt, the time of day is recorded, the scale balance displacement determined, and the rat's head-in or head-out status noted. When a head-

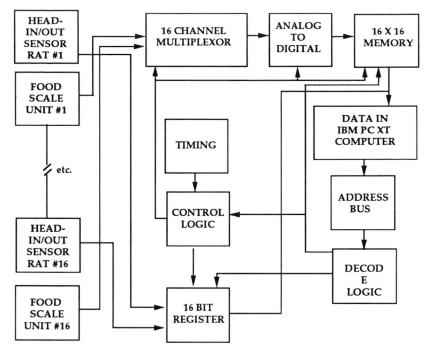

FIG. 2 Block diagram of electronic circuitry and logic enabling continuous monitoring of spontaneous food intake in caged rats.

out condition is detected following a head-in signal, a 12-sec delay is initiated, and then the scale balance value is recorded. The delay was found empirically to compensate for minor oscillations in the balance following removal of the rat's head from the food dish. During periods of feeding inactivity (absence of any head-in signals) the scale balance value is determined every 6 min to compensate for any minor balance drift.

Two exclusion criteria are applied to all the data accrued. The balance exclusion criterion eliminates data from any rat when the difference between daily food intake from strain-gauge and observer weighing exceeds 2.1 g. The photocell exclusion criterion eliminates data in which head-in duration exceeds 600 sec. Using these criteria, there is a high correlation ($r = 0.99$) between daily spontaneous food intake as verified by observer weighing and as measured by the strain-gauge balance (Fig. 3). The frequency of such exclusions is generally less than 5% of the total database.

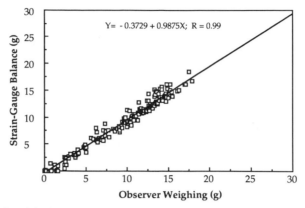

FIG. 3  Relationship between observer and measured daily food intake generated after the application of exclusion criteria for spurious values.

## Definitions of Terms

Concepts and terminology were developed only after extensive visual observations of the rats and subsequent inspection and correlative review of the continuous recordings of the raw eating data. When the raw data of free-feeding Fischer 344 rats were represented graphically and examined (see Fig. 4), the periodicity of eating activity during a 12-hr cycle could clearly be identified. Feeding activity was characterized by groups of rat accesses to the food cup, some of which included food consumption and some of which did not. Those episodes of head-in, no food eaten, head-out were visually noted to be associated with both olfactory and vibrissal exploration of the food cup; these have been termed sniffs. Episodes consisting of head-in, food eaten, head-out have been termed bites. Feeding activity is further characterized as a sequential group of accesses resulting in food consumption and including (a) a finite duration, (b) consumption of food, and (c) a variable number of sniffs.

By visual inspection of data such as that shown in Fig. 4, the gaps between accesses consisted of both inter- and intrameal intervals. The intrameal intervals were of shorter duration than the intermeal intervals. To select a minimum interval that would appropriately separate sequential grouped accesses into more clearly defined meals, a curve was constructed by taking a series of specific intermeal intervals and, for each, determining from raw data how many individual "meals" per 24 hr would have been eaten if that particular intermeal interval was used as the criterion. The specific intermeal intervals ranged from 1 to 180 min. To differentiate the effects of both the

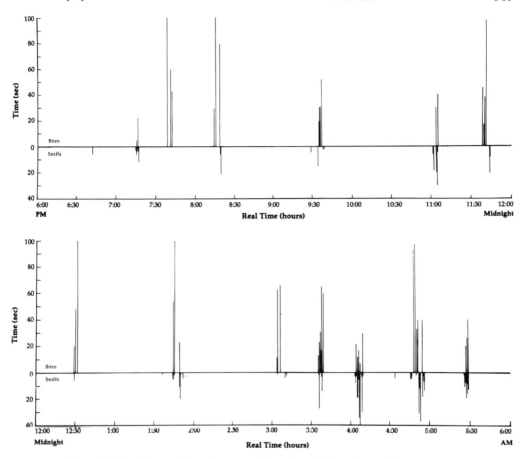

FIG. 4   Visual recording of one typical 12-hr dark period (sniffs and bites). Both the meals and their microstructure are clearly recognizable.

inter- and intrameal intervals on meal number, the curve was approximated by a two-exponential equation (11):

$$Y = a_1\, e^{-b_1 t} + a_2\, e^{-b_2 t}$$

The influence of the intrameal intervals rapidly decreased as the minimum of intermeal intervals was increased. At a 5-min intermeal interval, the influence of intrameal intervals (2%) was less than the usually assumed biological error of 5%. Hence, 5 min was taken as the minimum intermeal interval that appropriately separates sequential grouped accesses into clearly defined meals. Thus, a meal was defined as a bite or a group of bites preceded and followed by at least 5 min of feeding inactivity.

Other terms, based on the above definitions of a meal and measured using the ACREM, for either 24 hr or for each 12-hr light/dark cycle, include the following: food intake, amount of food (g) consumed; number of meals, total number of meals in a measured period; meal size, total amount of food consumed per total meal number (g/meal); meal sniffs, total number of sniffs occurring during all meals; meal duration, total duration of all the meals per total number of meals eaten (min/meal); eating activity, total number of bites and sniffs per total meal number; and intermeal sniffs, total number of sniffs occurring between meals. A calculation based on meal size and meal duration is the food consumption rate, which equals the total amount of food consumed per total meal time (g/min), also in either a 12-hr cycle or a 24-hr period.

## Statistical Treatment of Feeding Data

From a statistical perspective, data collected using the ACREM and TPN model present many of the problems common to longitudinal studies and repeated measures designs. There is usually some error variability, both between and within rats, in feeding indexes, and this error variation may well violate the strict sphericity and homogeneity assumptions of repeated measures and split-plot analyses of variance. It is typical for a fraction of the anticipated data values to be missing, occasionally because of equipment failure in early studies, but usually because of the illness or death of a rat. These problems often make it difficult to apply traditional analysis of variance methods.

Our strategy has been to analyze ACREM data with a curve-fitting approach, modeling the time course of the means of each feeding index with a relatively simple curve. The type of curve-fitting technique is chosen to suit each specific experimental design, but two methods have been particularly effective: nonparametric regression, as shown in Fig. 5 (12), and polynomial regression splines (see Fig. 6). Between- and within-group comparisons of feeding patterns are made in terms of the mean values of the fitted curves, rather than in terms of the observed means of the feeding indexes. This method provides a form of trend analysis custom-tailored to the design of each experiment, and it offers several advantages beyond producing a convenient graphical summary of the data.

First, fitting curves tends to smooth away a good deal of the noise inherent in the data, so that underlying trends can be more easily detected. A corollary is that tests which compare fitted curves are often more powerful than the corresponding tests which compare observed means; this is because the

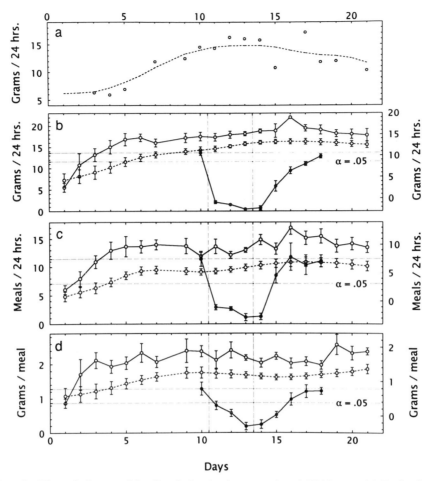

FIG. 5 Three indexes of feeding behavior in control and TPN rats. (a) Daily food intake of a control rat between days 1 and 21 of an experiment reported by Meguid *et al.* (12). The open circles are the measured intake for this rat; the dashed curve is the result of smoothing the measured intake using nonparametric regression. The remaining plots summarize results for the following indexes: (b) daily food intake, (c) meal number, and (d) meal size; each plot follows the same format. The uppermost curve (open circles, solid line) shows the daily mean ± S.E. of the measured index for eight control rats. This curve has been shifted upward for clarity; the right-hand axis gives the correct units. The dashed curve marked by open circles shows the daily mean ± S.E. for the same eight rats after the data record for each rat is smoothed as in the top panel. The filled circles plot the daily mean ± S.E. of the measured index for a group of eight rats receiving TPN; the vertical grid lines mark the onset and offset of TPN. The upper horizontal grid line marks the day 10 mean of the TPN group. The lower horizontal grid line indicates the level required for any of the day 11–18 means to be significantly below the day 10 mean (simultaneous level of significance, 0.05).

318

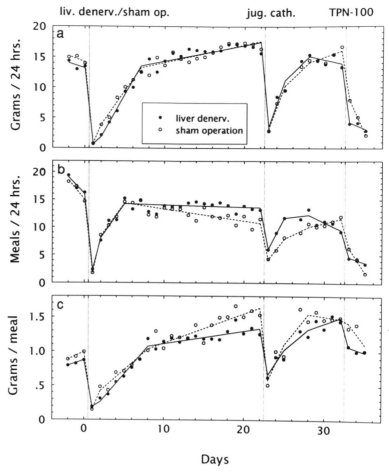

FIG. 6 Time course of three indexes of feeding behavior in liver-denervated and sham-operated rats: (a) daily food intake (g/24 hr), (b) meal number (meals/24 hr), and (c) mean meal size (g/meal). The filled circles represent the mean of the liver-denervated rats and the open circles the mean of the sham-operated control rats; the solid line is the regression spline for the liver-denervated group, and the dashed line is the corresponding spline for the sham-operated controls. The vertical grid lines show the three change points in the experimental design: liver denervation or sham operation, insertion of jugular catheter, and onset of TPN.

standard error of a fitted curve at a given time point is generally smaller than the standard error of the mean of the data collected at that same time point. The nonparametric regression approach provides this advantage without imposing any particular shape for the fitted curve.

Second, when appropriate, the regression spline method allows us to design the shape of the curve to impose a reasonable model of known or presumed physiological relationships. For example, it is known that infusing TPN inhibits food intake, and it is thus sensible to expect mean food intake to be at its lowest level at the end of the TPN period. In any given experiment, however, mean food intake may not decline monotonically during the TPN period, simply because of error variation. Nevertheless, it is generally easy to capture nearly all of the day-to-day variation in the observed mean food intake with a spline that does decrease monotonically throughout the TPN period. Another advantage of regression splines is that they are fit by ordinary multiple linear regression methods, so that familiar $F$- and $t$-tests are available. Also, the spline technique is practicable when there are too few data points to use the nonparametric regression approach.

Third, when necessary, the chosen curves can be fit to the data records of each individual rat. This provides a natural way to estimate missing data values for any given rat, should such estimates be required. A larger advantage of fitting curves to individual data is that computing the mean and its standard error at each time point in such a set of curves yields a picture of the time course of mean feeding patterns, along with—at each time point—error bars that require none of the stringent assumptions of standard analysis of variance methods for within-subjects designs.

## The Rat Total Parenteral Nutrition Model

To illustrate the application of the ACREM and our statistical analysis of the data generated, feeding patterns during and immediately after TPN in normal rats (12) and in rats after total liver denervation (13) are described. Adult male Fischer 344 rats, weighing about 250 g (Charles River, Cambridge, MA), are individually housed in a constant experimental environment of 12-hr light/dark cycles (0600 to 1800 hr and 1800 to 0600 hr), 26 ± 1°C temperatures, and 45% humidity. Rats have free access to water and to fresh, coarsely ground Purina Chow (No. 5008, Ralston Purina, St. Louis, MO).

Rats have a right internal jugular vein catheter inserted, which is exteriorized at the nape of the neck via a subcutaneous tunnel. The catheter is protected by a weighted spring and swivel device (Instech Corp., Plymouth Meeting, PA), which is sutured to the skin at the neck exit site. Catheter patency is maintained by constant 3 ml/hr infusion of normal saline with a calibrated Imed 960 volumetric infusion pump (Imed Corp., San Diego, CA). After 10 days, a TPN solution providing 100% of the rat's daily caloric need is infused for 3 consecutive days. TPN is then discontinued and normal saline infused. Control rats receive normal saline throughout the entire study period.

The composition of the TPN solution in these experiments is that which we use in clinical practice (14). The caloric ratio of glucose to fat (Intralipid 20% stock solution; KabiVitrum, Stockholm, Sweden) to a well-balanced amino acid solution (Novamine; KabiVitrum) is 50:30:20, and the ratio is kept constant throughout the study.

## Feeding Patterns with Total Parenteral Nutrition

An illustrative experiment using the ACREM to measure feeding patterns with TPN is reported in Meguid et al. (12). Figure 5a displays the daily food intake for one of the control rats. The day-to-day variation in food intake is quite typical of control animals in our experiments. The dashed line in Fig. 5a is the result of smoothing the plot of the food intake of this rat with a nonparametric regression estimator. The specific smoother used here is the kernel regression method described in Hart and Wehrly (15), but with the optimal Epanechnikov kernel function; other nonparametric smoothers produce very similar results. Note that several data points are missing for this rat (in addition to day 8) and that the smoother gives reasonable estimates of those missing values.

The remaining plots in Fig. 5 demonstrate other advantages of smoothing the control group data. In Fig. 5b, which shows the results for daily food intake, the uppermost curve (open circles, solid line) plots the mean ($\pm$S.E.) of the measured daily food intake for the eight control rats, that is, it summarizes eight sets of raw data, one of which is portrayed by the open circles of Fig. 5a. To improve clarity, this curve has been shifted upward by 3 g; the right-hand axis shows the correct units. The second curve (open circles, dashed line) plots the mean ($\pm$S.E.) of the smoothed food intake of the same eight rats, that is, the daily mean ($\pm$S.E.) of a set of eight nonparametric regression curves, of which the dashed line in the top panel is one member (the left-hand axis gives the correct units).

The overall trend in food intake is more readily discerned in the plot of the smoothed data, since much of the day-to-day noise has been minimized by the curve-fitting process. Furthermore, a good deal of within-day noise is also eliminated by the curve-fitting process; note that the error bars are much tighter for the smoothed means than for the raw means. Figure 5c,d displays, in the same format, results for two other feeding indexes, daily meal number (meals/24 hr) and mean meal size (g/meal). For each index, smoothing the data from individual rats prior to computing daily means and standard errors gives a clear reduction in both between- and within-day noise.

The filled circle in Fig. 5b–d plot the (raw) daily mean ($\pm$S.E) of rats for mean food intake, mean meal number, and mean meal size while receiving TPN for days 11, 12, and 13. The curve-fitting technique used with the control data is not useful for these data because the data record of each animal is divided into three segments, namely, days 10, 11–13, and 14–18, each of which is too short to apply a fully nonparametric smoother. An alternative might be to apply a semiparametric technique (e.g., regression splines) to the daily feeding indexes. However, that approach is unnecessary here because the effects of TPN are so clear-cut that they can easily be detected without any smoothing of the raw data.

Each of the graphs in Fig. 5b–d has two horizontal grid lines, the higher of which marks the day 10 mean for the TPN group. The lower horizontal gridline shows the level at which any of the day 11–18 means becomes significantly lower than the day 10 mean (simultaneous level of significance 0.05 for the eight daily comparisons with day 10). In other words, an analysis of variance contrast $t$-test comparing day 10 with any of the following eight daily means is significant, with simultaneous $\alpha = 0.05$, wherever a plotted mean (solid dot) falls below the lower horizontal grid line. Examining Fig. 5b–d, it is clear that each of the three feeding indexes shows an immediate and highly significant decrease with the introduction of TPN, a decrease that is maintained throughout the TPN period. Following cessation of TPN, each index begins to normalize, and by day 18 has returned to pre-TPN levels.

In short, during TPN rats decrease their food intake by eating fewer and smaller meals. Meguid et al. (12) also showed that mean meal duration, mean meal consumption rate, and mean meal and intermeal sniffs decrease with infusion of TPN. On stopping TPN, feeding patterns normalize in 3 to 4 days.

## Feeding Patterns with Total Liver Denervation and Total Parenteral Nutrition

A second illustrative experiment using the ACREM is reported in Ratto et al. (13). The liver is postulated to play a key role in regulating food intake by sending sensory information on nutrient availability, and possibly on nutrient type, to the central nervous system. The study of Ratto et al. (13) is one of a series of investigations from this laboratory on the possible role of the liver in mediating the changes in feeding patterns observed during TPN.

Total liver denervation (TLD) is performed as described by Bellinger and co-workers (14–21). Rats are placed into individual ACREM metabolic cages for postoperative convalescence. After 22 days, a right internal jugular vein catheter is placed using aseptic techniques. After a further 10-day recovery period, a TPN solution providing 100% of the rat's daily caloric needs is infused for 3 consecutive days. Figure 6 plots the time course of three feeding indexes: daily mean food intake, mean meal number, and mean meal size. The filled circles represent the eight TLD rats, the open circles the seven sham-operated (SO) control rats. The vertical grid lines mark the beginning of the three change points in the experimental design: TLD or SO only, insertion of jugular catheter, and onset of TPN.

A fully nonparametric curve-fitting approach (e.g., the kernel regression method) is practicable only when the data record for each rat contains a sizable number of data points obtained under homogeneous experimental conditions (e.g., the control data shown in Fig. 5a). Generally, however, our studies using the ACREM yield a rather short series of observations, subdivided into experimentally defined segments (e.g., the TPN data shown in Fig. 5b–d). In these cases, a natural curve-fitting approach is the method of polynomial regression splines (22), that is, polynomial segments joined end to end. Splines have the obvious advantage of allowing the shape of the fitted curve to switch dramatically at change points in the experimental design (e.g., at the onset of TPN). By constraining the polynomial segments in various ways (e.g., by forcing monotonicity or pseudoasymptotes), a regression spline requires rather few parameters to be estimated, and thus it can be used even when the data are sparse. The resulting curve (spline) retains many of the advantages of the nonparametric approach discussed earlier: trends in the data are more easily discerned, and spline-based tests comparing different experimental groups are generally more powerful than tests based on the observed group means. Figure 6 demonstrates the method using regression splines composed of a sequence of straight-line segments.

To each set of means in Fig. 6, we fit a regression spline consisting of 11 straight-line segments. The solid line is the regression spline for the TLD group, and the dashed line is the spline for the SO control group. The join points (or "knots") of these splines were chosen to permit the resulting curves to track the observed means accurately; the placement of the join points can be inferred from Fig. 6 (especially Fig. 6a). A spline with 11 straight lines requires 12 parameters to be estimated; as there are 38 daily means, this leaves plenty of degrees of freedom to test for lack of fit to the observed means. Lack-of-fit $F$-tests were nonsignificant in each group and for each feeding index.

Following liver denervation or sham operation, each feeding index begins an immediate recovery toward preoperation levels. By day 22, just prior to insertion of the jugular catheter, the two experimental groups are indistinguishable in mean food intake, averaging around 16 g/24 hr. However, the data of Fig. 6b,c suggest that this intake is achieved in different ways in the two groups. On day 22, mean meal number ($\pm$S.E.) is 13.1 $\pm$ 1.3 in the TLD group and 11.5 $\pm$ 1.2 in the SO control group (Fig. 6b); a $t$-test is not significant ($p = 0.18$). On the other hand, the mean ($\pm$S.E.) of the fitted splines at day 22 is 13.5 $\pm$ 0.5 in the TLD group, and 10.7 $\pm$ 0.5 in the SO control group; a $t$-test is now highly significant ($p < 0.001$). For mean meal size (Fig. 6c), the day 22 observed mean ($\pm$S.E.) is 1.26 $\pm$ 0.10 g/meal in the TLD group and 1.53 $\pm$ 0.13 g/meal in the SO control group; a $t$-test is marginally significant ($p = 0.055$). However, the mean ($\pm$S.E.) of the fitted spline at day 22 is 1.33 $\pm$ 0.06 in the TLD group and 1.64 $\pm$ 0.06 in the SO control group; a $t$-test is quite significant ($p = 0.001$). The ability of the regression splines to smooth away noise leads to small standard errors and hence intergroup comparisons that are considerably more sensitive than those based on the observed means.

Thus, the 22-day postoperation period provides evidence of the effect of total liver denervation. Although food intake was the same for TLD and SO rats, liver denervation altered feeding patterns, with TLD rats eating more meals of smaller size than the SO controls. This 22-day period is long enough to apply our other curve-fitting method, nonparametric regression, and doing so provides a different view of this phenomenon.

Figure 7 shows the result of smoothing the data record for each rat from days 1 to 22 of the study and then computing the mean ($\pm$S.E.) of the smoothed data for each day (i.e., the curves in Fig. 7 were obtained in exactly the same manner as the curves plotted with dashed lines in Fig. 5b–d). It is clear that total food intake was effectively identical in the two groups throughout the 22-day postoperation period (Fig. 7a). Moreover, mean meal number and mean meal size (Fig. 7b,c) are each nearly indistinguishable in the TLD and SO groups until about day 10 or 11, after which the two curves follow distinctly different paths. By day 22, the differences in mean meal number and mean meal size are significant ($p < 0.05$ in both cases).

These results demonstrate that neural communication between the liver and brain influences normal feeding patterns. It is especially worth noting that the effects of liver denervation on feeding patterns would not have been apparent without meal pattern analysis.

## Conclusions

The ACREM has been applied to address several different experimental questions in our laboratory. The results have been consistent and have shed light on some of the factors that are, and are not, involved in TPN-associated

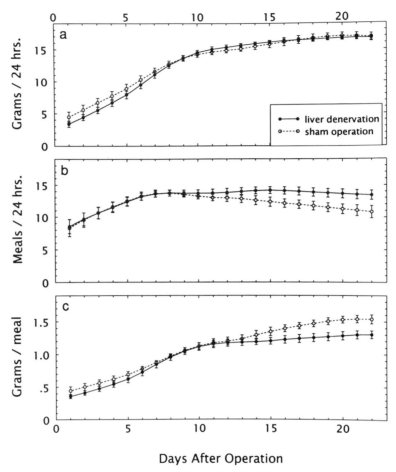

FIG. 7   Smoothed version of the data shown in Fig. 6 between days 1 and 22. Each curve was produced in the same manner as the plots shown with dashed lines in Fig. 5 b–d. The day 1–22 data for each rat were smoothed by applying the kernel nonparametric regression method. Then, for each of the two groups, the mean ± S.E. of the resulting smoothed data was calculated for each day.

anorexia. However, the application of the ACREM is not confined to TPN models. Indeed, we have used the ACREM to evaluate the changes in feeding patterns accompanying cancer-associated anorexia (23). Analysis of the feeding microstructure has emphasized discrete similarities and differences among models that may yield information on the physiological and learning mechanisms that regulate feeding.

## Acknowledgments

We thank Mrs. Darlene Thompson for editorial assistance.

## References

1. P. M. B. Leung and Q. R. Rogers, "Umami: A Basic Taste," p. 565. Dekker, New York, 1987.
2. M. G. Hulsey and R. J. Martin, *Physiol. Behav.* **50**, 403 (1991).
3. P. G. Clifton, I. N. Rusk, and S. J. Cooper, *Pharmacol. Biochem. Behav.* **33**, 21 (1989).
4. S. Hsiao and R. Spencer, *Behav. Neurosci.* **97**, 234 (1983).
5. T. W. Castonguay, D. E. Upton, R. M. P. Leung, and J. S. Stern, *Physiol. Behav.* **28**, 911 (1982).
6. M. L. Pelchat, H. J. Grill, P. Rozin, and J. Jacobs, *J. Comp. Psychol.* **87**, 140 (1983).
7. D. Fallon, *Science* **148**, 977 (1965).
8. S. Balagura and D. V. Coscina, *Physiol. Behav.* **3**, 641 (1968).
9. H. R. Kissileff, *Physiol. Behav.* **5**, 163 (1970).
10. M. M. Meguid, Y. Kawashima, A. C. L. Campos, *et al., Physiol. Behav.* **48**, 759 (1990).
11. S. D. Morrison, *Physiol. Behav.* **25**, 893 (1980).
12. M. M. Meguid, T.-Y. Chen, Z.-J. Yang, *et al., Am. J. Physiol.* **260**, E126 (1991).
13. C. Ratto, J. R. Gleason, Z.-J. Yang, R. Bellantone, F. Crucitti, and M. M. Meguid, *Physiol. Behav.* **50**, 607 (1991).
14. M. M. Meguid, M. Akahoshi, S. Jeffers, *et al., Arch. Surg.* **119**, 1294 (1984).
15. J. D. Hart and T. E. Wehrly, *J. Am. Stat. Assoc.* **81**, 1080 (1986).
16. L. L. Bellinger, V. E. Mendel, F. E. Williams, and T. W. Castonguay, *Physiol. Behav.* **33**, 661 (1984).
17. L. L. Bellinger, G. Trietley, and L. Bernardis, *Physiol. Behav.* **16**, 299 (1976).
18. L. L. Bellinger and F. E. Williams, *Physiol. Behav.* **26**, 663 (1981).
19. L. L. Bellinger and F. E. Williams, *Physiol. Behav.* **30**, 463 (1983).
20. L. L. Bellinger and F. E. Williams, *Physiol. Behav.* **43**, 527 (1988).
21. J. R. Martin, J. L. Beverly, D. B. Hausman, and L. L. Bellinger, *J. Nutr.* **120**, 893 (1990).
22. P. L. Smith, *Am. Stat.* **33**, 57 (1979).
23. M. M. Meguid, M. Muscaritoli, J. L. Beverly, Z.-J. Yang, C. Cangiano, and F. Rossi-Fanelli, *J. Parenter. Enteral Nutr.* **16**, 565 (1992).

# Section VI

## Other Behavior

# [21] Measurement of Illness Behavior: Review of Concepts and Common Measures

Samuel F. Dworkin and Leanne Wilson

## Introduction

The concept of illness behavior represents an attempt to account for response factors in illness (1). The term is used broadly to include not only readily observable behaviors related to illness (e.g., visits to the physician, bed rest, and taking medication), but the perceptions and appraisals that motivate these behaviors, that is, choosing between being compliant versus resistant to medical treatment or choosing to engage or not in behaviors (e.g., smoking or drinking) that place one at risk for illness. Thus, illness behaviors begin with physiological processes that potentially underlie symptom detection and extend to adaptive or maladaptive responses to these perceived signals.

Responses to illness are readily acknowledged to be both diverse and difficult to explain if the explanation must rely solely on pathophysiological changes in the physical structures of the body. Among the most perplexing aspects of illness are the following: (1) people commonly present symptoms which are not consistent with documentable physical change, and (2) individuals will maintain dysfunctional illness behaviors which bear greater relationship to their personally or culturally derived explanation of illness than to the biomedical model of disease (2). In addition to leading to mutual frustration on the part of both the physician and the patient, these illness behaviors are of critical interest to researchers in the area of health care utilization and delivery.

Engel (3) suggested that the biomedical model, with its preeminent focus on pathobiology, was too narrow a model to explain the onset and maintenance of disease. He proposed instead a biopsychosocial model which stresses that biological changes are not sufficient to explain the onset of disease. According to the biopsychosocial model, illness is the result of a complex interaction of biological, psychological, and social variables. Diversity in expression of any illness, including its severity, duration, and consequences for the individual, is accounted for by the interrelationship among biological changes, psychological status, and the social cultural contexts which shape the perceptions and responses of a patient to illness.

Implicit in the distinction between the biomedical and biopsychosocial models is a differentiation between disease and illness. Disease is most

*Methods in Neurosciences, Volume 14*

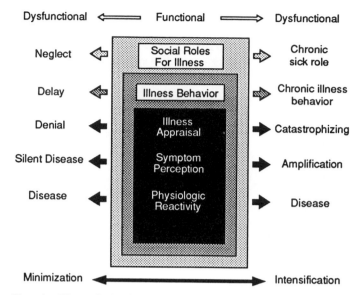

FIG. 1   Illness behavior: an integrated biopsychosocial model.

usefully defined as a biological event representing a disruption of a body structure or organ system as a result of anatomic and/or physiological change. The disease process may involve single or multiple organs or organ systems, may advance, regress, or remain dormant, and may or may not be clinically apparent. Illness is defined as encompassing the subjective experience of physical discomfort, emotional perturbation, behavioral limitations, and psychosocial disruption of activities and relationships. Illness behaviors constitute a class of events most usefully analyzed separately from their presumed physical bases (4).

The specific components of illness behavior as developed by Mechanic (1, 5), Engel (3), and others are depicted schematically in Fig. 1. The broad concept of illness behavior, then, subsumes a range of interrelated constructs, from the intraindividual—physiological reactivity, perception, and appraisal—to observable behaviors including the interpersonal and social—sick role behavior. As the model indicates, it is possible to inappropriately minimize or intensify each of these processes. Illness manifestations (6) can range from complete denial of the presence of a diseased structure, so that no aspect of the person's life is apparently disrupted, to complete dissolution of personal capacities to maintain responsibility for any aspect of daily life; from total avoidance or noncompliance regarding health care services to excessive utilization of health care, especially indiscriminate reliance on surgery and medications. Health care researchers are most often interested in the intensification of the dimensions of illness behavior, rather than in their minimization, and the measures available reflect this bias.

Generally, measures of illness behavior aim to assess only one of the dimensions, although some measures assess adaptation at multiple levels. The term "illness behavior" may be used to convey one or more of these constructs across different experimental contexts or disciplines. This chapter describes and evaluates measures for physiological reactivity, symptom perception, illness appraisal, and illness and sick role behavior. For the purpose of considering available measures, we combine the categories of chronic illness behavior with that of chronic sick role behavior to refer to two broad domains of behavior (7): (1) a decrease in activities associated with personal responsibilities and interactions at work, home, and socially, and (2) an increase in informal and formal help seeking in the pursuit of health remedies and medical intervention.

We have selected for review those measures that have been most widely used and validated and are most central to each of the dimensions of illness behavior, as depicted in Fig. 1. We discuss only briefly the construct of illness behavior at the physiological level, since this topic is covered in depth in the remaining chapters of this book.

## Physiological Reactivity

All illness behavior has as its central feature a concern over bodily functioning. We find it most useful to make the theoretical assumption that all such concerns are associated with physiological signals, transmitted in the nervous system as information to higher centers in the brain. Those signals have the potential for being perceived, appraised, and acted on as symptoms, indicating that something is wrong with the body, that is, the signals are negative, noxious, or aversive, most commonly painful, and have consequences for behavior.

It is important to note that the eventual perception and appraisal of physical symptoms does not require the presence of active pathology or pathophysiological processes. Both laboratory and clinical research confirms the possibility of eliciting symptom report in the absence of observed pathology (8–11). Conversely, the presence of discernible pathology is not necessary nor sufficient to yield illness behavior, as attested by studies of silent myocardial ischemia, hypertensive disease, and even pain-free rampant carious destruction of teeth.

Thus, the measurement of illness behavior at the physiological level is that of objective signs of structural change or organ dysfunction, a subject well beyond the scope of this chapter. However, the measurement of physiological reactivity per se is not the central challenge to the biopsychosocial model of illness behavior. The most perplexing challenge, far from being resolved, is to assess with reliability and validity the relationship of physical measures (1) to psychological measures of symptom perception and appraisal and (2) to behavioral measures related to coping with illness, such as adopting a sick role. In general, attempts to correlate physiological and behavioral information in the arena of illness behavior have not proved fruitful, and some workers have even challenged the usefulness of continuing to dichotomize symptom report, and by extension illness behavior, into physical versus psychological domains of measurement.

Consider, for example, the issue as applied to the study of chronic pain. In most settings in which human pain data are gathered, it is not possible to obtain information at the level of physiological (i.e., nociceptive) processes such as neurotransmitter concentrations, specificity of neural pathways being discharged, and so forth. In clinical settings, the equivalent of the nociceptive level of measurement is the physical level of recording objective pain-related clinical signs through physical examination and, to a lesser extent, through laboratory tests. Common physical measures include, for example, assessment of range of motion for musculoskeletal pains, use of radiographs and other imaging methods to detect abnormal structural changes, and laboratory tests for indications of painful systemic diseases such as rheumatoid arthritis or ischemic cardiac pain. In any case, it remains one of the major enigmas of pain assessment that measurements at the physical level, typically clinical signs, are often inconsistent with or inadequate to explain subjective reports of persistent pain and suffering frequently accompanied by dysfunctional chronic pain behaviors (12, 13).

## Symptom Perception

Much of the research on illness behavior has focused on the experience of physical sensations or symptoms reported by an individual. Somatization is the term that has been used most commonly to describe the perceptual aspects of illness behavior. Somatization refers to a process of reporting abundant physical sensations, often in multiple organ systems (14), in the absence of or in gross excess of observable pathophysiologic changes. Others have described somatization as the somatic expression of underlying psychiatric disturbance (6), particularly depression (15).

## Symptom Checklist 90-R Somatization Scale

### Description

A commonly used measure of physical symptom reporting is the Symptom Checklist 90-R (SCL-90-R; Ref. 16) somatization scale. The somatization scale is one of nine scales of the SCL-90-R, which measures dimensions of psychological adjustment and distress (obsessive–compulsive, interpersonal sensitivity, depression, anxiety, hostility, phobic anxiety, paranoid ideation, and psychoticism). The SCL-90-R is a 90-item self-administered questionnaire. Individuals report their level of distress arising from each of the items, using the response categories "not at all," "a little bit," "moderately," "quite a bit," and "extremely." The measure is designed to be used with "normal" individuals as well as those with medical or psychiatric disorders.

The 12 items on the somatization scale are headaches, faintness or dizziness, pains in heart or chest, pains in lower back, nausea or upset stomach, sore muscles, trouble with breathing, hot or cold spells, numbness or tingling, a lump in the throat, feeling weak in parts of the body, and heavy feelings in arms or legs. Information on scoring the SCL-90-R, normative data, and detailed information on psychometric properties are available in the manual by Derogatis (16).

### Reliability

Derogatis (16) indicates that the test–retest reliability ($r_{tt}$) of the somatization scale is 0.86, and the internal consistency reliability (coefficient $\alpha$) is 0.86.

### Validity

Derogatis (16) reports a correlation between the SCL-90-R somatization scale and the hypochondriasis scale of the Minnesota Multiphasic Personality Inventory (MMPI; Ref. 17) of 0.48. It was more highly correlated with MMPI Wiggins scales (0.62 with Organic Symptoms and 0.58 with Poor Health).

Several studies provide evidence for the construct validity of the SCL-90-R somatization scale. Schwartz and DeGood (18) reported that back pain patients whose drawings were rated as anatomically inappropriate scored significantly higher on the SCL-90-R somatization scale than those whose drawings were rated as appropriate. Similarly, Gil *et al.* (19) found that patients with sickle cell disease who were categorized as having pain drawings with sites inconsistent with the expected sickle cell disease patterns had somatization scores in the clinically significant range. In our work we found that a high score on the somatization scale was highly predictive of temporomandibular disorder pain that was widely dispersed, both on measures of self-report and on a clinical examination (20), and was significantly correlated with scores on the depression scale.

*Commentary*

The SCL-90-R somatization scale is an easily administered measure of physical symptoms. One possible problem with interpretation is that it was originally developed for psychiatric patients, not individuals with physical or medical conditions. Responses of pain patients to this scale, for example, may need to be interpreted with caution, since 5 of the 12 items are pain related. Another difficulty with the measure as one of symptom *perception* is that the wording of the questions confounds symptom detection with distress or discomfort about the symptom. Development of a longer scale that involves rating of the extent to which a variety of physical symptoms are present, rather than the extent to which they cause distress, would be a useful addition to the field.

# Illness Appraisal

Appraisal refers to the process of assigning meaning to perceived sensations or signals. Appraisal involves both cognitive and emotional processes and yields judgments about the origin of these signals, their significance to well-being, and decisions about action to be taken for self-coping or treatment seeking (21). Extreme negative appraisal of physical signals is considered to involve hypochondriasis, characterized clinically by intense preoccupation with or fear of illness, usually of severe or life-endangering illness, despite medical reassurance to the contrary. Hypochondriasis and somatization are overlapping concepts, but somatization is understood to involve perceptual processes, while hypochondriasis invokes a process of negative appraisal and fear regarding sensations or symptoms.

## *Illness Behavior Questionnaire*

### *Description*

Despite its name, Illness Behavior Questionnaire (IBQ; Ref. 22) is more accurately identified as a measure of illness appraisal than illness behavior in that it assesses self-reported hypochondriasis and other aspects of cognitive distortions and psychological processes assumed to be associated with maladaptive illness behavior. The 62-item IBQ is a self-administered questionnaire with a yes/no response format, and seven dimensions identified empirically through factor analysis: general hypochondriasis, disease conviction, psychological versus somatic perception of illness, affective inhibition, affective disturbance, denial, and irritability. The IBQ incorporates a previously developed measure of hypochondriasis, the 10-item Whiteley Index of Hypochondriasis. A manual describing a reference sample and scoring of the questionnaire is available (22).

### Reliability

Test–retest correlations for a sample of 42 patients yielded correlations ranging from 0.67 to 0.87 for the seven scales (22). With 71 subjects the 10 hypochondriasis items have been shown to have a test–retest correlation of 0.81 (22, 23).

### Validity

Pilowsky reported evidence for the validity of the 10 hypochondriasis questions from the Whiteley Index (23). Two studies are available on the validity of the full 62-item measure. Pilowsky *et al.* (24) reported that scores on six of the factors discriminated 38 patients with angina who responded to coronary bypass surgery from 12 patients who did not respond. Pilowsky *et al.* (25) developed a discriminant function equation from interviews with a sample of U.S. pain patients and general practice patients that was applied to similar patients in Australia. He reported a sensitivity of 97% and a specificity of 73.6% in classifying the Australian patients using the measure.

### Commentary

Factor analytic studies on the IBQ are considered somewhat weak, and validity has not yet been well established. However, the IBQ is the most widely used and validated measure presently available for assessment of hypochondriasis, disease conviction, and related constructs. As indicated earlier, it does not specifically identify or measure overt behaviors associated with illness.

## Behavior

Measurement of the behavioral responses to illness involve assessment of (1) changes in work, social, and recreational activity level and (2) increases in use of self-coping, home health remedies, and medical care.

## The Sickness Impact Profile

### Description

The Sickness impact Profile (SIP; Ref. 26) is the primary standardized, self-report measure used to assess the impact of significant illness on daily activities and behaviors. The measure can be self-administered or administered by an interviewer. It was designed to be applicable across a range of types and severities of illness. Subjects respond to each of 136 statements in 12 categories: Sleep and rest, Eating, Work, Home management, Recreation and pastimes, Ambulation, Mobility, Body care and movement, Social interaction, Alertness behavior, Emotional behavior, and Communication. The SIP is composed of statements which describe behaviors, such as "I spend much of the day lying down in order to rest" and "I have difficulty reasoning or solving problems." Respondents check those items which describe them on that day and are related to their health. The instrument can be scored to yield individual category scores, two-dimensional scores (physical and psychosocial), and an overall score. Manuals with the measure and its administration are available (27, 28).

### Reliability

Test–retest reliability for the SIP has been consistently high, from 0.88 to 0.92 for the overall score (26, 29, 30). Reliability for the 12 category scores was 0.82 (26). Cronbach's coefficient $\alpha$ for internal consistency was 0.94, although mailed self-administered questionnaires had a lower internal consistency $\alpha$ of 0.81 (26).

### Validity

Validity trials compared the SIP with subjective ratings made by the respondents, with clinical assessments, and with other functional assessment instruments for several illness populations (26). The SIP was found to correlate 0.69 with a self-assessment of limitation, 0.63 with a self-assessment of illness, 0.50 with a clinician's assessment of limitation, and 0.40 with a clinician's assessment of sickness (26). Deyo *et al.* (31) reported significant correlations between overall score and physical dimension score and several indicators of disease severity for a sample of arthritic patients.

### Commentary

The SIP has been carefully developed and appears to have good psychometric properties. It is applicable to a wide range of medical conditions, both acute and chronic. Few data are available on the relationship between this measure and other aspects of maladaptive illness behavior, such as excessive health care utilization or medication use. It seems most useful for characterizing patients experiencing significant manifestations of disease.

## Health Diaries

A comprehensive method for measurement of illness behavior is through use of a health diary. Whereas the SIP was designed and is most commonly used to assess illness-related changes in performance and function in those with significant disease, health diaries are used to assess a wide range of illness behavior in both physically ill and "well" individuals across an extended period of time. Health diaries are typically daily self-report records focused on the domains of symptom perception and health and illness-related behaviors. Researchers have used health diaries for daily report of symptoms or injuries, activity restrictions and other disability behavior, self-care actions for illness (taking medicine, consultation with family or friends), physician phone contact or visits, hospitalizations, and expenditures for health services. Although both the SIP and health diaries rely on self-report data, they have significant advantage over retrospective reports of illness behavior in that respondents are asked to report on behaviors from that day, rather than about events from several days or months prior, circumventing numerous problems related to memory and recall bias.

A detailed description of the administration and scoring of individual diaries is beyond the scope of this chapter. Verbrugge (32) provides an excellent review of the content, reliability, and validity of health diaries.

Pain researchers have been particularly interested in assessment of activity level; operantly based pain programs often target increased activity as well as decreased use of pain medications as treatment goals for patients. Follick et al. (33) have developed a daily activity diary for chronic pain patients which they have demonstrated to be a valid and reliable instrument for the assessment of time spent lying, sitting, or standing/walking, use of pain relief devices, time spent in pain relief activities, and use of analgesic medication.

Both general health diaries and pain diaries are considered self-observational techniques; respondents are asked to "observe" and report on their own behavior. Validity of these data is likely to be highest when the behaviors to be reported are clearly specified, when the number of behaviors and frequency of assessment are not excessive, and when respondents are reinforced for their efforts in recording information (34).

## Coping Measures

Coping involves both appraisal of symptoms and the individual's efforts to alleviate or minimize illness or discomfort. Although the health diary method can be used to include assessment of general illness-related coping strategies, in fact few measures have been developed specifically to assess strategies for coping with illness. Two exceptions are measures that have been developed to assess the use of cognitive and behavioral coping strategies for chronic pain. The Vanderbilt Pain Management Inventory (35) is a questionnaire developed to assess the use of active and passive coping strategies. Patients rate the frequency with which they use each of 18 strategies when their pain reaches a moderate or greater intensity on a five-point scale, from $1 =$ never do when in pain to $5 =$ very frequently do when in pain. Active strategies are those that involve an effort on the part of the patient to control pain or to function in spite of pain; passive strategies involve relinquishing control to others or allowing other areas of life to be adversely affected by pain. The active and passive scales have been found to be internally reliable, and validity for the measure comes from a study of rheumatoid arthritis patients in which passive coping was associated with higher depression scores, more functional impairment, and higher pain ratings, with the opposite pattern shown for active coping scores (35). A subsequent study indicated that passive coping predicted more pain flare-up activity, more depression, and decreased activity 6 months later (36).

Whereas the Vanderbilt Pain Management Inventory focuses heavily on behavioral responses to pain, the emphasis of the Coping Strategies Questionnaire (CSQ; Ref. 37), is on cognitive appraisal and use of cognitive coping approaches. The CSQ is a 42-item self-report questionnaire; patients rate how often they use each strategy on a scale from $0 =$ never to $6 =$ always. They also rate perceived control over pain and perceived ability to decrease pain on a six-point scale. Although the test was originally developed to include six scales, factor analytic studies have generally yielded two or three primary factors. In their original study, Rosenstiel and Keefe (37) identified three factors, namely, cognitive coping and suppression, helplessness (including "catastrophizing" thinking about pain), and diverting attention/praying, each of which were related to measures of disability and adjustment in patients with low back pain. In a study of sickle cell disease patients, Gil *et al.* (38) identified two factors: coping attempts and negative thinking/passive adherence. They reported that high scores for the latter were positively related to more severe pain, more activity reduction, and more psychological distress.

Both of these pain coping measures have been quite widely used and studied. They have been found to relate in important ways to other aspects of illness behavior among patients with a variety of painful conditions.

## *Observational Measures*

All of the measures described to this point involve self-report of behavior, and the validity of the measures rely on the veridicality of that report. Ideally, measures of behaviors such as ambulation, social activity, self-care, use of medication, and so forth should rely on objective, reproducible observations of the target behaviors. Studies of illness behavior in general have tended not to use observational measures, which are often impractical outside of an inpatient hospital setting. However, methods for assessment of both activity level and observable illness-related behaviors have been developed and studied in conjunction with research on chronic pain.

Sanders (39) developed an automated device for recording time spent standing or walking, which has been termed "up-time," for use with chronic pain patients. This type of device could be applied to measurement of rest versus activity in other types of patients.

Other motor behaviors, in addition to "up-time," are central to the assessment of illness behavior in chronic pain patients. Patients communicate pain both verbally, through pain complaints or sighing, as well as nonverbally, through guarded movement, body posturing, and facial behavior. Social and environmental factors are presumed to shape the frequency and intensity of these behaviors (34). Keefe and Block (40) developed an observation system for measuring five pain motor behaviors: guarded movement, pain avoidant posturing (bracing), rubbing the painful area, grimacing, and sighing. The coding system has been found to be reliable, and the frequency of observed pain behavior correlated both with patients' self-reported pain and with pain ratings made by naive observers. In a study of pain behavior observed while patients were undergoing a physical examination, rate of total pain behavior was most strongly predicted by medical status, which included examination findings and number of prior lumbar operations (41). Patients receiving disability or financial compensation payments or with disability decisions pending displayed a significantly higher level of pain behavior than those not receiving disability payments; however, disability status was not significant after controlling for the medical status variables. In a study of gait patterns of low back pain patients, Keefe and Hill (42) reported that gait pattern distinguished patients receiving disability payments or taking narcotic analgesics from those who were not.

## Other Measures

Several additional measures and measurement domains merit mention. Some of these have received little empirical study or are not easily administered. Others, while not central to the assessment of symptom perception, illness appraisal, or sick role behavior per se, do bear important relationsips to illness behavior.

Barsky et al. (43) have developed a measure of symptom perception. Barsky hypothesized that somatization and hypochondriasis occur as a result of symptom amplication, a dysfunctional predisposition to magnify perceived signals originating from physiological processes. He and colleagues have developed a brief self-report scale to assess this predisposition to amplify physiological sensations; respondents report the degree to which they tend to be reactive to sensations such as cold or hunger contractions. This theoretically important scale has yet to be widely used or validated.

The hypochondriasis (Hy) scale of the MMPI, another measure of symptom perception or report, has received a good deal of empirical study, particularly with regard to its relationship (along with other MMPI scales) to chronic pain. The hypochonriasis scale, however, is not intended to be administered or interpreted separately from the entire MMPI, which is long and time-consuming to complete.

Robbins and Kirmayer (44) describe the initial development of a measure of illness appraisal designed to assess the extent to which individuals interpret symptoms as somatic, psychological, or environmental (external) in origin. The Symptom Interpretation Questionnaire consists of 13 common symptoms such as those found on the SCL-90-R somatization scale. Subjects are asked to imagine that they were experiencing each symptom and to hypothesize its predominate cause. This measure represents a promising addition to the assessment of illness appraisal.

Research and measurement related to illness appraisal has primarily focused on cognitive aspects of appraisal; recent work in the area of negative affectivity shifts the focus to the role of emotions in the appraisal process. Negative affectivity refers to a tendency to experience aversive emotional states in response to environmental stimuli ranging from physical to interpersonal. A number of established measures of psychological distress are presumed to measure the construct of negative affectivity (45). Several researchers (46, 47) have demonstrated a relationship between negative affectivity and heightened reporting of physical symptoms. Thus, illness behaviors such as treatment seeking may be understood as partially reflecting a long-standing tendency to experience dysphoria and physical symptoms.

Finally, a well-constructed and validated measure, but limited to chronic pain, covers symptom report and appraisal as well as behavioral responses. The Multidimensional Pain Inventory (MPI), developed by Kerns *et al.* (48), is a comprehensive self-report inventory for the assessment of pain severity, interference due to pain, general activity level, responses of significant others to the pain condition, and psychological distress. The MPI represents a good model for the development of more generalized measures of illness behavior.

## Conclusion

Illness behavior encompasses a wide range of interrelated constructs, from symptom perception and appraisal to sick role behavior characterized by activity limitations and excessive use of health care. We have described the scales and measurement approaches available for assessment of illness behavior. Most of the measures, including those used to assess behavior, are brief, self-report inventories. It is clear from the preceding review that illness behavior has been extensively elaborated from a theoretical perspective. However, careful development of measurement instruments to quantify illness behavior have not received nearly so much systematic attention. Although numerous scales have been made available which touch on one or another of the components of illness behavior described earlier, typically they have not been subjected to careful psychometric analysis of their reliability and validity. More empirical work is needed to delineate clearly the dimensions of illness behavior and to establish appropriate measures for each.

# References

1. D. Mechanic, *Soc. Psychiatry* **1**, 11 (1966).
2. A. Alonzo, *Soc. Sci. Med.* **19**, 499 (1984).
3. G. L. Engel, *Science* **196**, 129 (1977).
4. D. Mechanic, *in* "Illness Behavior" (S. McHugh and T. M. Vallis, eds.), p. 101. Plenum, New York, 1985.
5. D. Mechanic, *N. Engl. J. Med.* **286**, 1132 (1972).
6. A. Kleinman, *in* "Illness Behavior" (S. McHugh and T. M. Vallis, eds.), p. 149. Plenum, New York. 1985.
7. T. Parsons, *Milbank Mem. Fund Q.* **53**, 257 (1975).
8. D. Giddon, *in* "Psychology and Dentistry" (W. A. Ayer and R. D. Hirschman, eds.), p. 35. Thomas, Springfield, Illinois, 1972.
9. H. Merskey, *Pain* **3**, 74 (1986).
10. S. F. Dworkin and J. A. Burgess, *in* "The Management of Pain" (J. J. Bonica, ed.), Vol. 1, p. 784. Lea & Ferbiger, Philadelphia, Pennsylvania, 1990).
11. T. L. Bayer, P. E. Baer, and C. Early, *Pain* **44**, 45 (1991).
12. R. A. Sternbach, "Pain Patients." Academic Press, New York, 1974.
13. S. F. Dworkin, *Can. J. Physiol. Pharmacol.* **69**, 662 (1991).
14. L. Kirmayer, *in* "Illness Behavior" (S. McHugh and T. M. Vallis, eds.), p. 111. Plenum, New York, 1985.
15. W. Katon, A. Berg, A. Robins, and S. Risse, *in* "Illness Behavior" (S. McHugh and T. M. Vallis, eds.), p. 355. Plenum, New York, 1985.
16. L. R. Derogatis, "SCL-90-R: Procedures Manual II." Clinical Psychometric Research, Towson, Maryland, 1983.
17. S. R. Hathaway and J. C. McKinley, "The Minnesota Multiphasic Personality Inventory Manual." Psychological Corp., New York, 1967.
18. D. P. Schwartz and D. E. DeGood, *Pain* **19**, 383 (1984).
19. K. M. Gil, G. Phillips, M. R. Abrams, and D. A. Williams, *Pain* **6**, 105 (1990).

20. L. Wilson, S. F. Dworkin, L. LeResche, C. W. Whitney, and B. G. Dicker, Meeting of the Society of Behavioral Medicine, Washington, D.C. (1991).
21. R. S. Lazarus and S. Folkman, "Stress, Appraisal and Coping." Springer-Verlag, New York, 1984.
22. I. Pilowsky and N. D. Spence, "Manual for the Illness Behavior Questionnaire (IBQ)." University of Adelaide, Australia, 1983.
23. I. Pilowsky, *Br. J. Psychiatry* **113,** 89 (1967).
24. I. Pilowsky, N. D. Spence, and J. L. Waddy, *J. Psychosom. Res.* **23,** 39 (1979).
25. I. Pilowsky, T. G. C. Murrell, and A. Gordon, *J. Psychosom. Res.* **23,** 203 (1979).
26. M. Bergner, R. A. Bobbit, W. B. Carter, and B. S. Gilson, *Med. Care* **19,** 787 (1981).
27. Department of Health Services, "The Sickness Impact Profile: A Brief Summary of Its Purposes, Uses and Administration." University of Washington, Seattle, 1979.
28. Department of Health Services, "The Sickness Impact Profile." University of Washington, Seattle, 1977.
29. W. E. Pollard, R. A. Bobbitt, M. Bergner, D. P. Martin, and B. S. Gilson, *Med. Care* **14,** 146 (1976).
30. W. E. Pollard, R. A. Bobbitt, and M. Bergner, *Soc. Indic. Res.* **5,** 279 (1978).
31. R. A. Deyo, T. S. Inui, J. D. Leininger, and S. S. Overman, *Med. Care* **21,** 180 (1983).
32. L. M. Verbrugge, *Med. Care* **28,** 73 (1980).
33. M. J. Follick, D. K. Ahern, and N. Laser-Wolston, *Pain* **19,** 373 (1984).
34. W. E. Fordyce "Behavioral Methods in Chronic Pain and Illness." Mosby, St. Louis, Missouri, 1976.
35. G. K. Brown and P. M. Nicassio, *Pain* **31,** 53 (1987).
36. G. K. Brown, P. M. Nicassio, and K. A. Wallston, *J. Consul. Clin. Psych.* **57,** 652 (1989).
37. A. K. Rosenstiel and F. J. Keefe, *Pain* **17,** 33 (1983).
38. K. M. Gil, M. R. Abrams, G. Phillips, and F. J. Keefe, *J. Consul. Clin. Psych.* **57,** 725 (1989).
39. S. H. Sanders, *Pain* **9,** 103 (1980).
40. F. J. Keefe and A. R. Block, *Behav. Ther.* **13,** 363 (1982).
41. F. J. Keefe, R. H. Wilkins, and A. C. Wesley, *Pain* **20,** 59 (1984).
42. F. J. Keefe and R. W. Hill, *Pain* **21,** 153 (1985).
43. A. J. Barsky, J. D. Goodson, R. S. Lane, and P. D. Clearly, *Psychosom. Med.* **50,** 510 (1988).
44. J. Robbins and L. Kirmayer, *in* "Illness Behavior" (S. McHugh and T. M. Vallis, eds.), p. 283. Plenum, New York, 1985.
45. D. Watson and L. A. Clark, *Psychol. Bull* **96,** 465 (1984).
46. P. T. Costa and R. R. McCrae, *Pers. Soc. Psychol.* **38,** 668 (1980).
47. D. Vassend, *Soc. Sci. Med.* **28,** 29 (1989).
48. R. D. Kerns, D. C. Turk, and T. E. Rudy, *Pain* **23,** 345 (1985).

# [22] Measurement of Maternal Behavior

Phyllis E. Mann

## Introduction

Maternal behavior in most mammals depends on the degree of development of the offspring at birth, which ranges from altricial to precocial. Thus, a wide variety of maternal behavior strategies to nurture the offspring have evolved. Rodent pups, for example, are usually extremely altricial at birth, depending on the mother to provide food, warmth, protection and even anogenital licking to stimulate urination and defecation. Ungulates, such as sheep, deliver precocial young that are able to follow the mother very soon after birth. Ewes form an exclusive bond with their offspring, and maternal care in that species includes licking the offspring, leading them, and displaying maternal aggression when necessary. In contrast, primate infants are semiprecocial. Some primate mothers continually carry their offspring, whereas, in other species, infants are able to cling to the parent or are left in a nest (e.g., humans). The ability to measure maternal behavior is essential for understanding the neurobiological basis of this behavior and to determine the effects of substances that affect the behavior. This chapter explains in detail how to measure maternal behavior in rats and presents an overview of maternal behavior assessment in other species.

## Rodents

The most commonly used species in the measurement of maternal behavior is the laboratory rat (*Rattus norvegicus*). Rats are relatively inexpensive, easy to handle, and are generally very good mothers. Rat young are altricial at birth and therefore require extensive caretaking behavior by the dams. The full array of maternal behaviors is readily observable and quantifiable, and can be inhibited or disrupted and, under certain conditions, stimulated or potentiated (1–5). Female rats do not display maternal behavior until right before or at parturition (6); therefore, the majority of studies that test for maternal behavior do so postpartum. The description of maternal behavior that follows will use the postpartum female rat as the model. Virgin, nulliparous female rats can be induced to display maternal behavior under a variety of hormonal regimens and/or exposure to foster pups for a period of time

(7). Details of this technique and how it differs from postpartum testing are described below.

## Conditions

### Strain and Age

The majority of research studies on maternal behavior in the rat use either the Sprague-Dawley (Crl:CS[SD]BR; 1–5) or Wistar strain (8), but others employ Long-Evans rats (9). The subject rats are usually between 90 and 150 days old when they are first mated to males. An important consideration is the parity of the female. Studies have shown that there are differences in behavior (including maternal behavior; 2, 3) and physiology (3) between primiparous (one pregnancy) and multiparous rats (several pregnancies); therefore, females with unknown parity should not be used. The majority of research has so far been performed on primiparous rats, and thus maternal behavior testing is performed after the female's first pregnancy and during her first lactation.

### Light Cycle and Temperature

Light/dark cycles used in most laboratories range from 12:12 to 14:10 hr (light:dark; 1–5, 9). A reversed light/dark schedule is sometimes employed when experimental conditions require that maternal behavior testing should be done during the dark cycle under infrared lighting (10). The temperature of the facility should remain constant (20–25°C), since changes in ambient temperature affect maternal behavior by influencing the amount of time the female stays on the nest (11–13).

### Cages

Rats are typically group-housed either in suspended wire-bottom cages or plastic "shoe-box" cages filled with contact bedding until several days before parturition. At that time they are singly housed in opaque polypropylene cages until the end of the experiment. Transparent polycarbonate cages may be used to facilitate the observation of the behavior, but they may cause added stress to the experimental animals. The size of the cage should be sufficiently large so that when pups are placed in the cage there is enough room to observe behavioral interactions. Standard laboratory cages (45 × 25 × 20 cm) are suitable, although some researchers use larger cages (50 × 40 × 20 cm; Ref. 10). Plexiglas partitions (3.8 cm high) may be placed in the cage to separate it into compartments so that the pups can be placed in different areas of the cage, thus preventing the pups from crawling to the female on their own. Specially designed maternity cages with clear plastic front panels are sometimes used to facilitate observations (14, 15). If maternal behavior testing is not performed in the female's home cage, sufficient time should be allocated to adapt the female to the novel environment (16). The bedding in the cage is usually wood shavings, and some researchers make paper strips available to the females for nest building.

### Surgical Procedures

If intracranial infusions of drugs are to be performed postpartum on the dams, cannulations should take place on days 13–15 of gestation (4). This allows the female a minimum of 1 week to recover from surgery prior to infusions. Intraatrial catheterizations are typically performed postpartum 1 or 2 days before drug administration (3).

### Procedure Postpartum

The day after parturition the litter is culled (e.g., to 6 pups), and the weight and sex of the pups are recorded. During lactation handling of the cages is restricted in order to reduce the amount of stress to the female and pups and to maintain controlled quantities of nesting material (usually 2 liters). Maternal behavior testing can be conducted anytime during lactation. If the females are scheduled to receive intracranial infusions before testing, for example, on day 5 postpartum, the animals are handled on days 3 and 4 postpartum in order to adapt them to the infusion procedure. If intraatrial catheterizations are required, this surgery is usually performed on day 3 or 4 postpartum. Maternal behavior testing starts between 0900 and 1000 hr on day 5 postpartum. The female's nest is rated (see below for description), and the positions of the nest, female, and pups are recorded. The pups are carefully removed from the cage and placed in a container with sufficient bedding to keep them warm. Some researchers place the container in a heated environment (34°C) to maintain optimal body temperature of the pups (17). The pups may be left in the testing room, although in certain situations, such as maternal behavior induction (see below), the test females should have as little contact as possible with the pups other than during the actual maternal behavior test. One hour later the pups are returned and the test session begins. During the hour before testing, infusions or injections should be performed. The interval between administration of the drug and testing depends on the nature of the drug.

## Maternal Behavior Test Session

The pups are placed in the cage away from the female. All six test pups can be placed together or spread around the cage. The behavior of the female is continuously observed for 15 min, and then spot-checks are performed at

15-min intervals up to 1 hr. If maternal behavior is not observed by that time, spot-checks are performed every hour until full maternal behavior is recorded.

### Nest Condition

The female's nest is rated on a four-point scale at the start of each test session. The ratings are as follows: 1 = poor, only a slight difference in the appearance of the shavings surrounding the female and the pups; 2 = fair, a nest is visible but the walls are not very high (less than 1 inch); 3 = good, the walls of the nest are at least 1 inch high surrounding the female and pups; 4 = excellent, in addition to very high walls surrounding the nest, all the bedding has been pushed toward the nest, leaving the rest of the cage empty of shavings. Some researchers include a fifth point, 0 = no nest at all (10).

### Latencies

The following behavioral latencies should be recorded starting when the pups are placed in the cage: (a) contact: when the female approaches a pup and sniffs it; (b) retrieval: when the female picks up (see Fig. 1) and places each pup in the nest; (c) grouping: when the female groups all the pups in the nest (sometimes the female may retrieve all the pups to the nest, but they may be scattered so that all the pups are not covered when the female crouches); and (d) crouching: when the female is crouching over all the pups (see Fig. 2). A crouch is an immobile posture over the pups in which the female's back is arched (ventroflexion) and her legs splayed over the litter.

### Full Maternal Behavior

Full maternal behavior is defined as retrieving, grouping, and crouching over all six pups in the nest within the 60-min test period. The number of females in each group that reach this criterion and their latencies to respond in the test session determine whether differences exist between the experimental groups.

### Analysis

Nonparametric statistics, such as Fisher's Exact Probability test or the Mann-Whitney $U$ test, are employed to determine significance between the groups on the basis of full maternal behavior. Analyses of variance and appropriate multiple comparison procedures may be used to determine if there are sig-

FIG. 1    Retrieval behavior in the laboratory rat.

nificant differences between the groups with respect to each behavioral measure.

*Variations*

Other behaviors and events in addition to the primary, pup-oriented, maternal responses may be recorded during a test session. Some researchers use computerized event recorders (e.g., Epson HX-2) to register the frequency and duration of various behaviors. For example, Stern and Johnson (9) record the number of times the females sniff, lick, and mouth the pups in addition to the time spent self-grooming, nest building, burrowing, exploring, and hovering over the pups, as well as the general activity of the females. It is also possible to distinguish maternal behaviors into motorically active (e.g., licking and retrieving pups) and motorically inactive (e.g., crouching) components (17). Crouching behavior can be differentiated into a "low" crouch where the back is slightly arched, as opposed to a "high" crouch when the back is definitely arched (17). Peterson *et al.* (18) record 15 different behaviors

FIG. 2   Crouching behavior demonstrating ventroflexion of the female's back.

distinguishing between active pup-directed, inactive pup-directed, and other behaviors, and they test subjects in a novel environment rather than in the home cage. Some researchers focus on retrieval behavior as the primary indicator of maternal behavior (10), whereas others may weigh the pups daily to determine if they are thriving (19). The amount of time spent continuously observing the animal's behavior may vary anywhere from 8 min (20) to 180 min (12), and the amount of pups used can also differ depending on researcher preference (21).

Methods for measuring the "intensity" of maternal behavior such as the T-maze can be used. A T-maze (entrance 35 cm long, each arm 40 cm) is connected to one end of the female's home cage before a test session. The pups are placed at the end of each of the arms of the T-maze, and the latencies to enter the maze and retrieve the pups back to the nest in the home cage are recorded. Previous studies have shown that it is possible to determine the motivation of females to act maternally using this apparatus (22–24).

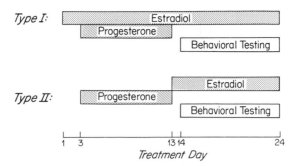

F<small>IG</small>. 3   Hormonal regimens employed in maternal behavior sensitization studies (25).

## Induction of Maternal Behavior

Female rats that have never had reproductive experience (nulliparous) are not spontaneously maternal when presented with foster pups, but they can be induced to display maternal behavior, also called sensitization or concaveation, by daily exposure to pups (7, 14). The sensitization model is useful to determine the neurohormonal and neuroanatomical changes that occur as a consequence of maternal behavior alone without the added involvement of all the changes that occur during pregnancy and parturition. In addition, this paradigm is utilized to discover which compounds may stimulate the onset of maternal behavior in nulliparous rats (1).

### Hormone Regimens

Two different hormonal regimens are routinely employed in our laboratory (see Fig. 3; 25). The type I regimen involves exposing the female concurrently to estradiol ($E_2$; Steraloids, Inc., Wilton, NH) and progesterone during most of the treatment time course, whereas type II exposes the females sequentially to progesterone followed by $E_2$. Both these regimens reliably induce maternal behavior in virgin rats, and although the type I regimen mimics pregnancy levels of the hormones more closely than type II, the latter has proved more useful in delineating the central role of hormones in regulating the onset of maternal behavior (25).

### Procedure

Females are ovariectomized 6 to 8 days before hormonal treatment. Hormone capsules are subcutaneously implanted under Metofane anesthesia according to the particular regimen employed. Beginning on treatment day 14 animals are tested daily for maternal responsiveness using foster pups. These pups are obtained from donor lactating female rats that have been mated for this express purpose in our colony. The donor females should be housed in a separate room from the subject animals, since previous studies have shown that the smell and sound of pups may reduce the latency to display maternal behavior (26). If drug administration is performed, this should occur before the test session.

### Test Session

The test session begins by placing three recently fed 3- to 8-day-old rat pups in each quadrant of the cage away from the female. A smaller number of pups are used in the sensitization procedure than in postpartum testing, since a small percentage (5–10%) of virgin females may initially cannibalize the pups. The maternal behavior test session is similar to the one described for postpartum testing except that, in addition to the measurements stated above, pup-killing responses are also recorded. If a female kills the pups, only one pup is used on the next test day. If a female kills any or all pups in two daily sessions, she is removed from the experiment. On test days 2–11 (treatment days 15–24), rat pups from the previous day's test session are removed and approximately 30 min later (depending on drug administration), a new set of recently fed test young are introduced into each test cage, thereby commencing that day's 1-hr test session.

Behavioral testing typically lasts for 11 days or until a female displays full maternal behavior (described above) on 2 consecutive test days, whichever occurs first (two consecutive days are used since a small percentage of animals are maternal on one day and not the next). The latency of the animal to exhibit maternal behavior is based on the test session in which the response is observed. For example, if an animal responds on test day 1, the latency to exhibit the response is 0 days. Animals failing to respond within the 11-day test period are assigned a score of 11. The latency, in days, to display full maternal behavior is the data used for experimental analysis. Contact, retrieval, grouping, and crouching latencies may also be used for analysis.

### Variations

Variations in the above procedure are found in the literature. Some researchers use varying number of pups and/or days of testing, or utilize different criteria to be classified as maternal. For example, Krehbiel and LeRoy (27) base their findings on the presence of retrieval only, whereas others include nest building and licking of the pups (e.g., 16). Steroid priming regimens also vary. Some researchers implant estradiol directly into the brain (28, 29), whereas others employ a pregnancy termination model (e.g., 30). Maternal or maternal-like behavior also can be induced in prepubertal male and female juvenile rats (31, 32). Measuring maternal behavior in juveniles is very similar to the procedure in adults, but whether the behavior display is actually "maternal" has been questioned since the patterning of behavioral units in "maternal" juveniles differs from that displayed by maternal adult females (32).

## Maternal Aggression Testing

During the last part of gestation a pregnant female rat will display aggression toward intruders, termed "home-cage" aggressiveness (33–36). After parturition maternal aggression becomes more pronounced and then gradually declines before weaning (35). Because the aggression displayed by the female occurs only during pregnancy or in the presence of pups after hormonal priming (36), this behavior is categorized as maternal.

### Animals

Adult male rats usually serve as intruders and are marked to differentiate them from the subject females. Previous studies have shown that using males from different rat strains does not necessarily influence the degree of aggression displayed by the female toward the male (34). Maternal aggression testing can be done on sensitized nulliparous females that have become maternal after hormonal priming and/or daily exposure to pups (see above) or on lactating females postpartum.

### Test Session

Testing is usually performed in the female's home cage with her pups present during the test session. The test starts when the male is placed in the female's cage and lasts for 10 min or until 10 attacks have occurred. An attack is defined as a quick lunge by the female toward the male, usually with an ensuing scuffle or with the male freezing in response. The following measures of maternal aggression are recorded: (1) first attack latency; (2) number of attacks; (3) presence of freezing by the male (more than 20 sec); and (4) length of test (if less than 10 min). The total aggression score can be calculated based on latency to attack and frequency of attacks (34).

## Other Rodent Species

Maternal behavior has also been measured in a number of other rodents including mice (37, 38), hamsters (39), gerbils (40, 41), and rabbits (42). Postpartum maternal behavior testing in mice is very similar to maternal behavior testing in rats, since they display very similar maternal behaviors. For example, Haney and Miczek (37) videotaped 5-min test sessions and recorded the latency, frequency, and duration of retrieval behavior, crouching over the pups, and licking or contacting the pups. They also record nest-building, rearing, walking, and grooming behaviors. The induction of maternal behavior in mice is also possible, but virgin mice, unlike rats, may be maternal spontaneously very soon after being exposed to foster pups. Therefore, a screening test is usually performed in mice to eliminate spontaneously maternal animals. Hormonal regimens are also employed to stimulate maternal behavior in mice, and the maternal behavior can also be induced by constant exposure to mice pups (43). Mice are more prone to display infanticide, and many studies have examined this phenomenon (37).

## Overview of Other Species

The majority of research on maternal behavior in larger mammals has focused on sheep and primates. The offspring of these species do not require as much caretaking behavior as rodent pups, but nevertheless easily quantifiable and observable maternal behaviors do occur.

### *Sheep*

Lambs are precocial at birth and are soon able to follow their mother, forming an exclusive attachment bond within 2 hr (44). The maternal behaviors displayed by ewes postpartum are easily distinguishable from the rejection behaviors displayed toward young by nulliparous ewes and can be recorded and evaluated. Nulliparous ewes will foster lambs and behave maternally toward them only if they are exposed to alien lambs for a period of days, receive hormone priming, and receive artificial vaginocervical stimulation which mimics the stimulation received at parturition (44, 45). Reproductive experience also plays an important role in sheep maternal behavior and should be taken into consideration when planning a study. Multiparous ewes are more likely to foster alien young and display maternal behavior than primiparous ewes, even while the subjects receive artificial vaginocervical stimulation (46).

#### *Animals*

Both multiparous and nulliparous adult ewes serve as subject animals in the majority of studies examining postpartum maternal behavior in sheep. Research that involves the induction of maternal behavior in nonparturient ewes, as measured by the acceptance of foster lambs, takes into account the reproductive history of the ewe, since parity may be one of the experimental variables. Again, multiparous ewes are more likely to foster lambs than nulliparous or primiparous ewes. At the start of an experiment estrous cycles are usually synchronized with vaginal sponges filled with progestins. The sponges are then removed, and injections of pregnant mare stimulating gonadotrophin (400 IU) are administered to elicit ovulation. Dexamethasone (16–20 mg/animal, i.m.) is injected on days 142–144 of gestation to induce parturition 36–48 hr later. Ewes and their lambs are then individually housed in indoor pens. Testing for maternal behavior may occur immediately after birth or several hours later with or without a period of separation between the ewe and the lamb, depending on the nature of the experiment (44–46). Artificial vaginocervical stimulation can be implemented before or during a test session to induce maternal behavior (46).

### Test Session

Both proceptive maternal behaviors and unreceptive "maternal behaviors" are recorded during a test session (46). Proceptive behaviors include low-pitched bleats, which females use to call the lambs, the female's acceptance of suckling behavior by the lambs, and licking and sniffing of the lamb by the ewe. Unreceptive maternal behaviors or aggression toward the lamb is indicated by head butts of the lamb or in that direction, withdrawal from the lamb, rejection at the udder by the ewe, and high-pitched bleats (protest vocalizations). The latency, frequency, and duration of these behaviors by the ewe are registered during the test session. The length of time the ewes are observed varies across studies and depends on the experimental variables. Maternal behavior is considered present if the ewe licks the lamb, demonstrates low-pitched bleats, and accepts the lamb at the udder.

## Primates

Maternal behavior in primates is similar to maternal behavior in other mammalian species in that it includes retrieving and grooming the infant. In contrast, however, most primate mothers do not make nests, spend a considerably longer time involved in maternal care, and are in close contact with the infant for substantial periods of time. In nonhuman primates the female's experience, whether undergoing pregnancy and parturition herself or observing it in others, is an important factor in the quality of maternal care displayed at first parturition.

*Monkeys*

Most early work on maternal behavior in monkeys involved rearing socially deprived infants that, as adults, lacked many normal social behaviors in addition to maternal care. Harlow's "motherless" monkeys displayed abusive behavior toward the infants after their first parturition, biting and sometimes killing them without allowing them to come near and nurse (48–50). Interestingly, after the second or third parturition of socially deprived females they became "adequate" mothers (defined as allowing the infant to nurse consistently so that it did not need to be fed by the experimenters). Maternal behavior in normal female monkeys was assessed by Seay (51) using feral, Rhesus monkeys. Mother–infant interactions were observed in specially built playpen cages for 165 days, starting 16 days after birth. There were two test sessions per day, 15 min in length, 5 days per week. Maternal behavior was categorized into two groups: positive and negative. Positive maternal behaviors were defined as cradling, retrieving, restraining, and grooming the infant and included signals by the mother for the baby to return. Negative behaviors included threat, rejection, and punishment.

More recent work describes maternal behavior in Japanese macaques (*Macaca fuscata*) and red-bellied tamarins. Troisi and D'Amato (52) examined maternal behavior in normal primiparous macaques compared to one abusive multiparous macaque. Mother–infant interactions were observed daily during 15-min test sessions starting at birth and continuing 6 days per week for 2 weeks. They recorded the following 10 behaviors by the mother toward the infant during successive 15-sec intervals in which that particular behavior occurred: (1) grooming the infant; (2) ventroventral contact between the mother and infant; (3) mother initiates ventroventral contact; (4) infant initiates ventroventral contact; (5) cradling the infant; (6) mother is in ventroventral contact with the infant while cradling; (7) mother is in ventroventral contact while grooming the infant; (8) restraining the infant if the infant is trying to get away; (9) mother rejects the infant by preventing it from making ventroventral contact with her; (10) abuse toward the infant. They also quantified indices of maternal warmth, protectiveness, rejectingness, and possessiveness by combining data from several of the above behaviors and obtaining ratios. For example, maternal warmth is the ratio of mother initiating ventroventral contact to the sum of mother initiating contact and infant initiating contact.

Pryce and co-workers (53) examined urinary estradiol levels in postpartum tamarin monkeys. First he defined the subject females as either "good" or "poor" mothers. These definitions were based on two criteria: whether the female had experience with siblings as an infant and, as an adult, how many of her infants survived at least 1 week of infancy. The females were maintained within their family group; thus, the father and other siblings remained with the female during the test session. The maternal behavior test session started immediately after birth and lasted for 2 hr. The observer stood behind a one-way viewing screen, which was 2 m from the cage. The following behaviors of the mother were recorded: (1) latency to carry one or more of the infants (tamarins usually give birth to twins); (2) the average time spent licking, carrying, and nursing the infant; (3) the duration of autogrooming; (4) the frequency of touch–stare (investigatory behavior); and (5) biting, pushing, and rubbing the infant off, active avoidance, and bared-teeth screaming responses toward the infant.

## Humans

Human maternal behavior can vary widely across cultures, from almost no handling when the infants are swaddled and placed on cradle boards to an extensive amount of carrying and tactile stimulation. The maternal behavior studies that are described below take in place in the United States and may portray maternal behaviors that are similar to those displayed by women in Western industrialized societies as a whole.

The assessment of maternal behavior immediately postpartum was performed by Schroeder-Zwelling and Hock (54) to determine if mothers with high-risk pregnancies have altered maternal behavior toward their infants. They used a Maternal–Infant Rating Scale to evaluate the behavior of both the mother and infant on a daily basis for several days postpartum. During the 15-min test session maternal behavior was observed for 30 sec then recorded for 30 sec. They assessed (1) contact with the newborn, (2) verbal interaction, (3) visual contact, (4) facial expression, and (5) sensitivity to feeding.

Mangelsdorf and colleagues (55) examined maternal behavior when the infant was 9 months old to determine if there was a relationship between maternal personality and infant temperament. Maternal behavior was assessed by two observers during several home visits and one laboratory visit by the mother and infant. The maternal behavior of mother was rated on the following scales: (1) quality of vocalization, (2) maternal expressivity, (3) positive regard during feeding, (4) attitude during feeding, (5) sensitivity, (6) delight, (7) supportiveness during play, (8) patience during play, and (9) attitude toward play. The maternal expressivity scale, for example, ranged from a score of 1 = no emotion is communicated to the child to a score of 9 = the mother uses expressions to communicate very much emotion to the baby. These scales are then used to arrive at a score for Maternal Warmth and Support.

Assessment of maternal behavior when the children were 2 years old was performed by Dunn and colleagues (56). They compared the behavior of the mother toward each sibling. Several 5-min test sessions were videotaped, and each revolved around a certain situation such as free play, play with a doll house, and a teaching task. During the doll house session, for example, the following scales were assessed: (1) intrusive/directive, (2) suggests pretend, (3) shares joy, and (4) responsiveness to the child's play.

## Conclusion

In addition to the animals mentioned above, the measurement of maternal behavior has also been performed in a variety of other mammals including goats, cows (57), and horses (58). Males in certain mammalian species also engage in parental behavior, and a considerable literature exists describing this phenomenon (see, e.g., Ref. 59). Finally, the use of standardized quantitative behavioral techniques such as those described above are invaluable in determining the causal relationships between biological factors and behavior. These procedures will aid us in understanding the function of the central nervous system and in identifying regulators of this important complex of behaviors.

## Acknowledgments

I thank Robert S. Bridges for comments and continuous generous support and Lisa Pellerin for help in the preparation of the manuscript. This work was supported by a National Institutes of Health Biomedical Research Support Grant (RR05852) award to Tufts University School of Veterinary Medicine.

## References

1. R. S. Bridges, R. DiBiase, D. D. Loundes, and P. C. Doherty, *Science* **227,** 782 (1985).
2. C. H. Kinsley and R. S. Bridges, *Biol. Reprod.* **39,** 270 (1988).
3. P. E. Mann and R. S. Bridges, *Brain Res.* **580,** 241 (1992).
4. P. E. Mann, C. H. Kinsley, and R. S. Bridges, *Neuroendocrinology* **53,** 487 (1991).
5. B. S. Rubin and R. S. Bridges, *Brain Res.* **307,** 91 (1984).
6. J. S. Rosenblatt and A. D. Mayer, *J. Comp. Physiol. Psychol.* **98,** 177 (1984).
7. J. S. Rosenblatt, *Science* **156,** 1512 (1967).
8. A. S. Fleming, U. Cheung, N. Myhal, and Z. Kessler, *Physiol. Behav.* **46,** 449 (1989).
9. J. M. Stern and S. K. Johnson, *Physiol. Behav.* **47,** 993 (1990).
10. M. Numan, J. McSparren, and M. J. Numan, *Behav. Neurosci.* **105,** 964 (1990).

11. J. E. Jans and B. C. Woodside, *Dev. Psychobiol.* **23,** 519 (1990).
12. P. Korda and J. Komorowska, *Acta Neurobiol.* **47,** 71 (1987).
13. M. Leon, L. Adels, and R. Coopersmith, *Dev. Psychobiol.* **18,** 85 (1985).
14. H. Moltz, M. Lubin, M. Leon, and M. Numan, *Physiol. Behav.* **5,** 1373 (1970).
15. J. S. Rosenblatt, *in* "Determinants of Infant Behavior" (B. M. Foss, ed.), Vol. 3, Methuen, London, 1965.
16. S. E. Fahrbach, J. I. Morrell, and D. W. Pfaff, *Horm. Behav.* **18,** 267 (1984).
17. J. M. Stern and L. A. Taylor, *J. Neuroendocrinol.* **3,** 591 (1991).
18. G. Peterson, G. A. Mason, A. S. Barakat, and C. A. Pedersen, *Behav. Neurosci.* **105,** 470 (1991).
19. J. Cohn and A. A. Gerall, *Physiol. Behav.* **46,** 333 (1989).
20. B. G. Orpen, N. Furman, P. Y. Wong, and A. S. Fleming, *Physiol. Behav.* **40,** 307 (1987).
21. S. Hansen, C. Harthon, E. Wallin, L. Lofberg, and K. Svensson, *Pharmacol. Biochem. Behav.* **39,** 71 (1991).
22. R. Bridges, M. X. Zarrow, R. Gandelman, and V. H. Denenberg, *Dev. Psychobiol.* **5,** 123 (1972).
23. J. Cohen and R. S. Bridges, *J. Comp. Physiol. Psychol.* **95,** 450 (1981).
24. J. M. Stern and D. A. Mackinnon, *Horm. Behav.* **7,** 305 (1976).
25. R. S. Bridges, *Endocrinology* (*Baltimore*) **114,** 930 (1984).
26. L. R. Herrenkohl and R. D. Lisk, *Physiol. Behav.* **11,** 619 (1973).
27. D. A. Krehbiel and L. M. LeRoy, *Horm. Behav.* **12,** 243 (1979).
28. S. E. Fahrbach and D. W. Pfaff, *Horm. Behav.* **20,** 354 (1986).
29. M. Numan, J. S. Rosenblatt, and B. R. Komisaruk, *J. Comp. Physiol. Psychol.* **91,** 146 (1977).
30. H. I. Siegel and J. S. Rosenblatt, *Horm. Behav.* **11,** 12 (1978).
31. C. H. Kinsley and R. S. Bridges, *Horm. Behav.* **22,** 49 (1988).
32. S. A. Brunelli and M. A. Hofer, *in* "Mammalian Parenting" (N. A. Krasnegor and R. S. Bridges, eds.), Oxford Univ. Press, London, 1990.
33. A. D. Mayer, H. B. Ahdieh, and J. S. Rosenblatt, *Horm. Behav.* **24,** 152 (1990).
34. A. D. Mayer, M. A. Monroy, and J. S. Rosenblatt, *Horm. Behav.* **24,** 342 (1990).
35. A. D. Mayer, S. Reisbick, H. I. Siegel, and J. S. Rosenblatt, *Aggressive Behav.* **13,** 29 (1987).
36. A. D. Mayer and J. S. Rosenblatt, *Horm. Behav.* **21,** 253 (1987).
37. M. Haney and K. A. Miczek, *Psychopharmacology* (*Berlin*) **98,** 68 (1989).
38. V. Soroker and J. Terkel, *Anim. Behav.* **36,** 1275 (1988).
39. L. J. Swanson and C. S. Campbell, *Behav. Neural Biol.* **26,** 364 (1979).
40. M. M. Clark, C. A. Spencer, and B. G. Galef, Jr., *Physiol. Behav.* **36,** 845 (1986).
41. D. Kleese and E. Hull, *Dev. Psychobiol.* **13,** 233 (1980).
42. M. L. Cruz and C. Beyer, *Physiol. Behav.* **9,** 361 (1972).
43. M. Koch and G. Ehret, *Physiol. Behav.* **45,** 771 (1989).
44. P. Poindron and P. Le Neindre, *Adv. Stud. Behav.* **11,** 75 (1980).
45. P. Le Neindre, P. Poindron, and C. Delouis, *Physiol. Behav.* **22,** 731 (1979).
46. K. M. Kendrick, F. Levy, and E. B. Keverne, *Physiol. Behav.* **50,** 595 (1991).
47. C. L. Coe, *in* "Mammalian Parenting" (N. A. Krasnegor and R. S. Bridges, eds.), Oxford Univ. Press, London, 1990.

48. G. L. Arling and H. F. Harlow, *J. Comp. Physiol. Psychol.* **64,** 371 (1967).
49. H. F. Harlow, M. K. Harlow, R. O. Dodsworth, and G. L. Arling, *Proc. Am. Philos. Soc.* **110,** 58 (1966).
50. G. C. Ruppenthal, G. L. Arling, H. F. Harlow, G. P. Sackett, and S. J. Suomi, *J. Abnorm. Psychol.* **85,** 341 (1976).
51. B. Seay, *Folia Primatol.* **4,** 146 (1966).
52. A. Troisi and F. R. D'Amato, *J. Nerv. Ment. Dis.* **172,** 105 (1984).
53. C. R. Pryce, D. H. Abbott, J. K. Hodges, and R. D. Martin, *Physiol. Behav.* **44,** 717 (1988).
54. E. Schroeder-Zwelling and E. Hock, *Res. Nurs. Health* **9,** 249 (1986).
55. S. Mangelsdorf, M. Gunnar, R. Kestenbaum, S. Lang, and D. Andreas, *Child Dev.* **61,** 820 (1990).
56. J. F. Dunn, R. Plomin, and D. Daniels, *Child Dev.* **57,** 348 (1986).
57. B. J. Buddenberg, C. J. Brown, Z. B. Johnson, and R. S. Honea, *J. Anim. Sci.* **62,** 42 (1986).
58. P. Duncan, P. H. Harvey, and S. M. Wells, *Anim. Behav.* **32,** 255 (1984).
59. M. W. Yogman, *in* "Mammalian Parenting" (N. A. Krasnegor and R. S. Bridges, eds.), Oxford Univ. Press, London, 1990.

# [23] Measurement of Exploratory Behavior in Rodents

Jaanus Harro

## Introduction

When reading numerous articles and notes on research that has exploited the study of exploratory behavior for various purposes, it may appear that just as many different techniques for the measurement of exploration exist as there are laboratories interested in it. Nevertheless, most of the techniques are more or less successful modifications or remodifications of a few original descriptions. In an attempt to bring some systematics to the measurement of exploratory behavior, this chapter describes some classic methods and their modifications. Several excellent reviews concerning the theoretical background of studies on exploratory activity as well as on the outcome of these studies are available and recommended (1–5). To reach not only statistically but also biologically significant results, one has to consider first what is being investigated using a particular test of exploratory behavior and why.

There are generally two main reasons to study exploratory behavior. First, it has been a routine approach to investigate anxiety; this approach is especially favored by psychopharmacologists, who seek a simple and drug-sensitive test for the assessment of spontaneous behavior. This approach implies that the test situation creates some anxiety in the experimental animal and relies on the fear of novelty. Second, the subject's normal pattern of exploratory behavior, guided by the exploratory drive, may be under investigation, and here the preference of novelty is expected. A third approach does exist in practice, when some feasibly observable component of behavior has been found to be useful for screening purposes: exploration tests have been exploited, for example, in studies on monoamine neurotransmission (6). In a strict sense, what is investigated in these cases is not exploratory behavior per se, since the adaptive value of behavioral acts for the subject is not evaluated. Techniques with emphasis on phenomenology are not discussed in this chapter.

Thus, if exploratory drive is studied, possible novelty-induced fearfulness must be minimized, or adequately controlled, if just the interference of exploratory and fear drives is under investigation. On the other hand, if neophobia and anxiety are the main issues to be characterized, one should

*Methods in Neurosciences, Volume 14*

still consider the curiosity factor. Rodents have a relatively poor repertoire of behavioral expressions compared to primates; however, this simplicity actually means that there is ground for misjudgment on the meaning of these expressions, making test design critical. In other words, if we attribute a test measure to a single variable, we should minimize the danger that the variable reflects activities in distinct behavioral systems in the test situation. One important point to bear in mind is that even among exploratory behavior tests of one type (e.g., forced exploration), there is frequently no correlation between activities displayed in different tests (7).

As a general rule, exploratory activity consists of three phases. First, the animal may be inhibited by the appearance of a novel stimulus, and, thus, the initial reaction could be avoidance (sometimes categorized as "freezing"). Second, as time passes, the animal will approach the stimulus. Third, as the stimulus becomes familiar, the rewarding effect of novelty diminishes and exploratory activity decreases. It is almost impossible to predict how long the first and second stages will last under a particular experimental setup, since this will depend on all the factors discussed below. Exploratory behavior, by definition, is the activity which facilitates familiarization with the environment, and an attribute of neotic behavior is the response decrement that occurs with repeated or continuous exposure to a stimulus. It follows that only behavior that diminishes on repeated presentation of the same test situation is exploratory (3); further, if not just the decline in exploration is under investigation, the subjects must be naive to the test situation.

## Tests of Forced Exploration

### Elevated Plus-Maze

Montgomery (8) examined the manner of exploration of open and enclosed arms in an elevated Y-maze and attributed the preference of rats for the enclosed arm(s) to differences in the extent to which the apparatus elicited fear and exploratory drives. This method has been subsequently introduced in a modified version (9). Because this test was pharmacologically, physiologically, and behaviorally validated as a measure of anxiety in the rat (10) and in the mouse (11, 12), it has become increasingly popular (13), making it necessary to present a critical evaluation of it in this chapter.

The X-shaped maze for rats consists of two open arms 45 × 10 cm, two closed arms 45 × 10 × 40 cm, and a central platform 10 × 10 cm. Two closed arms are opposite each other and have sidewalls and an end wall 40 cm high. The maze is usually elevated 50–70 cm above the floor. For mice, the arm dimensions could be 20 (30) by 5 cm. The arms are connected to one another by a central platform $5 \times 5$ cm, and the enclosed arms have 15 cm high sidewalls. The entire apparatus is elevated to a height of 40–50 cm above floor level.

The animal is placed at the center of the maze, facing one of the closed arms. In some studies, the animals are placed at the central square facing one of the open arms. In this case, few of the animals enter the open arm first, and the first entry preference can be added to the behaviors recorded; however, this seems to be of little value. Rats and mice spend a considerably greater amount of time in the enclosed arms. The measures commonly observed include the following: (a) number of open-arm entries, (b) number of closed-arm entries, and (c) time spent in open arms. Arm entry is counted when the animal has placed all of its four paws on it. The data used for the evaluation of exploratory activity are based on calculations that reveal the percentage of open-arm visits and time spent in open arms from the total number of arm visits and the total duration of the test, respectively.

Pellow and co-workers (10) suggested that the total number of arm entries would reflect unspecific locomotion, thus implicitly assuming that an increase in exploratory activity is reflected in a behavioral pattern in which additional open-arm entries are made on account of the number of closed-arm entries. A decrease in exploratory activity should lead to an increase in the number of entries into enclosed arms to satisfy the condition that locomotor activity is not diminished. The results from several laboratories suggest, however, that this is not necessarily the case (14). Significant increases in the total number of arm entries have been found after treatments that have increased the ratios, and anxiogenic challenge reduces not only the number of open-arm entries but also the number of entries into enclosed arms (9, 15). Moser (15) described rats moving about in one closed arm, but avoiding entries to the other arms, including the other enclosed one. Still, the animals displayed an interest in the other parts of the maze, since they frequently took up stretched-attend posture at the entrance of the closed arm. Furthermore, significant positive correlation exists between the validated open-arm exploration measures and the number of total arm entries (7), whereas there is no correlation between the sum of arm entries in the plus-maze and any locomotor activity-based measures in other tests. Moser (15) has advised that the total number of entries might better serve as an additional measure for anxiety, provided that appropriate control tests are performed to exclude sedation as the cause of decreased mobility. Certainly this measure is not valid as an index of unspecific locomotor activity, and preferably one can combine the plus-maze test with subsequent observation of the animal in another paradigm (e.g., open field).

In several studies, the number of open-arm entries made by mice or rats has been less than one per animal in 5 min (16, 17). If some of the animals do not make open-arm entries, the percentage measures become confusing, since the animals produce different numbers of closed-arm entries. Many researchers, however, have reported 10–15 arm entries in total, 10–40% of entries being made into the open arms. The percentage of time spent in the open arms is also a subject of considerable variability, ranging from 5 to 40% during the observation period.

Another overlooked point in the original method is that the central platform of the plus-maze is definitely more aversive than the areas protected with walls (14, 18). It has generally been ignored that both rats and mice normally spend a considerable amount of test time (up to 30–40%) in this small area. While remaining on the central platform, the animals show active rearing and peering around, and they obviously take a chance on exploring extramaze stimuli, in good accordance with the description by Halliday (19) of rat behavior in elevated mazes. If exploratory activity decreases in this forced exploration test (indicative of increased fear/anxiety), this occurs because of a decrease in both open-arm and central platform activity (18). Indeed, an anxious rat sometimes avoids the open areas by never placing both of its forelimbs on the central platform; however, it may walk around in the first selected arm and show a reasonable level of activity in a subsequent open field test. We have tried to alleviate the problem of low activity on this basis, dividing the open area (open arms and central platform) with lines into seven sections in a way that the central platform defines one section and both open arms are divided into three equal sections. The measures recorded in such a modified maze include, in addition to the customary measures validated by Pellow and co-workers (10), number of lines crossed with two forepaws when moving forward and time spent in the open area (20, 21). Based on the observation that diminished activity on the central platform contributes to the increase in the preference for the closed arms, this modification allows one to increase the resolution of the technique to some extent for the instance when baseline activity remains low.

The idea of dividing the open arms into sections and counting the line crossings has been used by another group (22), stemming from the observation that while rats explore the enclosed arms they enter in their full length, they rarely venture to the end after an open-arm entry. Thus, Costall and co-workers (22) divided the open arms into two halves and counted both open-arm entries and entries into the more distant half of the open arms. The latter measure is believed to be more sensitive. One additional measure that could be used is counting the number of approaches toward the central platform and dividing the total number of arm entries by this value, approximating more closely Montgomery's (8) suggestion that in an elevated maze one investigates an approach–avoidance conflict generated by simultaneous fear and exploratory drives.

Although it was stated initially that the activity of animals does not decrease over repeated exposure to the plus-maze (10), we, as well as others (18, 23), have observed less active exploration of open areas in the plus-maze if the experiment was repeated. Indeed, this would be more consistent with the nature of exploratory behavior, since the fear of open areas remains consistent (19) and the novelty factor diminishes, shifting the approach–avoidance conflict in the direction of avoidance. Open field exposure immediately before the plus-maze test has been suggested to upregulate exploratory activity; however, by no means have all laboratories replicated this finding. Prior exposure to a novel environment has been reported to reduce preference of novelty in a free choice test (24). Thus, an explanation for these discrepancies could be that, although neophobia may be reduced by this kind of behavioral treatment, the exploratory drive declines, and the relative impact of fear and exploratory drives in shaping the plus-maze behavior may not be the same in different experimental designs.

Modifications of the plus-maze apparatus have been described. The meaning of some of the deviations from the original design is hard to estimate (e.g., enclosed arms without end wall; similar arms placed at 90° angle to one another). However, several features are critical. Floor material used may be of importance (25). Care should be taken that the floor of the open arms should not be slippery, since this reduces open-arm exploration. Usually animals do not fall from the open arms; however, in one mouse study this was reported, and, thus, a slight lip 0.25 cm high was included on the open arms (18). Walls of the enclosed arms should be opaque, since in the case of Plexiglas walls the thickness may contribute to the behavioral outcome (7). Although the light levels did not modify open-arm exploration in one investigation (10), others have found that dim or red light facilitates visits to the open arms (18, 25).

An elevated plus-maze test offers a unique combination of features of forced exploration and still contains a choice measure. Under these conditions, one should be extremely careful in making interpretations of exploration scores, especially in cases when the treatment evaluated could be expected to influence several motivational systems (14, 26).

## Open Field

In the traditional meaning, an open field is a well-illuminated circular or rectangular area, several times larger than the home cage. The common dimensions have been in the range of $10 \times 20$ to $40 \times 40$ cm for mice and

$60 \times 60$ cm to $1 \times 1$ m for rats. The arena is crossed by photocell beams, electrostatic sensors, or lines drawn on the surface of the floor that define subdivisions (squares or concentric circles in a rectangular or circular field, respectively). Each beam crossing, sensor activation, or line crossed is scored. Additionally, the number of rears and the amount of time spent in sniffing can be measured. In the studies on emotionality, several other behavioral acts such as freezing, defecation, and urination can be quantified. The open field has also served as a basis for various tests that use detailed behavioral recording, in which all behaviors of subjects are categorized, quantified, and subsequently treated with multivariate analysis.

The most serious pitfall of the simplest open field paradigm is that it does not contain a choice measure. Furthermore, open field behavior is not a cluster of any particular behavioral domain. It is strongly affected by the factor of fear, which is assumed to decrease activity measures on the arena but under certain circumstances may increase them. Difficulties arise when one has to interpret the significance of concurrent behaviors, for example, line crossings versus sniffing, in the open field. It provides no way to estimate the animal's general ability to change location. The controversial nature of the open field test has been emphasized by several reviewers (3–5, 27, 28), and its use for any particular purpose has been strongly discouraged. Could it be, then, that the only reason for this test being so extensively used until now is, as proposed, its relative simplicity? (In fact, several computerized versions or highly detailed behavioral categorizations are anything but simple.) Probably the answer in part lies in the fact that the high sensitivity of the animal's behavior in the open field test to details in the design of the apparatus and procedure makes it possible for some laboratories to obtain highly reproducible patterns of ongoing behavior that do reflect one or another behavioral domain. The best characterization of the possibilities and pitfalls of the open field test is from Denenberg (29): "As long as I work within this context, I feel secure in making these particular interpretations. However, this does not mean that these same conceptual meanings may be attributed to open-field performance in other research contexts."

There are certainly some ways to modify the classic open field as a test of forced exploration. A short test period (2–3 min) in a brightly lit environment should be used to measure anxiety-related phenomena. Thigmotaxis (the tendency of rodents to stay in contact with objects or with perimeters in the environment) can be studied in the open field, measuring the duration of time which the animal spends near the walls (30). In the study cited, thigmotaxis was defined as the time in which the rat is in contact with, or within 2 cm of, any of the walls of the apparatus with most of its body. On the other hand, prolongation of observation time, dim light conditions, and a reasonably large arena (e.g., $40 \times 40$ cm or $1 \times 1$ m for mice and rats, respectively) can make the open field useful as a test for general locomotor activity. Animal activity close to the walls and in the central area can be recorded separately. In the case of a 1 × 1 m arena divided into 16 squares, the 4 central ones can define one response cluster and the remaining 12 the other. However, the animals may not have enough reason to explore the central area, which is clearly visible, unless it contains some objects of possible interest.

## Hole-Board

The hole-board apparatus was introduced for mice by Boissier and Simon (31), and it was subsequently adapted for rats and extensively characterized by File and collaborators (32, 33). In essence, the hole-board apparatus is simply a box, with floor dimensions of approximately 70 × 70 × 40 cm for rats and 40 × 40 × 20 cm for mice, that has holes in the floor. The number of holes in the floor is an important variable. It has been as high as 32 in some studies, but determinations of directed exploration are more reliable if the number of holes is kept low, usually 4, since the animals must have a chance to display locomotor activity without necessarily coming into contact with a hole. The holes have a diameter of 3.8 cm for rats and 3 cm for mice. Most frequently infrared beam interruptions have provided data for the automatic measurement of the number of head dips, time spent head dipping, and the mean duration of a head dip as a function of the two previous measures. However, a head dip can be scored by an observer using the definition that the eyes of an animal should disappear into the hole. In mice, repeated head dipping at the same hole is usually frequent and stereotyped. This should not be scored, instead making the limitation that the mouse must move away from the hole between consecutive dips (in case of automatic counting, the system should allow an adjustable refractory period). Repeated head dips occur more rarely in rats but may emerge after drug treatment.

Routinely, each animal is placed singly in the center of the arena. The use of large animal groups has been encouraged owing to the occurrence of a few individuals with highly deviant scores (32). The observation period has commonly been 10 min, and, during this time, 15–20 head dips for rats and 25–30 head dips for mice serve as reference values.

A practical suggestion that holds especially true for psychopharmacological studies is to place holes along the wall, 2.5 cm above the floor (1). However, the technique also has made it necessary to use the measure "hole visit," since a rat often produces a series of hole pokes while sniffing the surroundings of the hole. Analysis of the time course of hole poking has revealed the possibility of investigating two different aspects of exploration within this technique. Thus, during the initial test period, the number of hole pokes is maximal, whereas, when this measure decreases, the mean duration of hole pokes is increasing. However, in order to use the second phase for exploration studies, highly active strains of animals should be selected and the test conditions should be the least aversive.

For the purpose of increasing the curiosity factor, various objects can be placed under holes. This modification clearly increases the time spent in each head dip. Cheal (34) has developed an ethoexperimental method to study exploration in Mongolian gerbils, where some bedding from a cage of strange gerbils is placed under one of the five holes in the floor of a 30 × 45 cm arena. Since the animals are previously habituated to the apparatus and enter the arena voluntarily, this test is in principle more relevant to the novel object exploration tests described below. Gerbils spend a considerable proportion of the first minute exploring the novel object, and thereafter the behavior rapidly declines.

Finally, caution should also be used in noting multiple parameters in the "enriched open field" condition, since different behaviors considered exploratory (e.g., rearing and head dipping) compete with one another.

## Light/Dark Compartment Test

The natural tendency of mice to explore a novel environment is in opposition to the aversive properties of a brightly lit open field in the light/dark compartment test first described by Crawley and Goodwin (35). In their design, one-third of a polypropylene animal cage, with dimensions of approximately 45 × 20 × 20 cm, is separated with a partition containing a 13 cm long and 5 cm high opening, covered, and painted dark, whereas two-thirds of the cage is brightly illuminated. Photocells across the partition allow an electronic system to count automatically the transitions between the light and the dark chambers and to measure the amount of time spent in each. Total locomotor activity is measured simultaneously with the help of an Animex activity monitor on which the cage is mounted. The test time is 10 min.

In this model (35), mice tend to spend more time in the dark compartment; however, anxiolytic drugs increase the number of transitions between the dark and light chambers, which is taken as an index of exploratory activity. Others have suggested that the time spent in the light area would be more sensitive to anxiolytic and anxiogenic drugs (36). Subsequently, in the same model, another measure was suggested to reflect exploratory activity. Costall and co-workers (37) made an attempt to use the number of line crossings (the light area being divided into 9 squares) and rearings as the measures of neophobia. The relevance of this modification for exploratory behavior remains to be shown (13). Under the conditions described, mice have performed as many as 60–70 rearings in a 5-min test session, and this measure has been reported to increase further after various drug treatments to the level of one rearing per 3 sec, which is hardly compatible with normal exploratory behavior.

Misslin *et al.* (38) have modified the apparatus so that it consists of two boxes of equal size ($20 \times 20 \times 14$ cm) connected to each other by an opaque plastic tunnel ($5 \times 7 \times 10$ cm). A mouse is considered to be in a new box if it has all four paws on the floor of that box. Light intensity in the light half of the apparatus was demonstrated to be important, 2000 lux producing a significantly stronger avoidance of the light chamber than 200 lux. The choice of the starting box (i.e., light or dark) was found to be of negligible importance. In this design of the light/dark compartment test, both the number of transitions between the chambers and the degree of dark preference were affected by drugs that change the level of neophobia. However, only the first measure demonstrated a clear-cut habituation time curve. A significant decrease in transitions was observed between the first and second 5-min period, and a 5-min observation time was chosen for the method. During this period, 7–10 transitions occurred, and mice spent 70–80 sec on average in the light half of the apparatus, which is consistent with several other investigations.

The light/dark compartment test applies for mice but not rats since rats generally have a very strong preference for the dark chamber. If a higher baseline of exploration of the light area is desired for mice, a previous bright light housing of 1 hr is suggested (39).

## Four-Plate Test

The four-plate test was developed by Boissier and colleagues (40) to quantify suppression of the spontaneous activity of the mouse for the evaluation of the disinhibitory effect of anxiolytic drugs. Rat conflict models require a consummatory reward after previous deprivation (of water or food) to achieve punished responding. However, for mice the strong tendency to explore a novel environment is sufficient impetus to accept foot shocks, and anxiolytic as well as psychostimulant drugs increase the punished exploration (2).

The apparatus is a plastic box (approximately $20 \times 25 \times 15$ cm), the whole floor of which is covered with four identical rectangular metal plates separated from one another by a gap of 4 mm. At the beginning of the test, a mouse is gently dropped onto a plate, and 15 sec of free exploration is allowed. After this time, at each crossing between the floor plates, a brief foot shock is applied. The shock level should be established in each laboratory as the minimum level producing a consistent reduction in the number of crossings.

To allow either an increase or further decrease in the number of punished crossings, different shock levels can be used. In one study, 1 mA current in 60 msec resulted in an average of 4–5 crossings per minute, whereas 0.3 mA current of the same duration gave 2- to 2.5-fold higher values (11). Parallel studies of similar design (without delivering foot shocks) allow comparisons between punished and unpunished behavior. Typically, unpunished activity declines to 50% in 3–4 min, but it is important to note that punished exploration rates decline even more rapidly. Thus, measures should be taken during the first minute of punished exploration (if not just the habituation curves are being investigated). The first minute rate of crossings ranges between 12 and 20 or 4 and 10 in the case of unpunished or punished exploration, respectively. Relatively higher rates of punished crossings are achieved when each crossing is inevitably punished, since, in the beginning of the test, mice display flight reactions after foot shocks. In the original work this was counterbalanced with an observer-operated delivery of shocks: if a mouse continued running away, it received no new shock during the following 3 sec. In more recent work, this phenomenon has generally been ignored. After the animals have received one shock, they will cross from one plate to another with extreme caution. Thus, it is necessary to shock the mice only when they truly move from one plate to the other, and not when they tentatively paw or explore the next plate with their noses. Ten to twelve mice per group is generally sufficient to determine the differences between groups.

## Mirrored Chamber

An original idea that takes advantage of the approach–avoidance responses that many animal species exhibit on the novel placement of a mirror into the environment of an individual animal has led to the mirrored chamber test (41). Mice are placed in a 40 × 40 cm chamber with 30 cm high walls that contains a cubical inner mirrored chamber (30 × 30 × 30 cm). Thus, a corridor 5 cm wide runs around the mirrored chamber, which has three painted outer walls but one left open as the entrance to the wholly mirrored compartment. The focus of this method is to estimate the latency to enter the chamber of mirrors. Entry is registered when a mouse places all four feet on the mirrored floor panel, as the approach to the mirrored compartment consists of a series of partial entries with one to three feet placed on the floor panel separated by retreats to the corridor. The behavior shows rapid adaptation: once mice have spontaneously entered the mirrored chamber they leave and enter freely. The average latency to enter the mirrored compartment was approximately 1000 sec in BALB/c mice, whereas the average latency to enter a similar unmirrored chamber was 14 sec. The presence of mirrors on all sides of the inner compartment was found to be important. The obvious advantage of this technique is that it enables one to study exploratory activity of animals having a high baseline anxiety level (e.g., BALB/c mice).

## Tests of Novelty Preference: Choice Tasks

Modifications of forced exploration test apparatuses, albeit frequent, do not look so profoundly different from each other as do the novelty preference paradigms. However, the principles of choice tests are simpler to outline. They are based on either the possibility of choosing between a novel or familiar environment or on the measurement of exploration of a defined novel object in a familiar environment.

### Hughes' Box

A classic method to evaluate exploratory drive in the rat employs Hughes' box (42). In principle, the apparatus consists of two halves that can be interconnected via a removable partition. Both the familiar and unfamiliar halves of the box consist of three chambers, and the presence of a rat in any of the six chambers is sampled after each 10 sec in a 15-min period. Familiarization of one of the halves for rats is achieved by leaving the animals singly in that part of the apparatus for 24 hr with some food and a water bottle. Later modifications include dividing the floor of both the familiar and novel areas into squares in order to quantify locomotor activity. An open field with reasonable dimensions (1 × 1 m) can well be used together with an additional center wall equipped with a gate. Daily expositions of animals for 15–30 min to the chosen familiar half could be repeated 2–3 times. Otherwise, a single 1-hr confinement of each animal to one half of the exploration box with subsequent placement into the other can be used (43). It is wise to introduce half of the animals from each group to one and the other half of the group to the other half of the apparatus.

On the test day, the rat is placed on the familiar side of the apparatus, but the gate remains open. For 15 min, the number of squares (or chambers) visited in both the familiar and novel halves of the apparatus as well as the time spent in the familiar and novel compartments and the latency to enter the novel compartment are recorded. When given the opportunity to move around freely in simultaneously presented novel and familiar environments, rats spend a greater amount of time in the novel environment, and they achieve a higher score on activity measures in the novel area. One additional possibility is to attach a novel compartment to the home cage of individually housed animals and to record the latency to enter the unfamiliar area as well as activity measures in both the home cage and the novel compartment (44).

The design of Hughes' box has been modified by adding a third, starting chamber (45, 46). The modified apparatus is a rectangular box (approximately $120 \times 45 \times 30$ cm). Attached to the center of the longest wall is a start box ($9 \times 15 \times 15$ cm) with transparent walls. The entire apparatus is divided into two halves in such a way that the start box gives equal access to both. A rat is placed into one half of the apparatus for 1 hr, following which it is placed into the start box. The gate in the central wall is opened, and thereafter the rat is allowed to leave the start box (which is subsequently closed). In the next 10 min, the following recordings are made: (a) time taken to emerge from the start box; (b) side on which the rat emerged (novel or familiar); (c) location on either familiar or novel side recorded every 10 sec or, alternatively, total time spent on both sides; (d) number of squares entered in both sides; and (e) number of rears (with both forepaws off the ground and the back straight) in both familiar and novel sides. In the original studies, no preference of side was observable, but the number of rears was nearly 3-fold higher in the novel compartment (46).

## Exploration of Novel Objects

Berlyne (47) suggested that one must differentiate between inquisitive and inspective exploration, that is, between the exploration of the environment as a whole versus the exploration of particular items. Thus, behavior that brings an animal into a position giving it a better overview of the environment, such as locomotion and rearing, would be inquisitive exploration. On the other hand, inspective investigatory behavior would consist of sniffing, touching, or manipulating a particular object. Inquisitive exploration should not be confused with locomotor activity. Inspective exploration has the potential of being recorded out of stereotypic behavior, including components of displacement behavior. A simple way to differentiate stereotypic and exploratory behavior is to examine the habituation: whereas exploratory behavior tends to decrease rapidly, true stereotypic activity may increase when the situation becomes more familiar.

Berlyne also designed test conditions to measure object exploration, and a slightly modified method was used by Robbins and Iversen (48). The apparatus consists of two compartments. The main box has floor dimensions of approximately 60 × 45 cm, and, linked to one of the shorter sides is an alley or alcove 20 × 10 cm or bigger that serves as a place where objects to be explored are placed. For automatic quantification, photocells detect entries to the alcove; additional photocells in the main box provide data concerning general activity and/or exploration of the environment. Additionally, the duration of object exploration is measured by recording the time the animal spends sniffing, licking, or biting the object and, as defined by Robbins and Iversen, the time the animal observes the object, with its head 1 inch or less away from the object. The objects used could reflect an experimenter's fantasy, or simply what happens to be handy in the laboratory. What is important is that the approach response depend on the relative familiarity with the apparatus and the relative novelty of the objects. Both familiar and novel objects may be placed into the apparatus, and approaches to both may be quantified and compared.

## Controlling Variables

Exploration as a type of spontaneous behavior can be strongly influenced by several environmental as well as intrinsic variables. Some variables may exist that we are not aware of, but it is of primary importance to consider certain recognized variables that strongly interfere with exploratory behavior and to control or manipulate them. Even if some of the following may sound primitive to the experience researcher, too often several of these principles remain completely ignored in laboratory practice.

### Data Collection

Automatic quantification has always been a goal for designers of animal tests. The advantage is neutral, uniform, and comfortable measurement of some kind of movement. However, behavior is too complex to be simply measured with photocells, and, despite the availability of several advanced computerized techniques, the role of the human observer has not significantly diminished. Automatic quantification is appropriate if the experimenter is familiar with the common patterns of behavior in the setup. In the introductory phase of automatization, one should go through the tedious but seminal period of comparing the computer output with the direct observations. In several cases, however, exploratory behavior must be directly observed by the experimenter, whose full attention should be given to the animal behavior going on in the apparatus and to the variables that appear to influence that behavior. To make the presence of the observer less disturbing to the animals, video monitors are most desirable; alternatively, some researchers use mirrors placed on the observation field. One must bear in mind, however, that these means do not eliminate completely the influence of contact with the experimenter, since in most cases the animals have to be manipulated before they can be observed.

It is relatively simple to provide open field-like apparatuses with automatic movement registration devices. An accurate estimation of locomotor activity, however, requires a relatively dense network of photocells, since the distribution of the activity over the field can change. One must also consider whether the placement of beams corresponds to the measures defined. For example, if the arena is crisscrossed by photobeams, the animal active at the center interrupts more beams when changing location than another that moves along the walls.

Of primary importance is the order of animals being tested. This must be randomized with respect to the different groups compared. Otherwise, sensory stimuli, including odors left by previous animals in the apparatus, can strongly affect the data collected, and, if the experiment lasts several hours, daily rhythmicity also becomes important.

## Daily Rhythmicity

Exploratory activity shows great variation throughout the day. Most laboratories use a 12-hr light/dark cycle in their animal houses. Frequently, a reversed cycle is used, but by no means in all laboratories, and, moreover, the onset of a dark/light cycle varies to some extent (6–9 a.m.). It is widely known that rodents are more active during the dark phase, but exploratory behavior varies over lights-on periods as well. For head dipping, exploratory activity has been demonstrated to be the lowest in the middle of the light period (49). It is the author's experience that, during the light period, exploratory activity is highest in the hours before it ends. The daily rhythmicity of exploration not only determines baseline activity, but also modifies considerably the effect of manipulations (e.g., drug treatment) that in certain cases can be observable only in the afternoon if lights are turned on in the morning (50). However, for Crawley's version of the light/dark compartment test, no diurnal fluctuation in behavior has been reported (5).

## Sex

In a choice test, both female and male rats preferred the novel compartment to the familiar one, but female rats had higher exploratory activity in the novel half of the apparatus in a 15-min sampling period, as described by

Hughes (51). However, the latter phenomenon could have occurred because of the decrements in locomotor and exploratory activity in males that were observed over this relatively long observation time. Others have repeatedly found that female rats also emerge faster from their home cages than males, and they spend significantly more time in the novel compartment (3). Also, in forced exploration tests such as the open field and the elevated plus-maze test, females are more explorative (52). Sex differences in open field behavior appear around puberty, with males showing a postpubertal decrease in activity but females retaining their prepubertal levels. However, in the hole-board test, the sex of mice and rats has not been found to affect head dipping significantly (5). Overall, female rats usually provide higher scores in most of the exploration measurements. However, this can be caused by a greater vulnerability of male rat exploratory behavior to the complexity of environmental stressors.

## Hormonal Status

Ovarian status has no gross effect on exploratory behavior in female rats; however, lactating rats shortly postpartum are more active in forced exploration tests. Ovariectomy also does not have much effect, but it can modify the effect of drugs (53, 54); here, the phase of the estrous cycle can also be important (54).

## Hunger

Short deprivation of food (in the range of several hours) reliably increases exploratory activity in the forced exploration test (49) and, in the author's experience, the sensitivity of exploration to drug treatment is lost. Deprivations considerably greater than 24 hr, however, can lead to a decrease in exploration. Both hunger and thirst lead to an increased neotic approach in preference tests (3).

## Housing

Social isolation, especially if started at early age, can have a moderate diminishing influence on exploratory activity both in mice and rats (5, 16, 38, 55, 56). Animals reared in social isolation can be characterized as more fearful and less exploratory, since they generally display high activity in the most unspecified forced exploration tests (open field) but low activity in the plus-maze test, light/dark compartment test, hole-board test, and in novelty preference tests. Occasionally, social isolation can lead to heightened activity in tests of forced exploration other than the open field, such as the plus-maze test (56), and even in novelty preference tests (3). The discrepancies found remain unexplained but deserve consideration in all investigations using singly housed animals.

## Strain

Inbred mouse and rat strains show substantial variability in baseline exploratory activity (17, 57). As an example, in the light/dark preference test the maximal interstrain difference of exploration measures was more than 5-fold when five mouse strains were compared. Even more importantly, the baseline exploratory activity predicted the maximal anxiolytic-like effect of diazepam (57). Similar results have been obtained in a modified version of this test (38). Relatively high values can be obtained using NMRI and C57BL/6 mice and also Swiss albino mice, whereas DBA/2 and CF-1 mice and especially BALB/c mice strongly prefer the dark compartment. In BALB/c mice, even anxiolytic drugs often show no effect in several forced exploration tests. In most rat strains, differences in spontaneous behavior may not be so strong but become more pronounced after challenge (e.g., drug treatment). However, most common laboratory strains are different from wild rodents, since the process of domestication has resulted in several profound behavioral changes, including reductions in neophobia. Thus, in the laboratory, wild strains show low exploratory activity not only in intentionally fear-provoking forced paradigms, but even in novelty preference tests (3).

## Interindividual Differences/Home Cage Hierarchy

Spontaneous behavior can show large interindividual variability among animals from the same home cage. This also holds true for exploratory activity, especially in forced exploration paradigms (21, 58). Sometimes this forces the experimenter to use relatively large groups of animals. A group size of at least 10–12 individuals is suggested, since several treatments can yield splitting of the population into subpopulations of subjects with either an increase or decrease in exploration scores.

## Previous Experience

Considering the concept of exploratory behavior as novelty determined, it should not be surprising that not only does the performance in an exploration test change on reexposure, but the effect of a drug is also influenced by exposure to the test conditions even in the nondrugged state (13). On the other hand, previous drug experience alters exploratory behavior (5). Moreover, not only does the test situation experience affect the performance at reexposure, but the general level of fearfulness or motivation of the animal, determined by its past history, will also interfere (5). High ambient levels of sensory stimulation can make the novelty as a stimulus less reinforcing (3). Among other possible influences, handling is one of the most routine ones and has recently become a subject of investigations in its own right. With regard to exploratory behavior, however, the findings are controversial, with both increased and decreased activity in the forced exploration tests being reported, even in the same laboratory, without any intentional change of experimental design (59). As a rule, the direction of change depends on the baseline level of behavioral activity, which can, in some cases, even determine the effect of less vaguely defined treatment such as drug administration (58).

## Summary

One can easily note that several tests of forced exploration include more features of choice tests than others. The elevated plus-maze, light/dark compartment, and mirrored chamber tests offer the subject the possibility of evading more aversive areas, whereas the hole-board test gives the animal a chance to explore despite being in an aversive situation. On the other hand, even if working with novelty preference tests, one must bear in mind that any changes in the environment have the potential to produce neophobia.

The external situation and the internal state serve as the sources of a variety of stimuli that influence exploration. Thus, after selecting the appropriate technique, the investigator should carefully consider which behaviors appear to be salient under the particular circumstances, and should make an attempt to record all these acts. It is important to remember, however, that the behaviors will be competing. One could measure both the frequency and duration of different responses and keep the record of the temporal pattern of these responses at the initial stage of introducing an exploration paradigm in the laboratory. Nevertheless, to reduce noise, the crucial factors should be determined after the investigator has gained some experience.

## Acknowledgments

I remain grateful to Professor Lars Oreland and the Medical Faculty of Uppsala University for financial support and access to databases during the writing of this chapter, as well as to the Library Service at the Uppsala University Biomedical Center for the kind help that made this work possible.

# References

1. A. E. Kelley, M. Cador, and L. Stinus, *in* "Neuromethods" (A. A. Boulton, G. B. Baker, and A. J. Greenshaw, eds.), Vol. 13, p. 95. Humana, Clifton, New Jersey, 1989.
2. J. N. Crawley, *Neurosci. Biobehav. Rev.* **9**, 37 (1985).
3. D. T. Corey, *Neurosci. Biobehav. Rev.* **2**, 235 (1978).
4. R. N. Walsh and R. A. Cummins, *Psychol. Bull.* **83**, 482 (1976).
5. S. E. File, *Neurosci. Biobehav. Rev.* **9**, 45 (1985).
6. T. Ljungberg and U. Ungerstedt, *Pharmacol. Biochem. Behav.* **8**, 483 (1978).
7. L. A. Hilakivi and R. G. Lister, *Behav. Neural. Biol.* **53**, 153 (1990).
8. K. C. Montgomery, *J. Comp. Physiol. Psychol.* **48**, 254 (1958).
9. S. L. Handley and S. Mithani, *Naunyn-Schmiedeberg's Arch. Pharmacol.* **327**, 1 (1984).
10. S. Pellow, P. Chopin, S. E. File, and M. Briley, *J. Neurosci. Methods* **14**, 149 (1985).
11. D. N. Stephens, B. S. Meldrum, R. Weidmann, C. Schneider, and M. Grutzner, *Psychopharmacology* (*Berlin*) **90**, 166 (1986).
12. R. G. Lister, *Psychopharmacology* (*Berlin*) **92**, 180 (1987).
13. R. G. Lister, *Pharmacol. Ther.* **46**, 321 (1990).
14. J. Harro and E. Vasar, *Neurosci. Biobehav. Rev.* **15**, 473 (1991).
15. P. Moser, *in* "Behavioral Pharmacology of 5-HT" (P. Bevan, A. R. Cools, and T. Archer, eds.), p. 373. Lawrence Erlbaum Associates, Hillsdale, New Jersey, 1989.
16. S. C. Stanford, V. Parker, and A. Morinan, *J. Affective Disorders* **15**, 175 (1988).
17. C. Gentsch, M. Lichsteiner, and H. Feer, *Behav. Brain Res.* **25**, 101 (1987).
18. C. Lee and R. J. Rodgers, *Psychopharmacology* (*Berlin*) **102**, 507 (1990).
19. M. S. Halliday, *Q. J. Exp. Psychol.* **19**, 254 (1967).
20. J. Harro, M. Póld, and E. Vasar, *Naunyn-Schmiedeberg's Arch. Pharmacol.* **341**, 62 (1990).
21. J. Harro, R.-A. Kiivet, A. Lang, and E. Vasar, *Behav. Brain Res.* **39**, 63 (1990).
22. B. Costall, M. E. Kelly, and D. M. Tomkins, *Br. J. Pharmacol.* **96S**, 312P (1989).
23. R. J. Rodgers, C. Lee, and J. K. Shepherd, *Psychopharmacology* (*Berlin*) **106**, 102 (1992).
24. A. B. Sheldon, *Q. J. Exp. Psychol.* **67**, 516 (1969).
25. S. Morato and P. Castrechini, *Braz. J. Med. Biol. Res.* **22**, 707 (1989).
26. S. L. M. Widgiz and C. H. M. Beck, *Behav. Brain Res.* **40**, 109 (1990).
27. J. Archer, *Anim. Behav.* **21**, 205 (1973).
28. T. W. Robbins, *in* "Handbook of Psychopharmacology" (L. L. Iversen, S. D. Iversen, and S. H. Snyder, eds.), Vol. 7, p. 37. Plenum, New York, 1977.
29. V. H. Denenberg, *Ann. N.Y. Acad. Sci.* **159**, 852 (1965).
30. D. Treit and M. Fundytus, *Pharmacol. Biochem. Behav.* **31**, 959 (1988).
31. J. R. Boissier and P. Simon, *Therapie* **17**, 1225 (1962).
32. S. E. File and A. G. Wardill, *Psychopharmacologia* **44**, 47 (1975).
33. S. E. File and A. G. Wardill, *Psychopharmacologia* **44**, 53 (1975).

34. M. Cheal, *Physiol. Behav.* **21,** 299 (1978).
35. J. Crawley and F. K. Goodwin, *Pharmacol. Biochem. Behav.* **13,** 167 (1980).
36. T. Kilfoil, A. Michel, D. Montgomery, and R. L. Whiting, *Neuropharmacology* **28,** 901 (1989).
37. B. Costall, C. A. Hendrie, M. E. Kelly, and R. J. Naylor, *Neuropharmacology* **26,** 195 (1987).
38. R. Misslin, C. Belzung, and E. Vogel, *Behav. Processes* **18,** 119 (1989).
39. L. Singh, M. J. Field, C. A. Vass, J. Hughes, and G. N. Woodruff, *Br. J. Pharmacol.* **105,** 8 (1992).
40. J.-R. Boissier, P. Simon, and C. Aron, *Eur. J. Pharmacol.* **4,** 145 (1968).
41. P. L. Toubas, K. A. Abla, W. Cao, L. G. Logan, and T. W. Seale, *Pharmacol. Biochem. Behav.* **35,** 121 (1990).
42. R. N. Hughes, *Anim. Behav.* **13,** 30 (1965).
43. R. N. Hughes and L. A. Syme, *Psychopharmacologia* **27,** 359 (1972).
44. T. W. R. Hansen, T. Sagvolden, and D. Bratlid, *Brain Res.* **424,** 26 (1987).
45. B. J. Sahakian, T. W. Robbins, and S. D. Iversen, *Anim. Learn. Behav.* **5,** 193 (1977).
46. P. Winn and T. W. Robbins, *Neuropharmacology* **24,** 25 (1985).
47. D. E. Berlyne, *J. Comp. Physiol. Psychol.* **48,** 238 (1955).
48. T. Robbins and S. D. Iversen, *Psychopharmacologia* **28,** 155 (1973).
49. S. E. File and S. Day, *Anim. Behav.* **20,** 758 (1972).
50. M. F. O'Neill, C. T. Dourish, and S. D. Iversen, *Eur. J. Pharmacol.* **193,** 203 (1991).
51. R. N. Hughes, *Anim. Behav.* **16,** 92 (1968).
52. A. L. Johnston and S. E. File, *Physiol. Behav.* **49,** 245 (1991).
53. D. Bitran, R. J. Hilvers, and C. K. Kellogg, *Behav. Neurosci.* **105,** 653 (1991).
54. G. G. Nomikos and C. Spyraki, *Neuropharmacology* **27,** 691 (1988).
55. D. Einon and N. C. Tye, *Psychopharmacologia* **44,** 83 (1975).
56. L. A. Hilakivi, M. Ota, and R. G. Lister, *Pharmacol. Biochem. Behav.* **33,** 371 (1989).
57. J. N. Crawley and L. G. Davis, *Brain Res. Bull.* **8,** 609 (1982).
58. L. Valzelli, W. Kozak, and O. Giraud, *Methods Find. Exp. Clin. Pharmacol.* **9,** 437 (1987).
59. N. Andrews, A. Zharkovski, and S. E. File, *Br. J. Pharmacol.* **102S,** 305P (1991).

# [24] Measurement of Grooming Behavior

Maria Ester Celis and Enrique Torre

## Introduction

The biological significance of grooming behavior is far from clear, although different functions have been proposed. It is used by animals to care for their skin, fur, and feathers, but it can also be used to relieve stress and pain, to dearouse, and to reduce the incidence of sexually transmitted diseases, and has been associated with thermoregulation and social communication as well. Grooming is a group of autonomic endocrine somatomotor responses associated with adaptive behaviors which ensure the survival of the animal.

Grooming or its functional equivalents must be evolutionarily ancient because it is very nearly universally represented in animal taxa. In more familiar animal forms, including mammals and birds and many insects, grooming usually takes the form of moving the extremities over the body and of mouthing the body and the extremities, although sand-bathing is another common form of grooming in birds and many mammals.

The sequence of grooming acts in rodents and the apparent independence of some subroutines of grooming, for example, scratching, from others have also suggested a hierarchical organization to grooming. There are varieties of grooming; genital grooming in rats is usually the last event in a grooming sequence. In adult rats there is a clear cephalocaudal progression of grooming bouts. Grooming most commonly starts with nose wipes by the forepaws and most often ends with anogenital grooming, with the sequence of acts by no means being random between beginning and end either. However, scratching of the head by the hind leg can appear at any stage of the sequence, suggesting that scratching belongs to a different branch of the hierarchy than does the rest of the grooming sequence. The cephalocaudal progression of grooming in adults has now been described by numerous investigators in diverse species, including cat (1), mouse (2), walaby (3), squirrels of several species (4), cricket (5, 6), and giant rat (7).

Grooming behavior is a good tool for evaluating the effects of several peptides. Many peptides, among them adrenocorticotropin/$\alpha$-melanocyte-stimulating hormone (ACTH/$\alpha$-MSH), endorphins, and bombesin, can induce excessive grooming behavior. This is a convenient way to study the locus in the brain at which the peptides act and the neurotransmitters involved in the behavior (8–17). It is possible that each of these peptides administered

*Methods in Neurosciences, Volume 14*

centrally increases stress in animals. Grooming is considered to be a specific rodent behavioral response to stressful situations (18).

In the rat, this behavior follows a specific sequence. At the beginning, vibrating movements of the forelegs are generally observed when cleaning of the face and head are to follow. The face wash begins with the animal licking its forepaws and then brushing them over its nose with a series of brief horizontal movements, followed by wider ones which include its whole head. Grooming then continues along the rest of the body, comprising successively the flanks and back, hind legs, and genitals. Grooming is frequently interrupted by scratching with one of the hind legs. Another component of grooming behavior is shaking of the whole body, from head to tail, which is similar but less intense than the "wet-dog shakes" observed in neuropharmacological studies of the abstinence syndrome (9, 18).

In birds and mammals an increase in grooming behavior has been observed in response to situations associated with a new or with a conflictive environmental stimulus (19). These same stimuli activate the hypothalamic–adrenal–pituitary system (21), which suggests the involvement of this system in the induction of grooming behavior. Nevertheless, hypophysectomy, gonadectomy, and adrenalectomy do not affect the increase in grooming behavior associated with either novel stimuli or the injection of peptides (9, 21, 22). This suggests that the expression of this behavior may be independent of the endocrine system and may depend exclusively on the activation of central mechanisms.

This idea is sustained by two lines of evidence. (1) Intracerebroventricular administration of antibodies against $\alpha$-MSH or ACTH reduce the increase in the grooming response to novel stimuli (23). (2) Excessive grooming (EG) as well as the stretch and yawn syndrome (SYS) can only be induced by a central administration of the peptides. Intraperitoneal injections of ACTH or $\alpha$-MSH, even at high doses, do not modify this behavior.

Several authors agree that EG, induced by direct injection of $\alpha$-MSH/ACTH peptides into the brain, would not be a stereotyped expression of grooming behavior, since the animal performs the normal sequence of movements which comprise this behavior (see above), without increasing the number of episodes but increasing their length (9, 24). This response is clearly different from the stereotyped grooming behavior observed after treatment with dopaminergic agonists or amphetamines, when animals generally exhibit one or very few elements of this behavior, concentrated in a specific area of the body (17, 23).

EG does not appear to be affected by environmental factors either. The effect is comparable if the study is made in an enriched environment, in an open field, or in a restricted stimulation room (22, 25). Only high degrees of water or food deprivation or electric shock can interfere with EG induced

TABLE I    Coordinates for Cannula Implantation Sites [a]

| Area | Anterior | Lateral | Vertical |
|------|----------|---------|----------|
| Third ventricle (ACV) | 7.0 | 0 | −0.6 |
| Substantia nigra pars compacta (SNc) | 2.0 | 1.4–2.0 | −3.2 |
| Lateral substantia nigra (SNL) | 2.0 | 2.0–2.5 | −2.9 |
| Ventral tegmental area (VTA) | 2.0 | 0.5–1.2 | −3.2 |

[a] Coordinates for the implantation of cannulas into specific areas of the central nervous system are based on the atlas of Konig and Klippel (27).

by the peptides $\alpha$-MSH/ACTH (22). Colbern *et al.* (17) demonstrated that EG could be induced in a given animal during a period of 10 consecutive days by daily intracerebroventricular injection of ACTH $_{1-24}$. The levels and characteristics of the response during the 10 sessions were similar. Later, Jolles *et al.* (26) demonstrated that if a second injection of the peptide was made within 10 hr of the first, it did not modify normal grooming behavior. Between 10 and 18 hr after the first injection, the response began to increase until it reached the levels of the first, between 18 and 24 hr. In addition, cross-tolerance developed between $\beta$-endorphin, morpine, and [D-Phe$^7$] ACTH$_{4-10}$ (9).

## Experimental Procedures

### Animals

Male albino rats weighing 200–250 g are used. The colony room is maintained on a 14-hr on–off light cycle with the lights turned on at 0600 hr, with food and water provided *ad libitum*.

### Surgery

The animals are anesthetized with tribromoethanol (0.2 mg/kg body weight) and placed in a stereotaxic frame. The scalp is incised, and a stainless steel frame is attached to the skull. Holes are then drilled at appropriate rostral–caudal coordinates, according to the atlas of Koning and Klippel (27). Coordinates for the various implantation sites are shown in Table I.

Stainless steel cannulas, 14.5 mm long and 0.56 mm OD, are implanted intraventricularly or bilaterally into the substantia nigra. Guide cannulas are placed 1.5 mm above the nucleus in order to preserve it from the eventual damage caused by chronic implantation (12–14). The skull screw and the cannulas are cemented into place with dental acrylic, and the exposed portions of the cannulas are sealed with a piece of plastic tubing. After surgery, animals are individually housed and given food and water *ad libitum*.

## Injections of Artificial Cerebrospinal Fluid or Peptides

Intracerebral microinjections are made into unanesthetized animals using a 1-$\mu$l Hamilton syringe connected by PE-10 polyethylene tubing to a 30-gauge needle which extends 0.75 cm beyond the guide cannula. The volume for microinjections is 0.5 $\mu$l. Using this volume it is estimated that diffusion takes place into a sphere of tissue approximately 1 mm in diameter.

## Behavioral Test

EG induced by intracerebral administration of $\alpha$-MSH/ACTH is a behavioral expression in which the length of each grooming episode is increased without substantial alteration of the frequency of the episodes. This behavior can be measured by two methods: (a) direct timing of the grooming period and (b) assigning a value to the time and intensity of the response. Gispen and co-workers (21) designed a technique for the second approach which gave similar results to those obtained by timing methods (correlation coefficient, 0.9). Even though this methodology was developed to quantify EG, it can also be used to estimate other behavioral expressions such locomotion or elements of SYS (12). This is very important since behavioral expressions such as motor activity, which could be affected by pharmacological manipulation, can be evaluated along with alterations in grooming behavior, thus determining the effect on other functions controlled by these cells.

Another advantage of this method is that results are reproducible even when the same animal is injected on several consecutive days (provided the interval between injections is not less than 20 hr), which allows the use of every animal as its own control (9, 17). On the other hand, the discrimination of the various behavioral elements is very simple since they are evident expressions that leave no place for doubt, thus avoiding any confusion in the measurement of this behavior. Under optimal conditions, that is, using closed television circuits, computers, etc., over 12 animals can be monitored at the same time using this methodology.

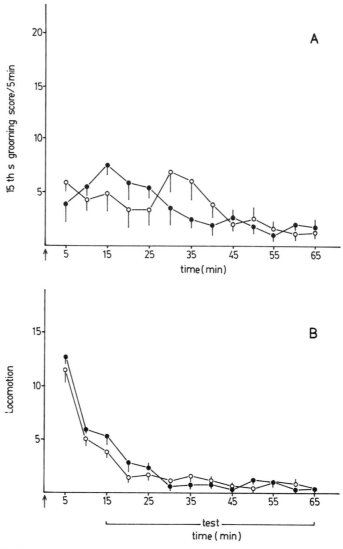

FIG. 1   (A) Time course of the grooming response following injection of 1 μl ACSF into the third ventricle (ACV) or ventral tegmental area (VTA). Values are the mean behavioral scores ± SEM and were calculated at 5-min intervals using the 15-sec sampling technique. (B) Time course of the locomotive response following injections of 1 μl ACSF into the ACV or VTA. Behaviors were scored using the 15-sec sampling technique (observations over 50 min, beginning 15 min after injections). The values are the mean behavioral scores ± SEM (*n* = 11 in all cases). •—• ACV; ○—○ VTA.

TABLE II   Behavioral Changes Induced by
Intracerebrospinal
Administration of Artificial
Cerebrospinal Fluid[a]

| Activity | Behavioral score |
|---|---|
| Grooming | 35.50 ± 4.80 |
| Rearing | 12.60 ± 3.70 |
| Locomotion | 11.30 ± 2.80 |
| Stretching and/or yawning | 1.20 ± 0.80 |

[a] Results are means ± SEM of 11 animals.

The steps followed in a typical test are described. Animals are used in the behavioral tests no sooner than 7 days after implantation. Behavioral tests are never carried out before this time, which allows animals to recover from surgery effects that could interfere with the tests, such as inflammation. Tests are carried out at 22°C in an isolated room, illuminated by fluorescent light. The rat is placed in a cage (40 × 40 × 25 cm) with transparent walls in a low noise environment.

The tests are run for 65 min. In the original design, as described by Gispen and co-workers (21), results from the first 15 min are not taken into account since in this period the grooming behavior is usually increased as a result of the new environment. Later this activity decreases, until it is almost nil by the end of the trial. In our experimental scheme, different behaviors are recorded throughout the 65 min of the trial, but only values registered in the final 50 min are computed for analytical purposes. This allows us to analyze the time course of $\alpha$-MSH from the moment the animal is placed in the observation box. Once there, the appearance of the behavioral components of EG, SYS, or motor activity are registered every 15 sec. Thus, the maximum

TABLE III   Effect of Injection of Artificial Cerebrospinal Fluid in
Different Brain Areas on Grooming Behavior[a]

| Area | Grooming | Rearing | Locomotion | Stretching and/or yawning |
|---|---|---|---|---|
| VTA | 39.00 ± 7.30 | 9.20 ± 1.70 | 11.50 ± 1.60 | 2.10 ± 0.85 |
| SNc | 40.25 ± 5.17 | 10.88 ± 4.10 | 9.75 ± 2.82 | 1.80 ± 0.70 |
| SNL | 34.10 ± 5.17 | 11.40 ± 4.20 | 13.10 ± 4.40 | 1.50 ± 0.65 |

[a] Grooming behavior, rearing, locomotion, and stretch and yawn were analyzed in every case. Behavioral scores are means ± SEM of 11 rats.

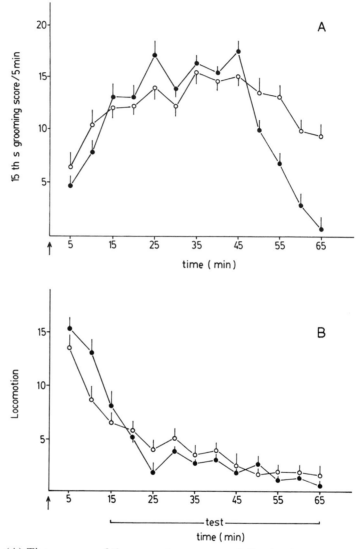

FIG. 2  (A) Time course of the grooming response following injection of 1 μg/μl of α-MSH into the VTA or ACV. Values are the means ± SEM and were calculated at 5-min intervals using the 15-sec sampling technique. (B) Time course of the locomotive response following injection of 1 μg/μl of α-MSH into the VTA or ACV. Behaviors were scored using the 15-sec sampling technique (observations over 50 min, beginning 15 min after injections). The values are the mean behavioral score ± SEM ($n$ = 10 in all cases). •—• ACV; ○—○ VTA.

number of positive responses is 200. The behaviors evaluated and the elements which comprise it are as follows:

*Grooming behavior:* vibrational movements of the forelegs, washing of the forelegs and the head, cleaning of the hind legs, body, tail, and genitals, wet-dog shaking, scratching

*Stretch and yawn syndrome:* stretching of the body with the back arched or crutched, yawn, simultaneous stretch and yawn

*Motor behavior:* locomotion: crossing the cage; rearing: raising forelegs, either resting them on the cage walls or not

One positive score is given when any of the components of the three behaviors is registered in a 15-sec period. For example, if an animal scratches itself, then walks to the other end of the cage and finally stretches and yawns, the score will be EG, 1; SYS, 1; Locomotion, 1. At the end of the test the animal is returned to its cage.

## Behavioral Response Induced by Injection of Artificial Cerebrospinal Fluid into the Brain

Rats injected with artificial cerebrospinal fluid (ACSF), either in the mesencephalon or intracerebroventricularly, and later placed in a new environment such as the observation box develop exploratory activity (locomotion, rearing, smelling) and grooming behavior in the initial phase of the observation period. This activity decreases until the animal usually falls asleep during the rest of the trial. The animal seldom yawns in the 65 min following the injection, and simultaneous stretching and yawning are even less frequently observed (Fig. 1; Table II). Similar results are obtained when ACSF is injected in other areas of the CNS such as the ventral tegmental area (VTA), substantia nigra pars compacta (SNc), or lateral substantia nigra (SNL) (Table III).

The analysis of the time course of EG induced by the injection of MSH into the third ventricle (ACV) or VTA shows that after the animal is introduced into the observation cage, scores for this behavior increase rapidly, until they become significantly different from the mean values shown by animals treated with ACSF 15 min after the injection, reaching a maximum 10 min later. The response begins to decline between 45 and 50 min later; the decline is slow if the peptide is injected in the VTA, and fast if it is injected intracerebroventricularly (Fig. 2A).

Table IV   Episodes of Stretching and
Yawning Induced by Peptide
Administration in Specific
Brain Areas[a]

| Area | Stretching and yawning | |
| | $\alpha$-MSH | ACSF |
| --- | --- | --- |
| ACV | 14.00 ± 2.5 (7)[b] | 1.8 ± 0.8 (11) |
| VTA | 2.5 ± 0.6 (9) | 2.10 ± 0.85 (7) |
| SNc | 1.00 ± 0.10 (15) | 1.8 ± 0.70 (8) |
| SNL | 2.20 ± 1.35 (6) | 1.5 ± 0.65 (10) |

[a] Behavioral scores are means ± SEM, with the number of
rats in parentheses.
[b] $p < 0.001$ when compared with all the other results.

Figure 2B also shows the time course of the response of locomotion in
animals injected in either the ACV or VTA. It can be seen that values
registered in treated animals are slightly higher over the entire observation
period; they are high in the initial periods but decrease soon afterward,
coinciding with an increase in grooming behavior, though they do not disappear, and show fluctuations that suggest a strong competition for other
behavioral components.

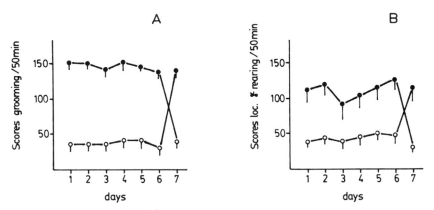

FIG. 3   Effects of repeated injections into the VTA on (A) grooming and (B) motor
behavior. Two groups of animals implanted bilaterally were injected for 7 consecutive
days. One group was given ACSF and the other $\alpha$-MSH for 6 days. On the seventh
day, animals in the first group received $\alpha$-MSH and those in the second group ACSF.
Results are mean behavioral scores ± SEM. ○—○ ACSF; •—• $\alpha$-MSH.

Finally, we could show that SYS was evident only if $\alpha$-MSH injections were made intracerebroventricularly (Table IV). The expression of episodes of stretching and yawning begins with a latency of 5–10 min after the injection, continuing in many cases beyond the end of the experimental period.

To evaluate whether repeated doses injected in the VTA could modify the behavior of the animals, two groups of animals implanted bilaterally are injected for 7 consecutive days. One group is given ACSF and the other $\alpha$-MSH for 6 days. On the seventh day, animals in the first group receive $\alpha$-MSH and those in the second group ACSF. Behavioral tests are performed every day. Figure 3 shows that there are no alterations in the performance of the animals, suggesting that the experimental design satisfies the requirements for reliability. The results presented also illustrate that in order to compare data from different authors, injections must be made in the same areas. In addition, some behaviors only appear if injections are made in a certain pathway.

All the techniques described in this chapter provide new knowledge and insight into the meaning of the presence of various peptides in the brain.

# References

1. R. M. Swenson and W. Randall, *J. Comp. Physiol. Psychol.* **91,** 213 (1977).
2. J. C. Fentress, *in* "The Biology of Behavior" (J. A. Kiger, Jr., ed.) p. 85. Oregon State Univ. Press, Corvallis, 1972.
3. E. M. Russell and D. C. Giles, *Behaviour* **51,** 19 (1974).
4. J. Ferron and L. Lefebre, *Behaviour* **81,** 110 (1982).
5. L. Lefebre, *Anim. Behav.* **29,** 973 (1981).
6. L. Lefebre, *Anim. Behav.* **30,** 1020 (1982).
7. R. F. Ewer, *Z. Tierpsychol.* **24,** 6 (1967).
8. A. R. Cools, V. M. Wiegant, and W. H. Gispen, *Eur. J. Pharmacol.* **50,** 265 (1978).
9. W. H. Gispen and R. L. Isaacson, *Pharmacol. Ther.* **12,** 209 (1981).
10. W. H. Gispen, J. Buitelaar, V. M. Wiegant, L. Terenius, and D. DeWied, *Eur. J. Pharmacol.* **39,** 393 (1976).
11. W. Ferrari, G. L. Gessa, and L. Vargiú, *Ann. N.Y. Acad. Sci.* **104,** 330 (1963).
12. E. Torre and M. E. Celis, *Neurochem. Int.* **9,** 85 (1986).
13. E. Torre and M. E. Celis, *Life Sci.* **42,** 1651 (1988).
14. E. Torre and M. E. Celis, *Acta Physiol. Pharmacol. Latinoam.* **39,** 49 (1989).
15. S. R. Barioglio, N. Lezcano, and M. E. Celis, *Peptides* **11,** 29 (1990).
16. J. P. Ryan and R. L. Isaacson, *Physiol. Psychol.* **11,** 54 (1983).
17. D. L. Colbern, R. L. Isaacson, E. G. Green, and W. H. Gispen, *Behav. Biol* **23,** 381 (1978).
18. J. C. Fentes, *Science* **179,** 704 (1973).
19. R. J. Bolles, *J. Comp. Psychol.* **53,** 306 (1960).

20. D. De Wied and J. Jolles, *Physiol. Rev.* **62,** 976 (1982).
21. W. H. Gispen, V. M. Wiegant, H. M. Greven, and D. De Wied, *Life Sci.* **17,** 654 (1975).
22. J. Jolles, J. Barendregt-Rompa, and W. H. Gispen, *Behav. Neural. Biol.* **25,** 563 (1979).
23. I. H. Ayhan and A. Randrup, *Psychopharmacologia* **29,** 317 (1973).
24. A. J. Dunn, E. J. Green, and R. L. Isaacson, *Science* **203,** 281 (1979).
25. R. L. Isaacson and E. J. Green, *Behav. Biol.* **24,** 118 (1978).
26. J. Jolles, J. Barendregt-Rompa, and W. H. Gispen, *Horm. Behav.* **12,** 60 (1979).
27. J. F. R. Koning and R. A. Klippel, "The Rat Brain." Williams & Wilkins, Baltimore, Maryland, 1963.

## [25] An Animal Model for Cardiovascular and Behavioral Reaction

### Lennart Svensson

## Introduction

Blood pressure is largely influenced by emotional stress. Traditionally, the individual's personality and way of coping with stress have been implicated in the etiology of essential hypertension. Both naturally occurring and laboratory-produced stressful stimuli have been used to study the influence of behavior on the establishment of hypertension. In animal models like the spontaneously hypertensive rat (SHR), several attempts have been made to link the abnormal regulation of the animal's circulation to its behavioral characteristics. It has been suggested that a genetically determined central nervous system (CNS) hyperreactivity may contribute to the development of high blood pressure in this rat strain (1). It is now well established that SHRs do deviate from their genetic controls, the normotensive Wistar-Kyoto rats (WKY), also with respect to behavioral reactions (2, 3).

In the study of the relationship between emotional stress and blood pressure, it seems important to choose a stressful stimulus that is relevant to the animal. One such stimulus is the acoustic startling stimulus, which is frequently experienced by the rat in its normal environment. Rats are nocturnal animals that orientate and protect themselves mainly by means of tactile, olfactory, and acoustic stimuli. Therefore, it seems reasonable to assume that the acoustic stimulus evokes distinct, alerting, but at lower stimulus intensities, not anxiogenic responses in the rat. The behavioral response to the acoustic stimulus is mainly dependent on stimulus intensity, the interval between stimuli, and the background noise level (4). By an adequate arrangement of these parameters it is possible to have a relatively stable baseline response level. Using the acoustic startle response, evidence was recently obtained for a dissociation between cardiovascular and behavioral reactivity in the SHR. These rats showed an exaggerated pressor response in contrast to a decreased startle response in comparison with WKY animals (5). Further evidence for a lack of correlation between blood pressure and behavior in SHRs has been presented, and the importance of behavioral reactivity in the etiology of hypertension in this rat strain has been questioned (6, 7).

In this chapter a method is described that permits simultaneous recording of acoustic startle and blood pressure in the rat. Somewhat different methods

*Methods in Neurosciences, Volume 14*

for simultaneous measurement of these parameters have also been described (8, 9).

## Blood Pressure

Blood pressure in the rat is rather conveniently monitored by means of an indwelling catheter. The carotid or the caudal tail artery is most often selected for catheterization. The caudal tail artery is accessed from the ventral side of the proximal end of the rat's tail. Light anesthesia is induced by, for example, Brietal (50 mg/kg, i.p.) and a longitudinal incision is made in the tail. Warming the animal by means of an ordinary lamp causes vasodilatation of the tail artery and helps dissection of the artery. A pair of forceps may be inserted under the artery to keep it stretched in open air. The artery is closed downstream with a suture. Upstream a small incision is made with a pair of scissors, and a 5 cm long plastic PE-25 catheter with the penetrating end cut at an angle is introduced into the artery. The catheter is fixed with sutures and flushed with saline containing heparin to make sure that there is a free passage and to prevent from clogging of blood. A few stitches are used to close the incision, and surgical tape is used to cover the wound. The distal part of the catheter is left free, with the cavity closed with a stopper. The rat is then left for postsurgical care, during which time precautions should be taken to prevent the animal from biting its tail.

Use of a short-acting anesthetic like Brietal permits the rat to be used in behavioral experiments a few hours after surgery (5). After a period of 4 to 6 hr, when the animal is fully awake and shows no evidence of distress, the free end of the catheter is attached to a PE-50 extension tubing that is connected to a commercial pressure transducer (e.g., Statham P23Db). The system is flushed again with saline, and recording can be initiated. A Grass 7B polygraph equipped with a 7P1 preamplifier and a 7P44 tachograph triggered from the pressor pulse can be used for continuous recording of blood pressure and heart rate, respectively. Alternatively, a conventional microcomputer equipped with an analog-to-digital input/output board may be used. In this case the input/output board should be programmed to sample the transducer output at a frequency of 10 Hz or higher in order to achieve satisfactory signal resolution. Because the rat tail artery regulates body temperature, the rat should be kept in a constant temperature environment during recordings. A constant temperature above room temperature is preferable since such an arrangement promotes tail artery blood flow.

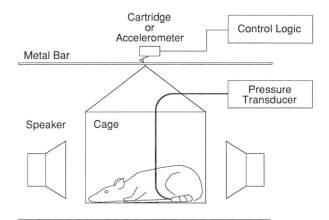

FIG. 1   Diagram of setup for simultaneous recording of acoustic startle and blood pressure in the rat. The cage containing the rat is suspended under a metal bar that is firmly fixed at both ends. Cage movement causes a bar displacement, the amplitude of which is picked up by the cartridge or accelerometer mounted on top of the bar. The acoustic stimulus is presented to the rat by means of a pair of high-frequency loudspeakers. The arrangement is housed in a sound-attenuated box, and the control logic and pressure transducer are positioned outside the box.

## Acoustic Startle

To measure the acoustic startle response some special equipment is needed (Fig. 1). The rat should be kept in a small cage during the test. It is the author's view that the weight of the cage should be small compared to the weight of the animal and that the cage should be suspended at the top, hanging free to move above the underneath surface. The latter arrangement has been found to facilitate the startle response (L. Svensson, unpublished results). A possible explanation for this effect is that the rat has to focus on its balance in the moving cage and thereby increases its muscular tone. The startle test should also be carried out inside a dimly lit, sound-attenuated box if possible. Cage displacement can be monitored by means of an accelerometer (e.g., Entran Devices EGAL-125-10D) or by means of a standard electrodynamic phonograph cartridge. In case of the latter arrangement the animal cage can be suspended under the center of a horizontal metal bar which is firmly fixed at both ends. The cartridge is then placed with the stylus resting on the bar. Cage movement thus causes a displacement of the bar, the amplitude of which is converted to an electrodynamic signal by the cartridge.

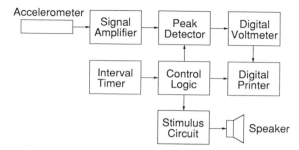

Fɪɢ. 2   Diagram showing the potential electronic circuitry needed to record the acoustic startle response.

A schematic diagram of the potential electronic circuitry needed to record the startle-elicited response is presented in Fig. 2. The startle-induced signal is fed to a voltage peak detector that should be set to sample the voltage for a 100–200 msec time period with start at the time of presentation of the startle-eliciting stimulus. The rat acoustic startle response latency is approximately 15 msec, and the longer the voltage sampling period, the greater the risk for measuring artifacts. By means of a digital voltmeter, the peak voltage can be presented visually or on a digital printer. Alternatively, the analog-to-digital input/output board used to sample the output of the pressure transducer can also be used to sample the startle response signal. The use of a microcomputer requires a certain amount of programming, but the changes that are recorded here are slow enough to permit high-level language programming (e.g., BASIC or Pascal) in most cases, which simplifies this task.

The acoustic startle stimulus should consist of a burst of white noise, which produces a more evenly distributed sound level than a single tone. The acoustic stimulus is presented to the animal by means of a high-frequency loudspeaker. A 100–105 dB (A) burst of noise of 20–40 msec duration is enough to elicit a startle response in rats of most strains. Response amplitude can be varied by adjusting the stimulus intensity. The background noise level is also important. A moderate level of background white noise has been shown to increase startle response amplitude (4). There is a certain degree of response habituation to successive stimulus presentations, which is highly dependent on the time period between the presentations; the shorter the time period, the faster the decline in response amplitude. To achieve a sufficiently stable response amplitude, the time period between stimulus presentations should not be shorter than 30 sec.

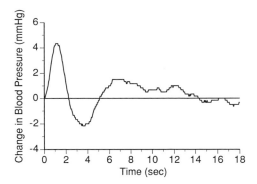

FIG. 3   Startle response-elicited changes in blood pressure in normotensive Wistar rats. The averaged response of 10 consecutive startle stimuli recorded from seven animals is shown. The acoustic stimulus (a 105 dB burst of white noise) was presented at time 0. The time between stimulus presentations was 30 sec.

## Signal Averaging

Moderate stimulus intensities evoke only minor changes in blood pressure and heart rate in the rat (5). It may therefore be necessary to use some form of signal averaging to verify the underlying changes in cardiovascular activity. One way of doing this is to use several stimulus presentations and to incorporate the evoked responses into one averaged response. The blood pressure data are displayed on a chart recorder by the Grass 7B polygraph, and these data need to be digitized before any further analysis can be performed. By

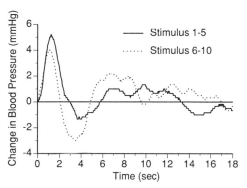

FIG. 4   Startle response-elicited changes in blood pressure in seven normotensive Wistar rats. The averaged responses of the first 5 versus the last 5 consecutive startle stimuli in a series of 10 are shown.

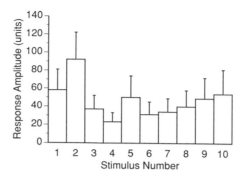

FIG. 5   Startle response amplitude of 10 consecutive acoustic startle stimuli [105 dB(A)] presented to seven normotensive Wistar rats. Data are represented by means ± SEM.

means of a digitizer (e.g., GTCO DIGI-PAD) these data can be transformed to digital data and stored on a microcomputer. Using the GTCO DIGI-PAD, this procedure allows a blood pressure resolution as low as 0.3 mmHg and a time resolution of 0.2 sec.

Because each digital data value represents both a response amplitude and time point, signal averaging is simplified. The first data value recorded after onset of the acoustic stimulus is subtracted from every data value in the recorded sequence of values covering a time period of 10–20 sec. This procedure standardizes the first data value to 0 and at the same time represents time 0. The data value representing the next time point is calculated as a mean value based on the second data value in all data sequences recorded for that animal, and so on. Finally a grand mean can be calculated by applying this technique for all animals tested. These calculations are time-consuming if done by hand, but computer algorithms for the calculations are conveniently programmed using any high-level language.

The averaged blood pressure response of 10 consecutive acoustic startle stimuli presented to seven normotensive Wistar rats is shown in Fig. 3. The startling stimulus caused an initial pressor response in the animals. This response was immediately followed by a depressor response, which probably reflects an opposing reaction to the sudden increase in blood pressure. These responses were then followed by a slow onset, longer lasting pressor response. Previous studies indicate that this response is a consequence of poststimulus arousal (5).

Figure 4 illustrates habituation of the cardiovascular response. When comparing the averaged response of the first 5 startle stimuli with the response of the last 5 stimuli, only minor modifications of the blood pressure response can be noted in the rats tested. It is also obvious from Fig. 5 that these animals show little habituation to the acoustic startle response under the experimental conditions, which might explain the lack of cardiovascular habituation.

# References

1. B. Folkow, *Physiol. Rev.* **62,** 347 (1982).
2. E. Hård, S. G. Carlsson, S. Jern, K. Larsson, A.-S. Lindh, and L. Svensson, *Physiol. Behav.* **35,** 487 (1985).
3. D. C. Tucker and A. K. Johnson, *Neurosci. Biobehav. Rev.* **5,** 463 (1981).
4. M. Davis, *Neurosci. Biobehav. Rev.* **4,** 241 (1980).
5. L. Svensson, C. Harthon, and B. Linder. *Physiol. Behav.* **49,** 661 (1991).
6. R. N. Leaton, J. V. Cassella, and D. Whitehorn, *Physiol. Behav.* **31,** 103 (1983).
7. J. E. Ledoux, A. Sakaguchi, and D. J. Reis, *Brain Res.* **277,** 137 (1983).
8. E. Borg, *Acta Physiol. Scand.* **100,** 129 (1977).
9. R. Rettig, M. A. Geyer, and M. P. Printz, *Physiol. Behav.* **36,** 1123 (1986).

# [26] Assessment of Cardiovascular Responses as Behavior

Bernard T. Engel, David E. Anderson, and Mark I. Talan

This chapter describes various experimental methods for modifying circulatory activity by characterizing cardiovascular responses as behaviors and then utilizing behavioral techniques to produce specific effects. A number of behavioral terms are defined to enable investigators to access the appropriate scientific literature. Separate sections deal with techniques for classical (Pavlovian) and operant (Skinnerian) conditioning methods for producing specific cardiovascular effects. A final section describes behavioral models for producing pathological cardiovascular effects.

## Introduction

The purpose of this chapter is to outline a number of techniques that can be used to assess cardiovascular responses from a behavioral perspective. In particular, we describe a number of experimental designs and analytic procedures that are appropriate for assessing behavior, and we note some of the pitfalls in interpreting the findings. We also review some common behavioral terms. We do not review the extensive literature on behavioral studies of the circulation; however, we cite articles that include relevant methodologies. The interested reader is referred elsewhere for scientific reviews of behavioral studies (1, 2). Thus, this chapter focuses on the following questions: How does one measure behavior? How does one interpret the findings from a behavioral perspective? The reasons why one should consider the circulation from a behavioral perspective are not discussed.

Traditionally, psychologists have defined behavior as the interaction of an organism with its environment. Thus, behavior, by definition, is a response to an environmental stimulus. We have modified this definition slightly to make it more congenial with physiological concepts; however, we believe that our definition maintains the traditional sense of the term.

Behavior means the interaction of an effector system with its environment. By effector system, we mean the glands or muscles that physiologists usually characterize as effectors, as well as the central and peripheral nerves which modulate these effectors. By environment, we mean the internal and external stimuli that precede or follow the responses of interest. Obviously, effector systems other than the circulatory system behave; however, we do not consider noncirculatory behavior in the present analyses.

There are two terms that are fundamental to any behavioral analysis: the first is punishment, and the second is reinforcement. Punishment is a process by which responses are inhibited, that is, their likelihood of occurrence is decreased. The stimulus that mediates punishment is called a punisher: its occurrence is a necessary condition for punishment to have occurred. For example, electric shock is often used as a punisher in animal studies; verbal punishers or loss of money have been used in studies with humans. Negative reinforcement is not a synonym for punishment. Reinforcement is the process by which responses are strengthened, that is, their likelihood of occurrence is increased. The stimulus that strengthens behavior is called a reinforcer. A reinforcer always follows the response it modifies. Positive reinforcement occurs when the animal responds in such a way as to increase the likelihood of occurrence of a reinforcer. Negative reinforcement occurs when the animal responds in such a way as to decrease the likelihood or continuation of punishment.

Food or water are common positive reinforcers, and avoidance of, or escape from, electric shock is a common negative reinforcer. Thus, reinforcement, whether positive or negative, will increase the probability of a response. A common synonym for a reinforcer is a reward, and reinforcement is always rewarding. It should be noted that in order for food or water to be reinforcing, the subject must be hungry or thirsty. A required antecedent condition to positive reinforcement is deprivation; thus, the astute reader will recognize that one could conceptualize positive reinforcement as negative reinforcement (e.g., eating is an "escape" from hunger). Nevertheless, there are important differences between punishment and reinforcement, and one needs to consider very carefully which mode of training is appropriate for the specific research (see Ref. 3 for a detailed discussion of the differences between reward and punishment on performance).

There are a number of other terms that we use. In general, these will be defined below as they are introduced. However, two additional terms are salient here: one is learning, and the second is conditioning. Learning is a systematic process of behavioral change in which the response that is learned is shown to occur in some predictable fashion in relation to antecedent and consequent stimuli. The antecedent stimuli are variously called cues, discriminative signals, or conditional signals, and the consequent stimuli are labeled unconditional stimuli, rewards, or punishers. It must be emphasized that not all responses that are elicited by stimuli are learned; for example, reflexes are not learned. However, even reflexes can be modulated with experience. Conditioning refers to any of several, formal processes through which responses are trained. Psychologists usually refer to conditioning as classical or Pavlovian when the training involves the establishment of an orderly relationship between an antecedent stimulus and the response. Operant or Skinnerian conditioning occurs when the training involves the establishment of an orderly relationship between a consequent stimulus and the response. It should be very clear that there are divisions among psychologists as to whether these two processes characterize different kinds of learning or only one kind of learning: there is no disagreement that the training procedures are different, only whether the behavioral mechanisms are different (see Refs. 4 and 5 for a discussion of some of these issues).

In the subsequent sections, we review and discuss some of the methods commonly used to study the behavior of the circulation. In the first section we discuss techniques by which cardiovascular effects are *elicited* by antecedent stimuli, namely, classical conditioning. In the second section, we consider studies in which cardiovascular responses are *emitted* to obtain consequent stimuli, namely, operant conditioning. The last section considers experimental designs in which the cardiovascular responses are acquired coterminously with other conditioned responses, permitting the study of interactions between elicited and emitted behaviors.

## Classical Conditioning

The hallmark of classical conditioning is the selection by the experimenter of a stimulus that reliably elicits a specific cardiovascular response and then pairing of this stimulus with another, neutral stimulus that does not evoke the same response. A typical eliciting stimulus (the unconditional signal) is one that reliably evokes a tachycardia or bradycardia, a change in blood pressure, or some other specific circulatory effect. The response elicited by the unconditional stimulus (US) is called the unconditional response (UR) and usually is characterized as a reflex, although it should be clear that it is often difficult to know whether a response is a reflex or a highly overlearned behavior. The (initially) neutral stimulus is called the conditional stimulus (CS). The goal of classical conditioning is to show that the CS can be made to elicit a conditional response (CR), which is similar to the UR. The main advantage of using the Pavlovian paradigm is that it provides the investigator with precise control of the stimuli (CS and US) and also allows one to specify the response, since the investigator selects the UR and CR. Thus, this paradigm is a useful model for studying central neural mechanisms of cardiovascular control.

In classical conditioning, learning occurs as a result of repeated pairings of the US and CS in the same temporal relationship. The CS is commonly a light or a tone, which initially does not reliably elicit the response. The

US might be an electrical shock or even an injection of a cardiovascular-specific drug. Although there can be considerable variations in the methodology, as noted below, there are certain procedural similarities across experiments. One is the temporal relationship between the CS and US. The procedure of choice is forward conditioning, in which the CS is presented prior to the US. There are two variants on this method: one is a delayed conditioning in which the CS is initiated and persists to the presentation of the US; the second is a trace conditioning in which the CS is initiated for a brief period of time and is terminated, and then, after a fixed interval, the US is initiated. Delayed conditioning is the most commonly used procedure in studies of the circulation. The delay between CS onset and US onset can range from a few seconds to 30 sec or more. A typical example of delayed conditioning can be seen in the cardiovascular responses to a cue signaling a subject on a treadmill that exercise will begin shortly. Physiologists have commonly labeled the response to this cue a "warm-up" effect, but it is a typical CR. Backward conditioning, when the US precedes the CS, has special interest only for scholars of learning and is not an appropriate model for studying the circulation per se.

Evidence for learning is inferred from several outcome measurements. First is the occurrence of the CR: specifically, the occurrence of the CR in the interval between the onset of the CS and the onset of the US. An important control measure of learning occurs when the US is withheld on specific trials. The response occurring at the time when the UR would normally be expected to occur is also considered to be a CR. Regardless which CR is measured, a major indication that learning has occurred is that there be some evidence of gradual acquisition; that is, if one were to graph the CR over trials, the magnitude of the response would be expected to change monotonically to an eventual plateau. There are also other, more complex measures of learning. If the US is consistently withheld, the CR should eventually diminish to control levels; this procedure is called extinction. If, on subsequent occasions, the CS–US pairing is repeated, then one would expect to see relatively rapid relearning, that is, faster than in the original training. Detailed discussion of the various outcome measures of conditioning can be found in specific articles or textbooks (6–8).

Several kinds of responses could occur in the CS–US interval: One is the desired CR, and another is any response elicited by the CS that is US nonspecific. In this sense, the CS is itself acting as a US. Responses elicited by the CS are usually called orienting responses (Pavlov called them the "what is that" response). Still another set of responses that could occur in the CS–US interval are ones that are *emitted* in anticipation of the US. These responses are often called adaptive or defense responses. Because orienting responses (OR) and defense responses (DR) could have cardiovascular components, which could be misinterpreted to be conditional responses, it is important to include experimental controls to separate them from one another.

The OR is characterized by several criteria: it is nonspecific to a stimulus; it does not increase with an increase in stimulus intensity; it decreases with repeated trials (extinguishes); and it reappears with any change of stimulus. The usual experimental control for orienting responses is a series of trials with the CS only. Defense responses are more difficult to characterize. However, a common experimental control (which also controls for OR), is to carry out a differential conditioning study. In this design, the subject is presented with an additional, discriminant stimulus (DS) (which is somewhat similar to the CS but different in some characteristic, e.g., a tone of different frequency or light of a different color) that will not be accompanied by the US or is accompanied by a different US. After sufficient repetition, the DS will elicit no CR, or it will elicit a different CR (9, 10). Inclusion of a DS has the added advantage that it enables the investigator to dissociate the cardiovascular component of conditioning from the somatomotor responses that might also occur during the experiment.

Several technical problems are likely to arise during the analysis of the experimental results. One such problem is the choice of a baseline measure, since there will always be "spontaneous" fluctuations of the responses of interest (spontaneous is used here in the sense that the responses are not contingent on the experimental protocol). An experimental procedure, which has been found to reduce variability, is to carry out the studies at fixed times of day under standardized conditions of feeding and watering. Statistical analyses of the results require the application of methods that are sensitive to repeated measurements.

Extensive reviews on classical conditioning of cardiovascular responses are readily available (11, 12). The usual targets for conditioning are heart rate and blood pressure; however, there are number of studies involving cardiac contractility (13), cardiac conductivity (14), and responses in different vascular beds (15–17).

## Operant Conditioning

In this section we consider training procedures that are designed to produce specific cardiovascular responses; that is, the cardiovascular response in question determines the occurrence of the reward or punishment. However, before describing the methods one can use to train subjects to emit specific cardiovascular responses, it might first be useful to consider some reasons why one would want to do so. Obviously, one reason is technical: if one were studying some particular aspect of the circulation, it would be useful to be able to produce that response, consistently and reliably. Students of the pulmonary system will certainly recognize that training for this purpose was common practice in pulmonary physiology for many years, and that much of the information about a variety of pulmonary reflexes in humans was obtained from the study of a few, highly trained subjects.

A second reason for wanting to train subjects to emit reliable cardiovascular responses is that it would permit one to investigate complex control systems under relatively natural conditions. It would seem very useful to have available subjects who could systematically modulate one or another cardiovascular response, so that one could study either the expression of such a response under specific control conditions, for example, the effect of voluntary lowering of heart rate during exercise under conditions of sympathetic blockade (18), or the impact of such voluntary control on reflex regulation of the circulation, for example, the effect of voluntarily induced increases and decreases of heart rate on baroreflex sensitivity (19). Finally, a third reason why one would want to be able to train subjects to modulate specific cardiovascular responses is that it might have therapeutic value (20–22).

The procedure that has been utilized to train a cardiovascular response is operant conditioning. The subject is introduced to an experimental setting, which is maintained under well-controlled conditions to rule out the possibility that fortuitous stimuli will *elicit* the response of interest. After some control period, a cue is enabled to signal the subject that the contingent stimulus is operative, and that the subject must *emit* the appropriate response to obtain the desired consequence. A typical example of this would be a study in which a monkey is trained to decrease or increase its heart rate in order to avoid an electrical shock to its tail (23). However, there are features of this training procedure that need to be elucidated further.

If the training procedure outlined above was initiated with a formal protocol, for example, if the monkey was required to slow its heart rate below resting level and was shocked every time it failed to perform successfully, the animal would receive a great many shocks, in part because it would be unlikely to emit the correct response and in part because the reflex reaction to tail shock would probably be a sustained tachycardia. Thus, while the animal might eventually learn the appropriate response, it would take a considerable amount of time to do so, and it would receive a great deal of punishment in the process. Because the learned response is the goal of the research, it is neither efficient nor humane to follow such a rigorous protocol from the outset. Rather, the experimenter should use what is called a *shaping* procedure, namely, a progressive technique that is designed to form or shape the behavior.

In the example cited above, the animal was initially trained to lower its heart rate by negatively reinforcing any lowering of heart rate, irrespective of the relationship of the absolute rate to the baseline rate. That was done by recording beat-to-beat variations in heart rate with a cardiotachometer, inhibiting the punisher whenever the heart rate moved in the correct direction, and restoring the punisher whenever the heart rate moved in the incorrect direction. As the animal's behavior, namely, its heart rate, became more appropriate, the criterion of successful performance was modified to include not only a change in the appropriate direction, but also a heart rate level that was nearer to the baseline level. In the final stage of shaping, when the animal could reliably slow its heart rate but could not consistently bring it to criterion within the experimentally desirable period (8 sec), the period was set for a longer interval (e.g., 16 sec) and then progressively reduced to the criterion interval. After about 40 shaping sessions, the animal was able to perform consistently and reliably: it lowered and maintained its heart rate below the baseline rate for 30 min. If it failed to perform successfully, it received a tail shock. Therefore, the animal would lower its heart rate well within the 8-sec interval, thus avoiding subsequent shocks (23).

There have been a large number of studies reported with human subjects in which the subjects were successfully trained to regulate heart rate, blood pressure, skin blood flow, or skin temperature. These studies, which have been generally grouped together under the rubric biofeedback, have been reported in a number of books and journals (24). There have been considerably fewer studies using animal models. However, there is clear evidence that monkeys can learn to modify heart rate (23), dogs can be operantly conditioned to modify coronary blood flow (25), and baboons can be trained to modify blood pressure (26).

There is one last technical issue to consider, namely, the relative value of negative and positive reinforcement in training. All of the examples cited above used negative reinforcement training procedures. That is no accident. In general, one finds that negative reinforcement produces much more reliable behavior than does positive reinforcement. In experimental studies, this translates into a smaller error variance, fewer tests, and fewer subjects. This is also true in the clinical setting. Most physicians are aware that one obtains better compliance from patients who have discomforting symptoms than from patients whose diseases are "silent." It is also well-known that compliance with a therapeutic regimen will be generally superior to compliance with a prophylactic regimen, that is, it is much easier to convince someone to "escape" from the consequences of an exacerbation of illness than it is to convince someone that abstinence will increase his/her well-being.

## Behavioral Models for the Study of Cardiovascular Pathology

The preceding sections outlined the formal procedures for studying how classical or operant conditioning procedures can be used to enable one to study how the nervous system modulates specific cardiovascular responses. This section describes experimental models for the investigation of cardiovascular regulation and the development of cardiovascular dysfunctions.

Objective behavioral analysis is founded on the presumption that activity of the individual is not initiated by "free will," but is regulated by the environmental consequences it generates. The individual responds to stimuli (cues) in the environment, and these responses result in the occurrence of other stimuli (rewards or punishers), which differentially affect those responses and increase (or decrease) the probability of their future occurrence. The emphasis on environmental feedback and the characterization of the cardiovascular responses as behavior differentiate this approach from that of traditional physiology in which behavior is conceptualized in terms of its form or topography. For example, exercise physiologists frequently criticize the concept of central command because they argue that there is no feedback signal by which cardiovascular responses can be modulated. However, in their analysis of "feedback," they ignore the environmental contingencies that initiate and maintain exercise. Many investigators believe that chronic cardiovascular disorders may evolve, in part, from habitual behavioral interactions whose consequences are more important than their form. A behavioral analysis of short-term and long-term cardiovascular regulation requires a set of concepts within which to analyze interactions of the individual with the environment.

Behavioral analyses are characterized by longitudinal observations of an individual in a controlled environment in which an independent variable (simple or complex) is systematically varied, and the results are described in terms of functional relationships between independent (environmental) and dependent (performance) variables. The fundamental distinction, which is relevant to behavioral studies of cardiovascular regulation, is between experimental procedures that strengthen or maintain performance and those that inhibit or suppress performance. Although most experimental studies have relied on aversive stimulation—in part because these stimuli usually elicit large changes in the circulation and in part (as noted earlier) because the stimuli produce reliable, stable performance—it should not be assumed that only noxious stimuli are relevant to the development of cardiovascular disorders. For example, laboratory animals will engage in operant behavior that provides access to ethyl alcohol, and the long-term effects can include chronic hypertension (27); also, they will engage in aggressive behavior characterizing the Type A personality, thought by some to increase the risk of heart attack (28), which may have been reinforced by the submissive behavior of others that it produced.

Two kinds of avoidance conditioning procedures have been widely used as behavioral models for the development of cardiovascular disease. Discriminated avoidance schedules elicit behavior in response to an explicit warning signal. Free operant avoidance schedules present no warning signal to the subject; instead, the passage of time since the previous aversive stimulus or the previous response (29) is the only cue. Discriminative avoidance schedules generate rates of responding that correspond to the frequency with which the warning signal is presented. Free operant avoidance generates steady rates of responding, which are a function of the intensity of the aversive stimulus even though the stimulus itself rarely occurs since the subject's behavior prevents its occurrence. Free operant avoidance schedules are experimentally very useful since they can maintain performance over considerable time in the apparent absence of an explicit stimulus. Direct observations of animals emitting well-learned avoidance behaviors reveal that performance is maintained with no observable emotional effects. Nevertheless, one can see that the cardiovascular effects of activation of the sympathetic nervous system are sustained over very long periods (30). Avoidance conditioning schedules are capable of potentiating the development of hypertension in animals on a high sodium diet (31). However, an understanding of the effective mechanisms requires attention to physiological adaptations between avoidance sessions, detailed later in this section.

Experimental procedures that inhibit or suppress performance may be even more useful for analyzing long-term cardiovascular regulation. When a response is punished, the result is a rapid suppression which persists as long as the punishment schedule remains in effect. When a stimulus that normally produces escape and avoidance is made contingent on a response which is concurrently maintained by a schedule of positive reinforcement, that response is variably suppressed, depending on deprivation and stimulus intensity parameters. Application of such conflict procedures have been shown to exacerbate the elevated blood pressure produced in spontaneously hypertensive rats (32) and in baboons that were also sodium loaded (33). When a stimulus that normally produces escape or avoidance is presented independently of the response, which is maintained by avoidance, an increase in response rate is often observed (34). Thus, the same aversive stimulus can increase or decrease the same performance, depending on other contingencies.

Behavior that is maintained by positive reinforcement may also be conditionally suppressed by presentation of a warning signal preceding an unavoidable aversive stimulus (35). This phenomenon of conditioned suppression may be particularly relevant to behavior of subdominant animals in social groups. For example, it has been shown that intermittent introduction of newcomers into monkey colonies temporarily upsets dominance hierarchies and increases vulnerability to the atherogenic effects of high cholesterol diets (36). Behavioral analyses are particularly relevant in such studies since the place of the individual in the dominance hierarchy determines not only its social behavior but also neuroendocrine factors that may be implicated in atherogenesis (37).

Sudden termination of discriminative stimuli, which prompts positively reinforced responding, might be an important determinant of cardiovascular disease. For example, experimental studies have shown that isolation of rodents from normal social interaction is sufficient to produce sustained experimental hypertension within periods of weeks (38). This mechanism may also be relevant to the cardiovascular response which develops progressively in well-trained animals immediately preceding sessions of free operant avoidance behavior. Under these conditions, the subject must inhibit responses that could reduce the likelihood that it will recognize and respond to the session onset. During such preavoidance periods, blood pressures of well-trained animals increase progressively due to peripheral vasoconstriction, while heart rate and breathing frequency decrease (39). During the avoidance sessions themselves, the rise in blood pressure is mediated by a rise in cardiac output; peripheral resistance actually falls. When avoidance sessions are scheduled at regular intervals, alternating patterns of cardiovascular responses associated with preavoidance and avoidance periods occur (40). Under conditions of high sodium intake, such schedules predispose dogs to sodium retention and the development of sustained hypertension (31).

# References

1. B. T. Engel and N. Schneiderman, *Annu. Rev. Physiol.* **46,** 199 (1984).
2. B. T. Engel, *Behav. Brain Sci.* **9,** 285 (1986).
3. B. A. Campbell and A. M. Church, "Punishment and Aversive Behavior." Appleton-Century-Crofts, New York, 1969.
4. H. D. Kimmel, *Pavlovian J. Biol. Sci.* **11,** 56 (1976).
5. R. A. Rescorla, *Am. Psychol.* **42,** 119 (1987).
6. R. A. Dykman, *Prog. Exp. Pers. Res.* **2,** 229 (1965).
7. G. Razran *Psychol. Rev.* **68,** 81 (1961).
8. J. A. Stern, *in* "Handbook of Psychophysiology" (N. Greenfield and R. Sternbach, eds.), p. 197. Holt, Rinehart, and Winston, New York, 1972.
9. T. W. Jarrell, C. G. Gentile, P. M. McCabe, and N. Schneiderman, *Brain Res.* **381,** 251 (1986).

10. D. C. Randall, T. L. Skinner, and G. E. Billman, *J. Auton. Nerv. Syst.* **13,** 125 (1985).
11. R. A. Galosy, L. K. Clarke, M. R. Vasko, and I. L. Crawford, *Neurosci. Biobehav. Rev.* **5,** 137 (1981).
12. D. H. Cohen and D. C. Randall, *Annu. Rev. Physiol.* **46,** 187 (1984).
13. D. C. Randall, G. E. Billman, and T. L. Skinner, *Physiologist* **25,** 263 (1982).
14. T. L. Skinner and D. C. Randall, *J. Auton. Nerv. Syst.* **12,** 23 (1985).
15. T. L. Billman and D. C. Randall, *Pavlovian J. Biol. Sci.* **15,** 93 (1980).
16. O. A. Smith, A. R. Hohimer, C. A. Astley, and J. D. Taylor, *Am. J. Physiol.* **236,** R198 (1979).
17. D. C. Randall, C. M. Conttrill, E. P. Todd, M. A. Price, and C. C. Wachtel, *Psychophysiology* **19,** 490 (1982).
18. B. T. Engel and M. I. Talan, *Physiol. Behav.* **49,** 373 (1991).
19. B. T. Engel and J. A. Joseph, *Psychophysiology* **19,** 609 (1982).
20. M. S. Glasgow, B. T. Engel, and B. C. D'Lugoff, *Psychosom. Med.* **51,** 10 (1989).
21. B. T. Engel and E. R. Bleecker, *in* "Cardiovascular Psychophysiology: Current Issues in Response Mechanism, Biofeedback and Methodology" (P. Obrist, A. Black, J. Brener, and L. DiCara eds.), p. 75. Aldine, Chicago, 1974.
22. R. R. Freedman, P. Ianni, and P. Wenig, *J. Consult. Clin. Psychol.* **51,** 539 (1985).
23. B. T. Engel and S. H. Gottlieb, *J. Comp. Physiol. Psychol.* **73,** 217 (1970).
24. J. P. Hatch, J. G. Fisher, and J. D. Rugh (eds.), "Biofeedback: Studies in Clinical Efficacy." Plenum, New York, 1987.
25. F. A. Ernst, R. K. Kordenat, M. S. Sandman, and C. A. Sandman, *Psychosom. Med.* **41,** (1979).
26. A. H. Harris and J. S. Turkkan, *Biofeedback Self-regulation* **6,** 11 (1981).
27. T. C. K. Chan and M. C. Sutter, *Life Sci.* **23,** 1965 (1983).
28. R. H. Rosenman, *in* "Biological and Psychological Factors in Cardiovascular Disease" (T. Schmidt, T. Dembroski, and G. Blumchen, eds.), p. 15. Springer-Verlag, London, 1986.
29. M. Sidman, *Science* **118,** 157 (1953).
30. R. P. Forsyth, *Science* **173,** 546 (1971).
31. D. E. Anderson, W. D. Kearns, and W. E. Better, *Hypertension* **5,** 286 (1983).
32. J. E. Lawler and R. H. Cox, *Pavlovian J. Biol. Sci.* **20,** 101 (1985).
33. J. A. Turkkan and D. S. Goldstein, *J. Hypertens.* **9,** 969 (1991).
34. M. Sidman, *Science* **132,** 61 (1960).
35. W. K. Estes and B. F. Skinner, *J. Exp. Psychol.* **29,** 390 (1941).
36. J. R. Kaplan, S. B. Manuck, T. B. Clarkson, and R. W. Prichard, *Adv. Behav. Med.* **1,** 115 (1984).
37. D. L. Ely and J. P. Henry, *Horm. Behav.* **10,** 156 (1978).
38. S. M. Gardiner and T. Bennett, *Med. Biol.* **55,** 325 (1977).
39. D. E. Anderson and J. G. Tosheff, *J. Appl Physiol.* **35,** 650 (1973).
40. D. E. Anderson *in* "Handbook of Hypertension, Volume 10: Behavioral Factors in Hypertension" (S. Julius and D. Bassett, eds.), p. 226. Elsevier, Amsterdam, 1987.

# [27] Assessment of Brain Electrical Activity in Relation to Memory and Complex Behavior

T. R. Vidyasagar

## Introduction

To understand the neuronal basis of higher brain functions, it is necessary to record the electrical activity of nerve cells in the appropriate part of the brain. Complex processes such as memory, attention, and motivation can be investigated at a functional level only by recording such activity in the conscious animal. Special procedures have been developed over the years to record single cell activity from the brains of awake animals without causing them detectable pain or discomfort. This chapter describes these methods, with particular reference to certain complex behaviors, such as memory tasks. Even though there is an appreciable amount of relevant literature on rats, the chapter focuses on monkeys. This choice not only provides a coherent and concise methodological review on one species, but also highlights the fact that the monkey is an easier animal to train on a number of complex behavioral tasks.

This chapter first provides a brief overview of the method of single cell recording from conscious monkeys before going on to describe the various experimental paradigms that have been applied in the study of memory and of some selected behavioral patterns at the neuronal level. The review is by no means exhaustive and is intended to act as a concise practical guide for the study of the neural basis of behavior through electrophysiological techniques.

## Techniques of Single Cell Recording from Awake Monkeys

Toward the end of the 1950s, the first recordings were made from the cortex of awake cats (1) and monkeys (2). The enthusiasm that these studies created was sustained by the extensive investigations and methodological improvements by Evarts at the National Institutes of Health during the 1960s (3–5). Evarts' work was acknowledged as a major advance, and the techniques have subsequently been applied by a number of laboratories around the world with certain minor but specific modifications. It is beyond the scope of this chapter to review all of these methods, and the reader is best advised to refer to the comprehensive review by Lemon (6), which sets out in great detail the many technical considerations in recording from conscious animals. Only the basic procedures are outlined here.

The technique involves several steps. First, under general anesthesia and aseptic surgical conditions, the skull is exposed. Metal devices for head restraint are then permanently implanted on the skull. A trephine opening is made over the area of interest in the brain, and a chamber is implanted over it. This chamber is closed with a screwable cap. When the animal has recovered from the operation, recordings are done for a couple of hours per day with an electrode on a lightweight microdrive which is mounted on the chamber. Some points related to the method are as follow.

## Stability

An important consideration in recording single cell activity in the awake animal is achieving the required mechanical stability and minimizing the movement between electrode tip and the neuron. The steps that are generally undertaken to this end are mechanical restraint of the animal, particularly its head, and the use of a stable microdrive that is mounted on the skull.

The monkey is usually seated in a primate chair, which it has been gradually made to accept over a period of time ("shaping"). Once the animal has learned to associate the chair with food, there is little difficulty for it in accepting the restraint of sitting in the chair. The operation on the skull to attach the artifacts necessary for restraining the head is usually done after a certain amount of training and sometimes even overtraining. Once the monkey becomes accustomed to the chair and performs various tasks to obtain a reward, it will also accept the head restraint after the operation. Apart from providing the stability for single cell recording, such restraint also helps to maintain the animal's head in a constant relation to the environment, especially to sensory stimuli produced by a visual monitor or projected on a screen and to task-related artifacts like the response lever. It also allows the experimenter to manipulate the spatial relation of the animal to the environment, say, by rotating the chair itself to see whether a response is specific to the direction of gaze or is object-centered.

The major factor in bringing stability to recordings from conscious monkeys has been the use of a micromanipulator, which is fixed in its relation to the skull. The micromanipulator is usually mounted on the implanted chamber at the time of recording and removed at the end of the recording session. The most commonly used types are either hydraulically driven (4) or driven by a stepping motor (7), both of which are remotely controlled. In the author's laboratory we have used the hydraulic drive with a miniature

X–Y table (David Kopf, Trent-Wells, Tujunga, CA) between the recording chamber and the microdrive, and at times we have used an additional special adapter between the chamber and the X–Y table to access structures not directly beneath the craniotomy. The wall of this adapter is angled to allow the electrode to penetrate the dura at a selected angle. The direction could be further altered by simply rotating the adapter. The weight of the entire setup is light enough to be mounted on the monkey's skull during the recording session. An alternative to the miniature X–Y table is using a system of eccentric (Starr) guides.

Figure 1 is a drawing of the apparatus used in Mountcastle's laboratory at the Johns Hopkins University for studies on the posterior parietal cortex. Apart from the electromagnetically driven stepping microdrive, the setup also features a number of salient points of the arrangement, such as the head-fixation apparatus, the reward tube, and the devices with which the monkey interacts to obtain reward. Figure 2 shows the arrangement in Lemon's laboratory at the Erasmus University of Rotterdam with a detailed illustration of the head-fixation and microdrive apparatus (8). The micromanipulator here is hydraulically driven.

Laboratories around the world that have resorted to chronic recording have introduced a number of modifications suited to meet special requirements. One of the unique designs has been developed by Porter and colleagues (8) in Australia. The training compartment is connected by a door to the home cage. The monkey soon learns to associate the training with food and readily enters the training cage, which is detached from the home cage and taken away for training and recording sessions. The restraining collar is easily fitted from outside. This method reduces handling of the monkey and facilitates transfer to the training cage with the least amount of stress to the monkey.

In Göttingen and Canberra, the author and colleagues have resorted to a relatively small head-fixation peg, whose base takes the contour of the skull. The X-ray pictures of the electrodes and guide tubes *in situ* shown in Fig. 3 give some idea of the size and shape of the fixation peg.

## Surgical Care

For ethical reasons and for efficient training as well as recording over a lengthy period (sometimes as long as 1 year or more), it is essential that the animal be kept comfortable and that all painful and stressful conditions be avoided.

In choosing implant materials it is important to avoid those likely to produce toxic reactions. Vitallium, which is an alloy of chromium, nickel, and

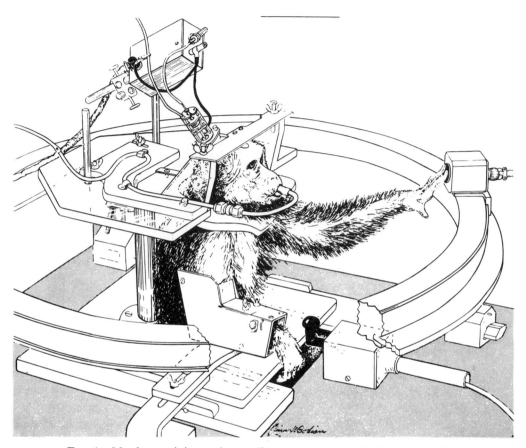

FIG. 1   Monkey training and recording apparatus used in Mountcastle's laboratory. The monkey is trained in a task to study the properties of neurons in the posterior parietal cortex. The right arm is restrained. When the target lamp mounted on the circular carriage is lighted, the animal has to press the lever with its left arm. After a variable period the light dims, and at this time the monkey has to release the lever and reach out and touch a panel next to the lamp. [From Mountcastle *et al.* (7), reproduced by permission of the American Physiological Society.]

magnesium, or stainless steel is the best material for appliances such as recording chambers and restraining devices implanted on the skull. When implanting metal electrodes into the brain, the tolerance of brain tissue to different metals should be carefully considered, especially when large currents are used for electrical stimulation (9).

All the surgical procedures are done under anesthesia with the appropriate pre- and postoperative care. Systemic administration of a broad-spectrum antibiotic is recommended in the week after the operation. It is important that the cage and the primate chair be kept scrupulously clean to minimize infections entering through the recording chamber or the tissue around the implant.

The recording chamber itself should be cleaned with sterile saline about three times a week. Routine aseptic precautions and regular cleansing are generally sufficient to keep infections down. At the first indication of infection with the dural surface appearing purulent, it is wise to add a few drops of an antibiotic (e.g., chloramphenicol) to the recording chamber at least three times a week after thorough cleansing. If the infection is persistent, even if mild, it is advisable to do a culture and sensitivity test to ensure that the appropriate antibiotic is administered. Persistent infection in the chamber can be either the cause or the result of chronic osteomyelitis around the chamber.

Bone necrosis can be a serious problem leading to loosening of the implants. This can be prevented by avoiding pressure on the bone from the nuts and bolts used for the implantation and by preserving as much of the periosteal cover over the bone as possible during the surgery.

One common problem faced by all investigators is the gradual thickening of the dura within the recording chamber. This may not be a serious problem in the case of deep recordings carried out with the aid of a guide tube, but for recordings from the surface of the brain just under the cranial opening, dural growth should be kept to an acceptable minimum. Preventing infection reduces growth of the dura considerably. Topical application of a corticosteroid is also helpful in reducing the dural thickening. Some investigators, including the author, resort to periodic scraping of the dura, say once a fortnight. If there are large blood vessels underneath the dura, it is necessary to apply a local anesthetic before scraping, because of the dense nociceptive innervation of the dura around major blood vessels. Others have used a thin stretched Silastic membrane to seal off the dura (10).

## Electrophysiology

The general principles of extracellular recording in nonrecovery or acute experiments all apply to the present situation. The most commonly used microelectrodes in chronic work are glass-coated tungsten and platinum–iridium ones, largely because of their strength. For deep recordings, we have used lacquer-coated tungsten electrodes (maximum outer diameter of about 0.35 mm and impedances between 3 and 10 M$\Omega$) inside stainless steel guide

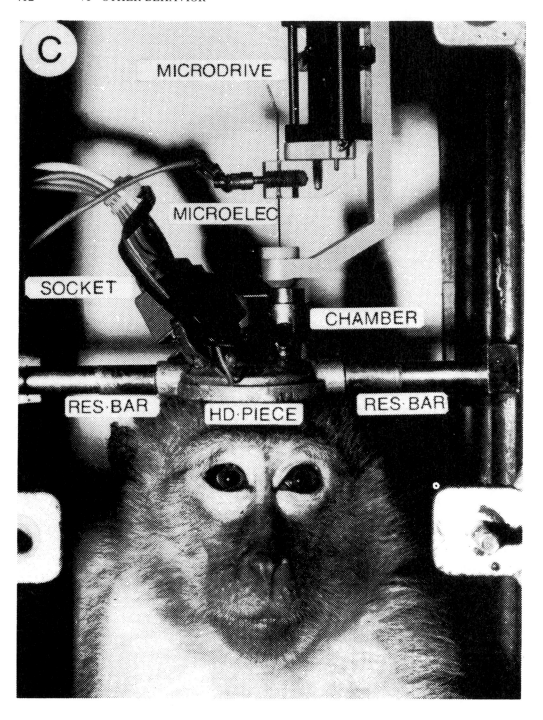

tubes. With a guide tube, there is little limitation on electrode possibilities, and glass micropipettes have even been used for intracellular recordings from the awake monkey (11). Iontophoretic experiments from deep brain regions have also been carried out using a multibarrel pipette with a long taper inside a guide tube (12).

A basic problem in chronic experiments, where repeated penetrations are done over many months, is identifying the exact location of the recorded neurons. In acute, nonrecovery experiments, electrolytic lesions can be done in each penetration and the tracks reconstructed histologically. This is obviously not possible with chronic experiments. Further, the initial dimpling created as the electrode penetrates the dura in chronic recordings makes the exact localization of cells recorded within the first couple of millimeters highly unreliable.

The localization of the general brain area is less difficult for recordings from the surface of the cortex directly beneath the craniotomy than in the case of deep recordings. Accurate notes made of stereotaxic coordinates prior to implanting, systematic records of the x–y coordinates of each penetration, and histological reconstruction of marker lesions made in the days prior to sacrifice of the animal all go a long way to achieving reliable localization of the neurons. However, the exact localization of neurons, for example, their laminar positions, remains unreliable.

For deep recordings, the best and the most common way of overcoming the problem of localization is to make X-ray pictures of the electrode *in situ* (13). The top half of Fig. 3 shows the frontal and lateral roentgenographic views of a lacquer-coated tungsten electrode used in the author's study of memory-related responses in the hippocampus and parahippocampus (14). Pictures of this type are taken with most penetrations. At the end of the series on an animal, a number of steel needles corresponding to the various angles of the penetrations over the series are inserted and left in place. They are then X-rayed (bottom half of Fig. 3), and the animal is heavily anesthetized and perfused with the needles in place. Histological reconstructions of the tracks of these needles, together with their X-ray pictures and those of the individual electrode penetrations, are studied to find the location of each recorded neuron. Relationships to permanently implanted electrodes in X

FIG. 2  Single unit recording apparatus in the conscious monkey, showing the implanted headpiece and recording chamber, the restraining bars, the hydraulic microdrive with the microelectrode, and the socket with multipin connector. The monkey is seated in a training compartment of the sort designed by Porter *et al.* (8). [From Lemon (6), reproduced by permission of John Wiley & Sons.]

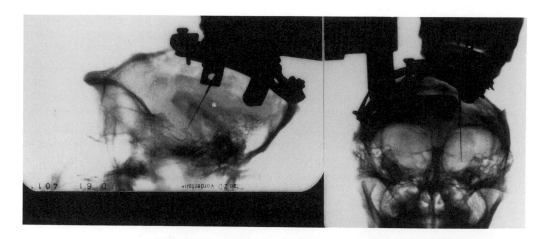

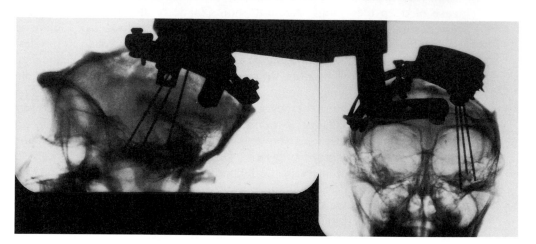

Fig. 3    Frontal and lateral X-ray pictures of the monkey's skull with a microelectrode in the top pair and three stainless steel needles in the bottom pair. Pictures like the top pair were made for each penetration at the end of a day's session. At the end of all experiments in a monkey and prior to perfusion, steel needles were inserted at the different angles of the penetrations and X-ray records made like those in the bottom pair.

rays and any electrolytic microlesions made during the last few penetrations are further aids in localization. In our study, this method gave us enough accuracy to localize most recorded neurons with confidence. However, we believe the precision cannot be more than a millimeter or two and that the laminar localization of neurons is relatively unreliable. Thus, in our study we were able to say with certainty that a neuron was in the hippocampus, but not whether it was from the CA1 region, CA3, or the dentate gyrus.

One useful guide for localization during deep recordings is when the electrode passes through structures whose identity is easily recognizable. For example, should the electrode pass through the lateral geniculate nucleus (LGN), it can be identified by the typical responses of the neurons to visual stimuli. Further, since the LGN has a clear topographical representation of the visual field, the receptive field position of the neurons in the track can be referred to topographical maps, and, using a stereotaxic atlas, the penetration can be taken as a reference point for future penetrations. Another notable example is the high maintained discharge (70 to 80 spikes/sec) of many neurons of the globus pallidus.

When using stimulating electrodes, it should be borne in mind that repeated pulses of even relatively small currents can cause local lesions (15). In certain regions, electrical stimulation can also lead to pain and distress.

The usual measures to eliminate electrical noise in nonrecovery, acute experiments all apply here. However, some additional factors such as muscular movements can contribute to noise in chronic experiments. To avoid picking up electromyographic signals, it is best to attach the indifferent electrode to a stainless steel screw driven into the skull or to one of the screws on the chamber if it can be done without affecting stability. It is important not to leave any loose electrical connection on the implants, since they can cause considerable noise during movements of the animal. It is best to leave the guide tube isolated from the ground, since there is often a capacitance between it and the electrode if it is grounded. On the other hand, the mouthpiece, through which the monkey obtains its reward, should be properly grounded to avoid lick artifacts. The exception is when the lick itself is used as the response to close a circuit to provide the reinforcement (16).

## General Training Strategies

In training monkeys to reach high performance levels in complex tasks, it is important to have an animal that is healthy, comfortable, and devoid of anxiety. Time is well spent at the outset in permitting the monkey to become accustomed to the laboratory environment and experimenters. The monkey has to be periodically weighed, and the total amount of food per week is kept at a level that avoids any loss of weight but does not lead to an excessive increase of weight above that expected from normal growth.

Partial food and water deprivation have both been used by investigators to achieve and maintain high levels of performance. It is the author's experience that only mild degrees of food deprivation are necessary to train cynomolgus monkeys in various visual discrimination and memory tasks. In the days before training (i.e., Sunday to Thursday), the monkey should get approximately 70 to 90% of the daily quota, and in the remaining days an extra amount should be added to the daily quota to bring the weekly intake to the required level. We found that an appropriate choice as reward—like orange juice, black currant juice, or even drinking chocolate—can greatly aid the speed of learning. This choice can differ from monkey to monkey. In most cases, our monkeys have unlimited access to water in their cages. This does not too seriously impair their motivation, although we have sometimes resorted to turning off the water in the cage a couple of hours before the training is due to begin. The usual daily quota of food is fed to the monkey after the period of training. This helps greatly in keeping the animal's motivation for doing the task high. In such a regimen, where there is tight control of food intake during the week, the monitoring has to extend into the weekend as well, since sudden access to large amounts of food and water at the end of the week can lead to excessive consumption. Occasionally this overindulgence leads to a "bloat syndrome," with much abdominal gas accumulation, that can end fatally (17). The risk of bloat can be greatly reduced by the administration of antiflatulent preparations containing aluminum hydroxide and simethicone, particularly when the risk is high toward the end of the week.

It is generally possible to train monkeys largely on rewards with little in the way of punishment. We have only used increased intertrial intervals as negative reinforcement for incorrect responses. One mild negative reinforcement for incorrect responses is the use of hypertonic saline, which is a sufficiently aversive stimulus to act as a deterrent (16).

Advances through the various stages of the task have to be made gradually as per the usual rules of operant conditioning. If big leaps are attempted, the monkey can be pushed into a "failure set." We have used a criterion of 90% correct performance at each level before advancing to the next stage in order to maintain reinforcement of the desired behavior.

## Specific Experimental Procedures

The behavioral tradition of investigators such as Pavlov, Thorndike, Skinner, and Konorsky has been to investigate higher brain functions with an impressive array of tests. Single cell activity has been studied during the performance of some of these tasks, especially those involving instrumental (operant) conditioning. This section describes some of the procedures commonly used in primate experiments, in particular those related to memory processes. Some procedures that are necessary as controls are also presented below. For example, a response that occurs during a memory task need not be specific to the task, but may be due to any of a number of factors unrelated to the mnemonic component of the task, like attention, motivation, expectation or consumption of reward, the behavioral response, or the processing of the sensory stimulus. Some of these are described, not so much to provide a practical list, but more to give a flavor of the approach. The section mostly focuses on tasks related to the visual system, but the principles can usually be generalized to other modalities.

## Visual Fixation Task

When performing a memory task using a particular sensory modality, say, vision, it is often important to rule out pure sensory responses that are not memory-dependent. It is useful to have the monkey trained on a fixation task, during which the visual field can be explored and any response to visual stimuli characterized. We have found a significant number of neurons in the hippocampus and parahippocampus responding during the presentation of the stimuli in a visual memory task (14). In these cases, we have attempted to plot circumscribed receptive fields or elicit responses to complex visual stimuli as the monkey was fixating on a spot. Such controls are essential to tell whether a response to a sensory stimulus is more than just a response to the physical dimensions of a stimulus. A neuron whose response to a visual stimulus is the same regardless of whether the stimulus appears as an integral part of a memory task or as a stimulus to which the monkey pays no attention cannot be termed as performing a specific mnemonic function.

The fixation task is usually performed in the following way. The monkey has to keep a lever pressed until the fixation point (not more than 2 mm in diameter) suddenly dims. The monkey is then required to release the lever within a certain response time (say, 600 msec) to obtain the reward of a few drops of fruit juice. The next trial is begun by the monkey pressing the lever again. A trained and well-motivated monkey leaves practically little intertrial interval; he presses the lever again as rapidly as possible after releasing it to obtain reward. Thus a fixation spot is almost continuously present on the screen, and the monkey's attention is focused on it.

## *Visual Discrimination Task*

When neurons respond during a typical memory task such as delayed match to sample (see below), it is necessary to test unit activity in a discrimination task to exclude responses that are due to nonmnemonic factors like reward, expectation, or the motor response. Thus, monkeys are trained to discriminate between two or more visual stimuli; they are expected to respond by pressing a lever to one set of pictures (GO stimuli) and by withholding the response when any of the other pictures (NO-GO stimuli) appear. The test can also be done with just two stimuli, one as GO and the other as NO-GO. For correct hits, the monkey is rewarded with fruit juice, and for wrong hits and misses, the next presentation is delayed. For correct rejection, the next picture appears after a relatively short intertrial interval. The use of this task as a control is illustrated in the next section dealing with a memory task.

## *Delayed Match to Sample Task*

The delayed match to sample (DMS) task is a true test of recognition memory, where the animal has to retain in short-term storage the memory of an object it has seen and then compare it with another object to decide whether the two are the same. This is also a specific test for human amnesics (18). The characteristic anterograde amnesia that occurs after bilateral damage to the medial temporal lobe (19–21) can be revealed by a test of delayed match to sample. Similar lesions in monkeys have also been shown to be sensitive to the same test (22).

In the classic versions of the DMS task (23–25), an object, the "sample," is shown, removed, and after a "delay" presented together with a second object. The task is to recognize the object seen earlier ("match"). Monkeys should be trained on the delayed matching task gradually. The procedure we apply (14), which is not quite similar to those used by other investigators, is the following. The animals are first trained on a visual dicrimination task, so they could, for example, discriminate between two like and two unlike pictures. Their response is to press a lever when two like pictures appear on the screen; when the pictures are different, the monkey has to restrain from pressing the lever. The two members of each pair are projected from two different projectors. The delay between the two pictures is then gradually increased until the pictures (which are members of a like or an unlike pair) appear one after another with delays up to 20 sec (Fig. 4). The standard sequence we use consists of 20 pairs of pictures, with the individual slides so distributed between the projectors to avoid any bias that can give unwanted cues to the animal. The pictures in this experiment (14) are not trial unique,

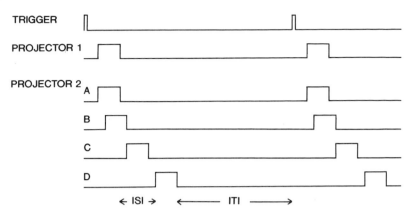

FIG. 4 Steps involved in training a monkey in a delayed match to sample (DMS) task. Two projectors were used to present pictures on a screen. When presented simultaneously (projector 1 + projector 2, A) they were either two like or two unlike pictures appearing side by side on the screen. The monkey, which had already been trained on a visual discrimination task, was now trained to identify the like pictures as GO and the unlike pair as NO-GO. Then a delay was introduced between the two pictures (projector 1 + projector 2, B), so that the onset of the two members of the pair was staggered. This delay was slowly increased until the image from the second projector appeared just after the image from the first projector was turned off (C) and eventually (D) after an interstimulus interval (ISI). The intertrial interval (ITI) was relatively long, so that the monkey did not get confused between the ISI and ITI. The trigger, usually an acoustic tone, is presented before the first stimulus and is a good reference point for the monkey to know when the trial begins.

but are familiar ones. Thus the task tests relative familiarity that involves a temporal discrimination in the remembrance of known stimuli. This element makes the task more difficult to learn than the classic DMS task that uses only trial unique stimuli.

Figure 5 shows the responses of a temporal lobe neuron. The responses on the left-hand side of Fig. 5 are for all the trials superposed. On the right-hand side of Fig. 5 the response is broken down to the four categories, based on the reaction of the monkey. For a well-trained monkey, most trials fall into the hits or the correct rejection category. Note the difference between the response to the first and second pictures for the hit trials. Even though the pictures are the same, the response to the second picture is suppressed. However, this does not provide conclusive evidence that the neuron is involved in the mnemonic part of the memory task, say, in "distinguishing" that the second picture is the same as the first. The action of the neuron could be related to the expectation of reward, to the reward itself, or even

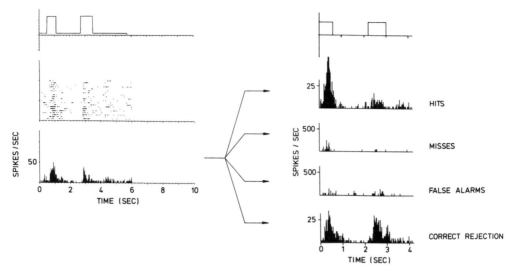

FIG. 5   Responses of a neuron in the temporal lobe during a delayed match to sample task. In the peristimulus time histogram (PSTH) at left, responses to 20 trials are superimposed. The dot display shows the spike activity for each trial. About one-half the trials presented like pictures and the other half unlike pictures. The PSTH is broken down to the four possible stimulus–response conditions on the right. Since different numbers of trials make up each histogram, depending on the number of correct and incorrect responses, the scale of the $y$ axis differs from one histogram to another. [From Vidyasagar *et al.* (14), adapted by permission of Elsevier Science Publishers.]

to the motor movements the monkey is making to obtain the reward. Controls need to be done to exclude these factors before assigning a possible mnemonic role for the neuron. One way of doing this is to test the neuron's response in a visual discrimination task. If a similar response occurs also in the visual discrimination task, the response in the DMS task cannot be attributed to a mnemonic factor.

This is illustrated in Fig. 6. Two neurons are recorded, both showing a response to the second stimulus for the GO trials in the DMS task. The neuron at left shows a specific memory-related response, since in a visual discrimination task no response was seen for the GO stimuli that elicited the lever-pressing response to obtain reward. On the other hand, the cell at right exhibits the response also for the GO trials in the discrimination task. In the case of this cell, its response in the DMS task cannot be termed as a response specific to the memory task. We have tested for nonmnemonic factors yield-

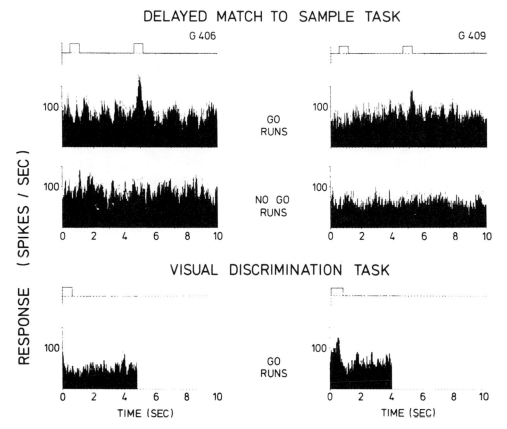

FIG. 6  Two neurons (possibly in the globus pallidus) that showed a response after the presentation of the stimulus in GO (hit) trials of the DMS task. Responses of the two neurons in a visual discrimination task are shown below. GO runs in the DMS task refer to the trials of hits. NO-GO runs are the correct rejection trials.

ing responses in the DMS task also by studying the response of a neuron under less formalized clinical conditions (see below).

## Delayed Nonmatch to Sample

The delayed nonmatch to sample is a variant of the DMS task, where the monkey has to respond if the two stimuli do not match (23). Since normal monkeys spontaneously prefer new objects, this task seems more dependent on short-term memory and less on long-term associative learning than the classic DMS task. Hence, the monkeys appear to learn this version more easily, but this task is sensitive to lesions of the temporal lobe in the same way as the classic DMS task (26).

## Serial Visual Recognition Task

The serial visual recognition task is another version of the DMS task that tests recognition memory, developed by Gaffan (27), and it is similar to a serial recognition task found to be impaired in human amnesics (28). The test has been widely used in electrophysiological experiments by Rolls *et al.* in Oxford (29). The monkey is shown each of several hundred visual stimuli twice per day. When a stimulus appears for the first time (novel), the animal is trained to withhold licking a tube to avoid aversive hypertonic saline. When it is shown for the second time (familiar), the monkey could lick the tube to obtain reward. Thus, in this task the monkey responds to whether the picture is novel or familiar. However, since a large number of pictures were being used, some familiar objects from one day were used as novel objects the next day. Thus in a strict sense, the task was testing relative recency.

## Delayed Response Task

Neuronal activity related to spatial memory tasks in the primate have been studied by measuring unit responses during a delayed response task in the prefrontal cortex (30, 31), cingulate cortex (32), and hippocampus (33). In all of these areas, many units exhibited responses during one or more phases of the delayed response task, which involved the monkey's remembering the place where a cue appeared and making a motor response to that place after a delay period. In this task, the monkey has to first press a lever and hold it for a few seconds until a cue light appears briefly at one of two choice keys (left or right). A delay period is then introduced, after which the monkey has to press the cued choice key to obtain reward.

When units show responses in this task, it is again essential to do the necessary controls to ensure that the response is not related to nonspatial or nonmnemonic factors such as reward.

## Object-Place Memory Task

In the object-place memory task, the monkey must remember both the object and the position in space where it appeared. Such a memory task is shown to be particularly impaired if hippocampal damage has occurred in the monkey

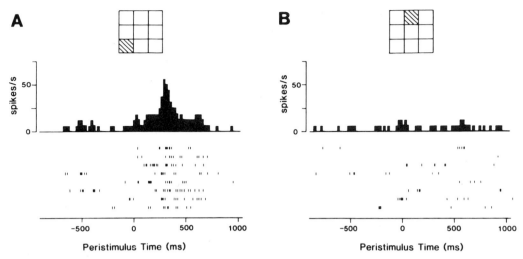

FIG. 7  Rastergrams and peristimulus time histograms of a hippocampal neuron during a 3 × 3 version of the serial object-place memory task. [From Rolls *et al.* (37), reproduced by permission of the Journal of Neuroscience.]

(34–36). Rolls *et al.* (37) investigated the responses of monkey hippocampal neurons during a serial multiple object-place memory task. In each trial, a stimulus was shown in 1 of 4 or in 1 of 9 positions on a video screen. Each stimulus was shown in any one position twice during a day's session. When a stimulus was shown the first time, the monkey had to avoid licking a tube; licking the tube led to a few drops of aversive hypertonic saline being administered. When the stimulus appeared for the second time at the same position on the screen, the animal could lick the tube to obtain reward. Figure 7 shows the response of a hippocampal neuron in this task. The cell fired more vigorously when an object was presented in the bottom left corner than anywhere else on the monitor. Nine percent of the neurons in the hippocampus and parahippocampus had spatial fields; they responded whenever there was a stimulus in some but not in other positions on the screen. Two to four percent of the neurons responded to a combination of spatial information and information about the object seen.

## Conditional Spatial Response Task

The conditional spatial response task requires the monkey to make one spatial response to one stimulus and a different spatial response to another stimulus, and it reveals impairment after hippocampal lesions (38, 39). In this task, when one stimulus is shown, the monkey has to press a panel three times to gain reward, and when a different stimulus is shown, it has to withhold pressing the panel for 3 sec to obtain reward. A new pair of stimuli is used every day, and the monkey has to learn the correct responses daily. This kind of learning is slowed after hippocampal lesions (39), and many single cells in the hippocampus of the normal monkey respond to combinations of a visual stimulus and the associated spatial response (40).

## Clinical Testing

Clinical testing refers to recording neuronal responses in an informal situation that does not involve a specific training paradigm. Even though such testing may be difficult to document and may appear anecdotal when presented, it has a definite place in the study of higher brain functions. Responses in a clinical setting can sometimes be used as a control to exclude various nonspecific factors contributing to responses in a specific trained behavioral paradigm. It is also particularly valuable to explore neural responses in untrained situations when investigating a new brain region. This can be done even before training an animal [e.g., the experiments on the parietal lobe by Hyvaerinen and Poranen (41) and Leinonen *et al.* (42)], and specific paradigms can be designed on the basis of the responses.

The value of clinical testing can be exemplified by the following account of our experiments (14). When we were studying the responses of hippocampal and parahippocampal neurons in a memory (DMS) task, we were often confronted with the finding that few neurons responded in a way that suggested that they could be involved in storing the information necessary for a memory task. Even though there were responses to the presentation of the stimulus that appeared context-dependent, there were few neurons whose response could be unequivocally related to the memory task per se. We then examined neural activity in many informal situations as the monkey was interacting with the experimenter in some way. It turned out that many units in both the hippocampus and parahippocampus gave vigorous responses under these conditions. Some were activated when the animal paid visual attention to the object, when some reward was expected, or when some fruit juice or solid food was consumed. Many of these situations also involved a human. We attempted to formalize some of these complex situations by means of a "raisin trial." Two neurons from the parahippocampal cortex (Fig. 8A,B) responded vigorously as the monkey was shown the reward, a raisin, but little during the other phases of the trial. Neither of these two neurons exhibited an appreciable response during the DMS task, nor did they have a visual receptive field that could be plotted during the fixation task. The responses were much less when the hand was shown without the raisin, but with the fingers apposed as if a raisin could be hidden, the response was better.

Figure 8C shows responses of a sample of neurons from the temporobasal cortex during the rasin trial. All of these neurons exhibited marked responses during some phase of the raisin trial. The responses were also not related to eye movements or angle of gaze. This was tested by monitoring eye movements (see top traces in Fig. 8C) using a noninvasive infrared system. Such monitoring of eye movements is often essential in chronic monkey experiments on higher cognitive functions. There have been many different methods used for this purpose. The most widely used and the most accurate (less than 1 min of arc) is a magnetic search coil system (43, 44). However, this technique involves surgical implantation of a coil on the sclera and careful calibration. We have used an infrared system developed by Bach *et al.* (45) that is noninvasive and is easy to mount and use. The accuracy with this method is around 5 min of arc.

Clinical testing often helped us to identify a response seen during the DMS task as one more likely to be due to nonmnemonic factors such as food reward or expectation of food rather than to the mnemonic aspect of the DMS task. The cell in Fig. 9A gave an enhanced response to the second stimulus of the match (GO) trials in a DMS task, followed by strong inhibition. However the excitation was seen also for the GO trials of a visual discrimination task, and the inhibition was simply related to the consumption of fruit juice. The inhibition seen after the second stimulus in the DMS task in the cell shown in Fig. 9B was also due to the fruit juice. One may be tempted to interpret the mild enhancement of activity seen during the latter half of the delay phase of the DMS task in Fig. 9C as due to a mnemonic component. However, this was probably related to the expectation of reward, since a similar enhancement was seen during the raisin trial as the monkey was expecting the raisin. Thus, our experiments did not provide direct support for the involvement of the medial temporal lobe in memory functions, even though at first many of the responses during memory tasks appeared to show some involvement. Clinical testing was invaluable in arriving at the conclusion that the neuronal responses in the temporobasal cortex of the monkey were more related to the internal state associated with a stimulus than to the mnemonic components of the memory task we used.

## Human Studies

In human clinics, electrodes are occasionally implanted in the brain, or recordings are done during neurosurgical operations. The commonest indication is epilepsy that is resistant to all medication, and surgical resection of the epileptogenic focus is undertaken after its identification using microelec-

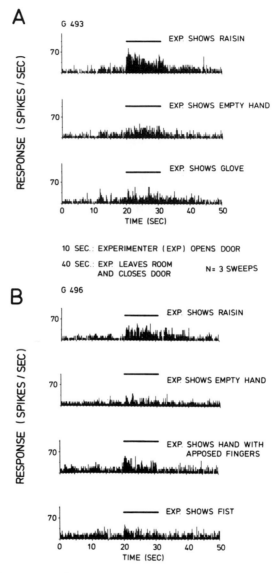

FIG. 8   (A, B) Responses of two parahippocampal neurons during the "raisin trial," along with a few controls. (C) Responses of four hippocampal (Hipp) and four parahippocampal (PH) neurons during the raisin trial, along with typical eye movement traces at the top. The different phases of the raisin trial are described at the bottom. In (A), response for showing the raisin is compared with showing an empty hand or a glove. In (B), the response is compared with different ways of exhibiting the hand. [From Vidyasagar *et al.* (14), reproduced by permission of Elsevier Science Publishers.]

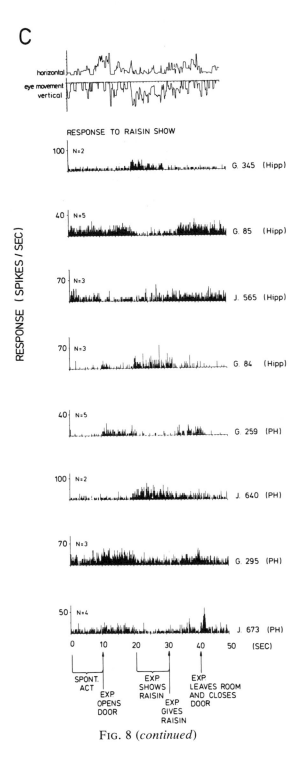

FIG. 8 (*continued*)

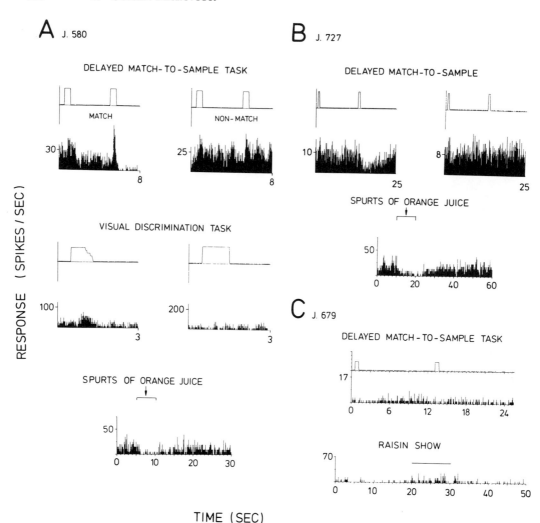

FIG. 9  Responses of three parahippocampal neurons during the DMS task and in other behavioral situations. (A, B) Match (GO) trials of the DMS task are on the left-hand side and nonmatch (NO-GO) trials are on the right-hand side. (C) Both match and nonmatch trials are superimposed. In (A), response in a visual discrimination task is also shown (GO trials at left and NO-GO at right). The change in discharge rate caused by administration of spurts of fruit juice is shown in (A) and (B). In (C), the response during the DMS task is compared with that during the raisin trial. [From Vidyasagar *et al.* (14), reproduced by permission of Elsevier Science Publishers.]

trodes. In such explorations, tests could be formulated to yield information useful both for clinical and research purposes. Studies have been done with implanted electrodes in the medial temporal lobe (MTL) of patients with temporal lobe epilepsy (46, 47). Based on clinical criteria and after informed consent, electrodes were implanted in the MTL to identify the epileptic foci. The MTL neurons in these experiments fired in a way suggesting that they contribute specific infomation to the cortex during retrieval of recent memories.

Single neurons have also been recorded during open temporal lobe surgery undertaken for the treatment of epilepsy (48, 49). The recordings were made under local anesthesia so that the patients were fully cooperative during the recording and testing. These experiments indicated that the activity of many neurons in the medial temporal lobe was usually related to the experience of a successful recognition and not to the memory match involved in the recognition task.

## Statistical Treatment of Data

Many factors are likely to introduce considerable variability in the neural responses recorded from conscious animals. It is therefore often useful to apply statistical methods to evaluate the results. To assess the significance of the changes in discharge rate in a peristimulus time histogram, Mann-Whitney $U$ tests or tests appropriate to the Poisson distribution of the spikes in the relevant bins can be compared to the discharge rate during a prestimulus period (16). Analysis of variance can be performed on the responses of a cell to different stimuli or to different phases of a particular task. If a significant difference between these responses is indicated, further tests (e.g., multiple $t$, Tukey, and Newman-Keuls' analyses) can be performed to determine how the different stimuli/situations differed in their efficacy in eliciting a response (50, 51).

## Conclusions

This chapter has summarized some of the methods available for assessing the electrophysiological correlates of memory functions and other higher brain processes. The technique of recording single cell activity from awake monkeys provides a powerful means to study higher brain functions and can be applied in a variety of behavioral situations. In probing the functions of various cortical areas, it is valuable to use a battery of behavioral tests, both formalized and clinical. The possible pitfalls and artifacts lurking in any one paradigm are many when working on higher functions in a conscious animal. While designing electrophysiological experiments and interpreting the data, it is also important to take into account data obtained through other methods like lesion experiments in animals, human clinical cases, anatomical connectivity studies, behavioral experiments, and relevant physiological studies of the structures in question done on acute, nonrecovery preparations or even on brain slices.

## Acknowledgments

All experiments by the author reported in this chapter were done in collaboration with Drs. E. Salzmann and O. Creutzfeldt. The author is grateful to Dr. G. H. Henry for a critical reading of the manuscript.

## References

1. D. H. Hubel, *J. Physiol.* (*London*) **147,** 226 (1959).
2. H. Jasper, G. F. Ricci, and B. Doane, *Electroencephalogr. Clin. Neurophysiol. Suppl.* **13,** 137 (1960).
3. E. V. Evarts, *J. Neurophysiol.* **27,** 152 (1964).
4. E. V. Evarts, *in* "Methods in Medical Research" (R. F. Rushmer, ed.), Vol. 11, p. 241. Year Book Medical Publishers, Chicago, 1966.
5. E. V. Evarts, *Electroencephalogr. Clin. Neurophysiol.* **24,** 83 (1968).
6. R. Lemon, "Methods for Neuronal Recording in Conscious Animals." Wiley, New York, 1984.
7. V. B. Mountcastle, J. C. Lynch, A. Georgopoulos, H. Sakata, and C. Acuna, *J. Neurophysiol.* **38,** 871 (1975).
8. R. Porter, M. McD. Lewis, and G. F. Linklater, *Electroencephalogr. Clin. Neurophysiol.* **30,** 91 (1971).
9. A. M. Dymond, L. E. Kaechele, J. M. Jurist, and P. H. Crandal, *J. Neurosurg.* **33,** 574 (1970).
10. E. E. Fetz and M. A. Baker, *J. Neurophysiol.* **36,** 179 (1973).
11. M. Sakai and I. Hamada, *Exp. Brain Res.* **41,** 195 (1981).
12. P. M. Cahusac, E. T. Rolls, and F. H. Marriott, *Neurosci Lett.* **124,** 39 (1991).
13. J. P. Aggleton and R. E. Passingham, *Exp. Brain Res.* **44,** 271 (1981).
14. T. R. Vidyasagar, E. Salzmann, and O. D. Creutzfeldt, *Brain Res.* **544,** 269 (1991).
15. H. Asanuma and A. P. Arnold, *Brain Res.* **96,** 103 (1975).
16. E. T. Rolls, M. K. Sanghera, and A. Roper-Hall, *Brain Res.* **164,** 121 (1979).
17. A. E. Turner and A. Cowey, *J. Instrum. Anim. Tech.* **22,** 181 (1971).
18. L. R. Squire, "Memory and Brain." Oxford Univ. Press, New York, 1987.
19. W. B. Scoville and B. Milner, *J. Neurol. Neurosurg. Psychiatry* **20,** 11 (1957).

20. S. Corkin, *H. M. Semin. Neurol.* **4**, 249 (1984).
21. S. Zola-Morgan, L. R. Squire, and D. G. Amaral, *J. Neurosci.* **6**, 2950 (1986).
22. S. Zola-Morgan and L. R. Squire, *Behav. Neurosci.* **100**, 155 (1986).
23. M. Mishkin and J. Delacour, *J. Exp. Psychol.* (*Anim. Behav. Proc.*) **1**, 326 (1975).
24. W. H. Overman and R. W. Doty, *Neuroscience* (*Oxford*) **5**, 1825 (1980).
25. C. L. Martin-Elkins, P. George, and J. A. Horel, *Behav. Brain Res.* **32**, 219 (1989).
26. J. Delacour, *Neuropsychologia* **15**, 681 (1977).
27. D. Gaffan, *J. Comp. Physiol. Psychol.* **86**, 1100 (1974).
28. E. K. Warrington, *Cortex* **10**, 289 (1974).
29. E. T. Rolls, D. I. Perrett, A. W. Cann, and F. A. W. Wilson, *Brain* **105**, 611 (1982).
30. H. Niki, *Brain Res.* **70**, 346 (1974).
31. H. Niki and M. Watanabe, *Brain Res.* **105**, 79 (1976a).
32. H. Niki and M. Watanabe, *Brain Res.* **110**, 381 (1976b).
33. T. Watanabe and H. Niki, *Brain Res.* **325**, 241 (1985).
34. M. Mishkin, *Philos. Trans. R. Soc. London* (*Biol.*) **298**, 85 (1982).
35. D. Gaffan and R. C. Saunders, *Q. J. Exp. Psychol.* **37B**, 61 (1985).
36. D. Gaffan, *in* "Cognitive Neurochemistry" (S. M. Stahl, S. D. Iversen, and E. C. Goodman, eds.), p. 45. Oxford Univ. Press, Oxford, 1987.
37. E. T. Rolls, Y. Miyashita, P. M. B. Cahusac, R. P. Kesner, H. Niki, and J. Feigenbaum, L. Bach, *J. Neurosci.* **9**, 1835 (1989).
38. D. Gaffan and S. Harrison, *Q. J. Exp. Psychol.* **36B**, 223 (1984).
39. N. M. J. Rupniak and D. Graffem, *J. Neurosci.* **7**, 2331 (1987).
40. Y. Miyashita, E. T. Rolls, P. M. B. Cahusac, H. Niki, and J. D. Feigenbaum, *J. Neurophysiol.* **61**, 669 (1989).
41. J. Hyvaerinen and A. Poranen, *Brain* **97**, 673 (1974).
42. L. Leinonen, J. Hyvaerinen, G. Nyman, and I. Linnankoski, *Exp. Brain Res.* **34**, 299 (1979).
43. D. A. Robinson, *IEEE Trans. Biomed. Eng.* **10**, 137 (1963).
44. S. J. Judge, B. J. Richmond, and F. C. Chu, *Vision Res.* **20**, 535 (1980).
45. M. Bach, D. Bouis, and B. Fischer, *J. Neurosci. Methods* **9**, 9 (1983).
46. E. Halgren, T. L. Babb, and P. H. Crandall, *EEG Clin. Neurophysiol.* **45**, 585 (1978).
47. G. Heit, M. E. Smith, and E. Halgren, *Nature* (*London*) **333**, 773 (1988).
48. G. A. Ojemann, O. Creutzfeldt, E. Lettich, and M. M. Haglund, *Brain* **111**, 1383 (1988).
49. O. D. Creutzfeldt, *in* "Vision, Memory and the Temporal Lobe" (E. Iwai and M. Mishkin, eds.), p. 187. Elsevier, New York, 1990.
50. J. L. Bruning and B. L. Kintz, "Computational Handbook of Statistics." Scott Foresman, Glenview, Illinois, 1977.
51. E. T. Rolls and G. C. Baylis, *Exp. Brain Res.* **65**, 38 (1986).

# [28] Measurement of Two Types of Classical Conditioning

Paul F. Chapman and Lonnie L. Sears

## Introduction

The diversity and complexity of mammalian behavior is an accurate reflection of the complexity of the mammalian brain. Just as neuroscientists have found it necessary to use simplified preparations to understand aspects of neurophysiology and pharmacology, behavioral neuroscientists have developed paradigms that allow accurate, objective, and replicable measurements of relatively simple behaviors. Among the most popular of these are a set of associative learning paradigms that employ classical (or Pavlovian) conditioning. We describe the measurement and analysis of two forms of classical conditioning: the conditioned nictitating membrane/eyelid response, an example of a discrete conditioned skeletal muscle response, and conditioned lick suppression, one of a number of ways to measure conditioned fear.

## Classical (Pavlovian) Conditioning

Although conditioned skeletal muscle responses and conditioned fear (or, more generally, conditioned emotional responses) differ in many respects, they also demonstrate several properties that are characteristic of conditioned behavior. These include (1) dependence on the temporal pairing of neutral and aversive stimuli, (2) adherence to a steep interstimulus interval function, and (3) sensitivity to the order of stimulus presentation. Successful measurement of conditioned behavior depends on an understanding of these variables.

One of the theoretically and practically important features of classical conditioning is that it can be either very simple or quite involved. In its most basic form, classical conditioning depends on the presentation of a neutral conditioned stimulus (CS), followed some time later by the unconditioned stimulus (US), which produces a reflexive unconditioned response (the UR). This simple formula can be elaborated by the addition of compound stimuli, for example, to produce higher order learning phenomena such as second-order conditioning, latent inhibition, and blocking. While these phenomena

*Methods in Neurosciences, Volume 14*

are interesting and important in their own right, a complete description of higher-order conditioning is beyond the scope of this chapter.

## The Conditioned Stimulus

The conditioned stimulus is one chosen to be "neutral" prior to pairing with the US, where the word neutral is generally taken to mean that the stimulus does not produce an overt behavioral response. This definition of neutrality is useful, and most conditioned stimuli used to condition vertebrates fall into this category. It should be noted, however, that successful conditioning can be achieved with stimuli that do produce small responses prior to associative conditioning. Most notably, this form of conditioning (sometimes called alpha conditioning) is used in the marine mollusk *Aplysia,* in which responses to tactile conditioned stimuli to the animal's siphon or mantle shelf are associatively enhanced by pairing with electric shock. Moreover, even when there is no detectable behavioral response, there is still evidence of reflex activation by some conditioned stimuli (1).

For the most part, conditioned stimuli are either auditory or visual stimuli. Auditory stimuli are usually either pure tones or white noise, although discrete clicks are sometimes used. When light is the CS, it is usually generated from a small light source that is turned on briefly in an otherwise dark conditioning chamber. Tactile stimuli are used less frequently in conditioning vertebrates, but vibration is an effective CS.

## The Unconditioned Stimulus

The US can be either appetitive (i.e., food or water) or aversive, although in practice aversive stimuli are usually favored. The advantage of aversive stimuli is that they often produce discrete, reliable, and easily quantified defensive responses. Tactile stimuli delivered to the face, for example, produce eyelid closure and contraction of other facial muscles that can be measured using electromyography (EMG). Mild electric shocks to either the forelimbs or hind limbs produces limb flexion, which can also be measured by either EMG or other methods (described below). In contrast, appetitive unconditioned stimuli, such as food or water, involve consummatory responses, which usually involve a constellation of behaviors such as the orienting of the head and body, approach of food or water source, and head and jaw movement, as well as physiological responses such as salivation. Moreover, for animals to generate consummatory unconditioned responses,

adjustment of their motivational state is required (i.e., food or water deprivation). Aversive stimuli are generally less dependent on motivational factors.

## Acquisition and the Interstimulus Interval

The essence of Pavlovian conditioning is that repeated presentations of the
CS paired with the US produces a change in the animal's response to the
CS such that it comes to resemble the response to the US. The term *paired
with* is generally taken to imply two features of stimulus presentation: (1)
the CS is presented prior to but within reasonable temporal proximity of the
US, and (2) the probability that the US will follow the CS approaches 1.0.
The time interval between the CS and US onset is called the interstimulus
interval (ISI) and is important for determining the strength and rate of acquisition of the CS–US association.

The optimal ISI varies considerably depending on the task. For eyeblink
conditioning in the rabbit, for example, the optimal ISI is between 250 and
500 msec (2). When conditioning taste aversions (by pairing a distinctively
flavored solution with gastrointestinal discomfort), an ISI of more than 1 hr
is effective. Whatever the optimal ISI for any particular form of conditioning,
it is a general characteristic of classically conditioned responses that they
follow a steep ISI function. Very short intervals (i.e., less than 50 msec) are
usually relatively ineffective at producing robust conditioned responses, as
are those considerably longer than the most effective value.

The simplest and most effective form of conditioning is called delay conditioning. In this condition, the CS is presented, and then after some time
delay (the ISI), the US is presented. The CS and the US then terminate at
the same time. In trace conditioning, the CS terminates before the onset of
the US. To form an association between the CS and the US, the subject
must retain a "memory trace" of the CS. This form of training produces
robust conditioned responses, but it typically requires more training. In
general, conditioning animals by presenting the US after the CS does not
produce learning. Backward conditioning can be detected under certain conditions, but for simple conditioning paradigms this extreme case of a short
(i.e., negative) ISI is not practical.

## Extinction

Once conditioned to asymptotic performance levels, the strength of the
conditioned association tends to remain relatively stable over time. If the
subject is then presented with stimulus conditions in which the CS is not an

accurate predictor of the US (i.e., US presentation is not made explicitly contingent on CS presentation), the strength of the association [and therefore the conditioned response (CR)] will diminish rapidly. This process is called extinction, and it is usually induced by one of two means. In CS-alone extinction, the CS is presented repeatedly without the US. In unpaired extinction trials, subjects receive equal numbers of conditioned and unconditioned stimuli, presented in random order. Evidence indicates that extinction is an active process, involving the useful adaptation to the condition where no association between the CS and US occurs.

## Skeletal Muscle versus Emotional Conditioning

Although there are many types of conditioning, we have chosen to focus on two general classes, and one example from each class. Aversive (or defensive) skeletal muscle conditioning has been used for decades to understand the neural circuitry and mechanisms underlying learning and memory. In particular, the conditioned eyeblink response, which was first examined in humans in the 1930s, has become increasingly popular since its initial description in rabbits in 1962 (3).

The conditioned eyeblink response offers several distinct advantages that are particularly appropriate for studying learning and memory. First, the response produced by the US is discrete, reliable, and easily measured. The acquisition of the response is rapid enough to permit the measurement of physiological correlates on a reasonable time scale, but not so rapid that learning curves are difficult to measure. Moreover, since the UR and CR are temporally distinct, it is possible to measure conditioned responses as they develop during repeated CS–US pairings. The contributions of nonassociative learning phenomena such as habituation and sensitization are negligible, suggesting that conditioned responses produced in the well-trained subject are the result of associative learning. The conditioned eyeblink is particularly well suited to examinations of the neuronal circuitry underlying simple associative learning. More detailed discussions of the use of the rabbit in classical conditioning (4) or techniques for neural recordings are available (1, 5, 6).

An important aspect of behavioral neuroscience focuses on the neural basis of emotion. Pairing a neutral CS with an aversive stimulus has proved useful for developing and testing animal models of strong human emotions such as fear and anxiety. Unlike the conditioned eyeblink response, however, both the unconditioned and conditioned responses are difficult to measure directly in any of the several paradigms used to test emotional learning. Consequently, the design of these experiments is somewhat more complicated than that used for skeletal muscle conditioning. Although the measurement of conditioned emotional responses requires distinct training and testing phases, the response is typically acquired rapidly, so the overall training and testing time actually compares favorably to what is required for conditioned eyeblink.

## Conditioned Eyeblink Response

An introduction to the biomechanics of the rabbit eyeblink is helpful for comparing and contrasting the different procedures used to measure the response. When an air puff US contacts the cornea, retractor bulbi muscles retract the eyeball into the socket (1). These muscles are innervated by cranial nerve VI (abducens) originating in the abducens and accessory abducens nuclei (7). Eyeball retraction results in a largely passive extension of the nictitating membrane (NM), which is a membrane composed of conjunctiva and cartilage located behind the inner canthus of the eye. In conjunction with eyeball retraction, eyelid closure occurs. Eyelid closure is controlled by muscles that are innervated by cranial nerve VII (facial).

## *Measurement of Nictitating Membrane Extension*

The eyeblink conditioned response has been measured using features of both NM extension/eyeball retraction and eyelid closure. Historically, the most popular technique has been to measure NM extension (3). Typically, a small loop of 6-0 nylon suture is placed in the NM by gently pressing on the cornea to produce extension and then suturing the loop at a point near the membrane edge. The loop should be in the outer membrane and away from the cornea to avoid eye irritation. A small hook can then be attached to the suture, which is attached by thread to various devices able to transduce NM movement into an analog signal.

A minitorque potentiometer with an arm attachment and a counterbalance converts NM movement into analog voltage changes. The potentiometer can be secured on the rabbit with a screw slot attached to the rabbit's skull with dental acrylic. This procedure is performed during surgery when recording electrodes are also cemented to the skull. Other methods for transducing the NM movement involve the use of a photosensitive diode in a Bakelite box. An arm attached to the NM with the suture and small hook produces rotation of an axle, varying the intensity of a light source measured by a diode. NM movement can also be measured without an NM suture. A light source can be directed at the cornea, and then reflected back to a shielded photocell. NM extension alters the reflected light that is detected by the photocell, which produces an analog signal. All of these techniques produce a signal that can be digitized and stored for subsequent analysis.

Several studies have utilized tailor hooks to retract the eyelids while measuring NM movement. This procedure is generally unnecessary, however, since similar NM measurements are obtained with unretracted eyelids. It may be important to retract the eyelids, however, when looking at specific behavioral or neural features related to the US. In a study of dorsal accessory olive US-evoked activity during conditioning, for example, eyelids were retracted to help differentiate neuronal changes related to learning from changes due to eyelid closure prior to air puff presentation (8).

## Measurement of Eyeball Retraction

A second measurement of conditioned responding is eyeball retraction. Because eyeball retraction produces NM extension, the two measures provide similar information. Retraction can be measured by placing a counterbalanced lever in contact with a polyethylene loop on the cornea. Movement is transduced to an analog signal by a potentiometer attached to the lever. Eyeball retraction is more difficult to measure than NM extension and is used less frequently.

## Eyelid Response Measurement

Measurement of conditioned eyelid responses can be accomplished with the counterbalanced arm of a potentiometer attached with a loop of suture to the upper eyelid. Although the eyelid response is produced by different musculature than eyeball retraction, measurements of the two responses yield comparable data regarding characteristics of this type of motor learning (9).

The musculature involved in the eyelid response is easily accessible for electromyographic recording. In this procedure, a recording electrode is implanted in the muscles of the upper eyelid. The electrode is typically a uninsulated 30-gauge stainless steel wire and is implanted during surgery. A 26-gauge hypodermic needle can be placed through the muscle just superior to the upper eyelid, with the stainless steel wire inserted in the needle. The needle can then be withdrawn, leaving the wire in place. The wire can be twisted, and a male Amphenol pin can be crimped over the wire for easy attachment to wire leads during the conditioning procedure. Output from the electrodes is amplified, rectified, integrated, and stored either digitally or on FM tape for analysis. EMG is more sensitive to changes in both the amplitude and latency of the conditioned response (10). Given these advantages, we recommend EMG recordings over the other described procedures for measuring behavioral responses during this type of motor learning.


## Data Analysis

With appropriate transformations, all of the above procedures provide data that can be analyzed in several ways to quantify learning. A frequently used measure of learning is percent conditioned responses, namely, the percentage of trials in which a CR was produced. The determination of what constitutes a CR must be specified according to both the latency and amplitude of the response. To distinguish conditioned responses from spontaneous blinks or alpha responses (reflex responses to the CS), a reasonable criterion would stipulate that the response must begin at least 80 msec after the CS onset, but before the onset of the US. Percent conditioned responses are frequently used to establish a criterion for asymptotic learning (e.g., 8 CRs within 9 consecutive trials).

Often the sensitivity of the apparatus for recording the eyeblink determines the amplitude at which a response can be reliably detected above baseline variations. A criterion of 0.5 mm of movement (eyelid, NM, or eyeball) is frequently used, although some equipment, such as EMG, may be able to reliably measure smaller movements. Another (less reliable) criterion that has been used is to define a CR as an eyeblink with an amplitude at least 10% of the UR amplitude.

The acquisition of associative learning can also be measured by the latency of the behavioral response. With repeated CS–US pairings, the onset of the eyeblink moves forward in time until the peak amplitude occurs simultaneously to onset of the US (2). Both onset and peak latency can provide important information to supplement the percent CR measure. Hippocampal or cerebellar cortical lesions, for example, can produce abnormalities in the timing of conditioned responses, leading to inferences about the involvement of these brain areas in the timing of learned responses (11, 12).

A third measure of conditioned responding is CR amplitude. As learning occurs, there is an increase in the amplitude of the eyeblink prior to US onset. Use of the amplitude criterion has an advantage over either percent CR or CR latency, in that there is no need to define a CR, since learning is determined by relative changes in amplitude. The CR amplitude can be a particularly sensitive measure of the ability of animals to perform the learned response, since several manipulations that may not affect percent CR disrupt

CR amplitudes (13). Also, the amplitude measure provides a useful means for assessing extinction of a conditioned response (2).

Finally, the magnitude of conditioned responses can be measured by calculating the area of the eyeblink during the CS period (14). The magnitude measure incorporates both amplitude and latency information while reducing across-trial variability in the CR measure. A number of software programs provide a magnitude measure as part of the data analysis options (15). A similar CR measure is the "efficiency ratio," which is the ratio of CR to UR amplitude and varies from 0.0 to 1.0 for a response of maximum efficiency (16). The efficiency ratio appears to be more sensitive to the initial phase of learning, compared to other measures, since relatively large changes in the ratio occur early in training, maximize, and rapidly stablize, thus yielding less additional learning information.

The determination of the preferred CR measure depends in part on the equipment used to record and analyze the behavioral response as well as the specific research question to be addressed. Because each feature of the conditioned response can be altered under different conditions, it is useful to analyze each CR index. If the various measures provide similar information, the data presentation can be simplified to a report of the percent conditioned responses.

## Trial Types

One distinct advantage of the rabbit eyelid conditioning paradigm is the opportunity it provides to control stimulus presentation. Three general trial types can be presented: CS–US paired trials, CS-alone trials, and US-alone trials. The presentation of CS–US paired trials is obviously essential for studying classical conditioning. By altering features of the stimulus presentation that occur in other trial types, the relative contributions of associative and nonassociative factors to responses can be distinguished.

Repeated presentations of unpaired CS- and US-alone trials can produce nonassociative learning (i.e., habituation, dishabituation, or sensitization). To differentiate changes in the behavioral response due to these nonassociative factors from learning due to the pairing of the stimuli, a control group that receives unpaired trials is important. If a procedure produces a change in any response measure as a result of paired but not unpaired stimulus presentations, changes in the characteristics of the behavioral response likely resulted from associative learning. The effects of nonassociative factors on rabbit classical eyelid conditioning has been well studied (2) and should be controlled for in any study of associative learning.

US-alone trials can also be used to determine if an experimental manipulation alters performance components of the eyeblink response. For example, cerebellar involvement in classical eyelid conditioning is supported by the observation that cerebellar lesions disrupt the CR but not the UR amplitude (17). Since the UR is not disrupted, the cerebellar lesion is not impairing performance of the eyeblink per se, rather the lesion affects an aspect of the associative pairing of the CS and US.

Conditioned responses that are sometimes undetectable in the presence of a US can be measured accurately on CS-alone trials. Early in training, the CR is relatively small, and its onset latency relatively is long, so that it is often overshadowed by the much larger UR. With training, the latency of the CR decreases and its amplitude increases; however, relying solely on paired CS–US presentations to measure learning would result in an overestimation of the number of pairings required to establish the association. Moreover, experimental manipulations that delay the onset of the CR in well-trained animals could be misinterpreted as impairing associative learning unless CS-alone probe trials are also presented.

It is also useful to monitor the spontaneous eyeblink rate of the rabbit in the absence of both the CS and US. Variations in spontaneous blink rate can alter the number of conditioned responses observed, since it can sometimes be difficult to differentiate the two types of responses. Spontaneous blinks that occur by chance during the CS period could be incorrectly scored as conditioned responses. During eyelid conditioning with a 350-msec ISI, a spontaneous eyeblink occurs during the CS period in approximately 2% of trials (2).

## Measuring Eyeblink Conditioning in Humans

Prior to its application to the rabbit paradigm, classical conditioning of the eyeblink response was studied in humans to better understand learning-associated phenomena (18). More recently, the paradigm has been used to make inferences about the neuropathology of various brain disorders (19, 20). Many considerations regarding stimulus presentation and data analysis described for rabbit eyeblink conditioning also apply to human subjects and are not repeated here.

Measurement of human eyeblink responses in recent studies has been accomplished using several different procedures. A potentiometer can be attached to headgear worn by the subject, with the counterbalanced arm connected to the subject's upper eyelid with a thread attached to a false eyelash. EMG can also be employed by attaching an active and an indifferent electrode above the eye, over the muscles controlling the eyeblink. Finally, in conjunction with a cap electrode arrangement used for EEG recordings, an electrooculogram (EOG) can be recorded. The EOG is produced by recording a potential between the cornea and eyeball which varies with eye movement. Because EMG is the most direct and sensitive measure of the eyeblink, it is recommended over the other methods. All of these procedures produce a signal which can be digitized and stored for subsequent analysis.

An additional requirement of classical eyeblink conditioning in humans is controlling for attentional and motivational factors. Prior to experimental procedures experimenters often provide instructions suggesting that the subject should relax and "do what comes naturally." Particularly with subjects who may be uncooperative due to age or a disease state, it is also helpful to provide a silent movie for viewing while conditioning trials are presented. In addition, when testing subjects who may have a brain disorder, it is important to assess hearing as well the UR of the subject to determine if auditory or motor output systems are compromised, thus making interpretation of conditioning data more difficult.

## Conditioned Lick Suppression

There are three general phases to measuring a conditioned emotional response. In the first phase, the animal establishes some baseline level of a measurable activity. This could be startle amplitude in response to a noise burst, rate of bar press for food reward, or rate of licking a water spout. In the second phase, Pavlovian conditioning occurs when a neutral CS is paired with an aversive US, usually uncontrollable foot shock. Finally, conditioned fear is measured by presenting the CS while the animal is engaged in the phase one activity.

### Measuring Baseline Lick Rate

In using lick suppression to measure conditioned fear, it is first useful to establish a baseline lick rate. This serves a dual function: it allows identification of individual differences in the drinking rate of animals, and it allows the animals to become adapted to the chamber and the drinking tube. This baseline phase may last for two to four sessions, depending on how readily the animals adapt. It is generally our impression that it is better to err on the side of caution and allow four baseline sessions.

To encourage animals to drink during the testing sessions, it is necessary to limit their water intake prior to the first day of testing. This is accomplished by removing all water 48–72 hr before the first session. This period of deprivation is sufficient to ensure that subjects will begin drinking shortly after being placed in the cage for the first time.

Testing is most commonly performed in a standard operant conditioning chamber, allowing access to the spout of a water bottle. Licks are measured in one of two general ways. Conductive lick counters detect continuity between the metal spout of the water bottle and a metal grid chamber floor, which serves as the ground. The animal completes the loop by standing on the floor and touching the water spout with its tongue, producing an amplified output signal. Note, however, that the animal can also complete the loop by touching the spout with its forepaw as it drinks, or even by touching the spout with a whisker. In either case, continuous contact could produce a spurious response, which would either over- or underestimate the real lick rate, depending on the system. This problem is less likely to occur if the drinking tube is recessed as much as possible in the cage wall, ensuring that the animal can reach it only with its tongue.

It is also possible to measure licks using a photobeam assembly. For this procedure, the tip of the water spout is recessed into the cage wall, and a beam of light (from a fiber optic light source) shines across the tip of the drinking spout to a photodetector on the opposite side. When the animal licks the spout, it produces a break in the photobeam, which can be used to generate a logic signal for a computer, chart recorder, or mechanical counter. In a 5-min drinking session, a water-deprived rat (48–72 hr) will drink almost constantly. Rats will take in enough water in a 10- to 15-min period daily to maintain body weight throughout the duration of the experiment.

## Delivering the Conditioned and Unconditioned Stimuli

The second phase of lick suppression training is the one in which learning occurs. This is a relatively straightforward example of classical conditioning; the CS, which is usually a tone, is repeatedly paired with an inescapable foot shock US. The conditioning phase for lick suppression is similar to that used for any other measure of conditioned fear [e.g., conditioned freezing (21) or potentiated startle (22)]. The interstimulus interval used for conditioned fear is as critical as it is in eyeblink conditioning, but the intervals tend to be longer. Stronger conditioning occurs when the ISI is 15–60 sec than when it is in the millisecond range. Although pure tone is often used as a CS, white noise is equally effective, and other auditory stimuli, such as clickers (23), have been used successfully. Mild foot shocks (0.5–1.0 mA; 500–750 msec) are effective unconditioned stimuli.

Five to ten CS–US pairings are sufficient to produce strong conditioning (21, 23, 24). The trials can be massed in one session or distributed over two. If, for example, drugs are to be administered, or reversible lesions performed during training, ten tone–shock pairings can be delivered in approximately 10 min, which would produce robust conditioning in control animals. An intertrial interval of approximately 1 min is appropriate, although (as with other forms of Pavlovian conditioning) it is usually best to randomize the exact ITI somewhat.

## Measuring the Conditioned Response

The principal procedural distinction between conditioned skeletal muscle responses and conditioned fear responses is that the fear CR is not identical to the UR to shock. While autonomic responses can be measured "in real time" during conditioning (21, 25), there is reason to believe that the neural circuitry underlying autonomic expression of emotional responses is different from that subserving behavioral expression (26). Therefore, the effect of CS–US pairings must be measured indirectly.

Each of several ways of measuring conditioned fear has advantages and disadvantages. One can measure fear (or anxiety) by comparing the amount of startle elicited by a burst of white noise in the presence versus absence of the CS. If the CS elicits fear in the animals, their startle amplitude (typically measured as cage displacement) will be potentiated. This preparation is advantageous in that it does not require that the animal perform any motivated behavior, such as drinking or bar pressing. The response is a simple reflex, and its neural circuitry has been well described (22). The measurement of potentiated startle requires some specialized equipment (e.g., for measuring cage displacement), and it may therefore be less practical for those who do not intend to use the paradigm regularly.

Freezing is the simplest and most easily measured conditioned fear response (21). After pairing of a CS and shock, animals are placed in a novel environment, where they begin to explore. The CS is then presented, and animals typically stop moving. Freezing can most easily be measured by observation. Observers who are blind to the animal's condition determine the amount of time between the onset of the CS and the first movement. In well-trained animals, this latency measure is higher than in untrained ones, although the extent to which the freezing is learning-specific can depend on the baseline rate of exploration shown by the animals. It is also possible to automate this procedure, however, using one of a number of commercially available activity monitors.

Conditioned lick suppression measures the extent to which learned fear interferes with an ongoing, motivated behavior, namely, drinking of water by thirsty rats. Although this does require animals to engage in a behavior at a steady baseline rate, the behavior is rapidly acquired, requiring no special training. Moreover, measuring the lick rate for several days prior to CS–US paired training can establish a stable baseline. The equipment required to measure the response can be quite modest, consisting principally of an amplifier circuit for measuring animal contact with the drinking tube. Furthermore, the correct learned response is a suppression of ongoing activity, so that treatments that impair learning do so by increasing activity rather than decreasing, thereby minimizing the confounding effects of motivational, arousal, or motor impairments.

Two measures of fear conditioning are used for lick suppression: the lick ratio, which assesses the number of times the animal licks during the CS period compared to the immediately preceding period, and the lick latency, which measures how long it takes the animal to make a lick after CS onset. Both can be measured within the same session. Animals are placed in the test cage and allowed to lick for a period of time (e.g., 30 sec) to establish a baseline. Because there is often some reluctance to drink when the animals are first placed in the test cage after CS–US pairing, the session should begin with the first lick, rather than when the animal is place in the cage. The latency to first lick can also be used as a measure of generalized fear conditioning, particularly if the training cage and test cage are similar.

Using a 15-sec ISI as an example, the lick suppression ratio would be calculated by taking the lick rate during the 15 sec prior to the CS onset (pre-CS), and dividing that by the sum of the pre-CS lick rate plus the number of licks during CS presentation. The resulting ratio will be 1.0 if the animal does not lick at all during the CS presentation period, and 0.5 if the number of licks in this period is equal to the pre-CS rate. In a 5-min session consisting of 20 15-sec bins, the CS can easily be presented six times, so that multiple measures can be averaged for each animal in each session. This minimizes the impact of single trial anomalies, such as when an animal fails to lick during the pre-CS period.

## Conclusion

Pavlovian conditioning can be very useful in the examination of learned behaviors and, particularly, for studying the underlying neural circuitry in the mammalian central nervous system. Although they cover a relatively small portion of the entire spectrum of associatively learned responses, conditioned skeletal muscle responses and conditioned emotional responses represent a reasonable sample of the paradigms used by behavioral neuroscientists to investigate the relationship between neural circuitry and behavior.

# References

1. C. F. Cegavske, R. F. Thompson, M. M. Patterson, and I. Gormezano, *J. Comp. Physiol. Psychol.* **90,** 411 (1976).
2. I. Gormezano, E. J. Kehoe, and B. S. Marshall, *in* Progress in Psychobiology and Physiological Psychology'' (J. M. Sprague and A. N. Epstein, eds.), Vol. 10. Academic Press, New York, 1983.
3. I. Gormezano, N. Schneiderman, E. Deaux, and I. Fuentes, *Science* **138,** 33 (1962).
4. M. M. Patterson and A. G. Romano, *in* "Classical Conditioning" (I. Gormezano, W. F. Prokasy, and R. F. Thompson, eds.), Vol. 3, p. 1. Lawrence Erlbaum, Hillsdale, New Jersey, 1987.
5. T. J. Gould, L. L. Sears, and J. E. Steinmetz, *Kopf Carrier* **29,** 1 (1991).
6. T. J. Gould, L. L. Sears, and J. E. Steinmetz, *Kopf Carrier* **30,** 1 (1992).
7. T. S. Gray, S. E. McMaster, J. A. Harvey, and I. Gormezano, *Brain Res.* **226,** 93 (1981).
8. L. L. Sears and J. E. Steinmetz, *Brain Res.* **545,** 114 (1991).
9. D. A. McCormick, D. G. Lavond, and R. F. Thompson, *Physiol. Behav.* **28,** 769 (1982).
10. D. G. Lavond, C. G. Logan, J. H. Sohn, W. D. A. Garner, and S. A. Kanzawa, *Brain Res.* **514,** 238 (1990).
11. S. P. Perrett, B. P. Ruiz, and M. D. Mauk, *Neurosci. Abstr.* **16,** 268 (1990).
12. R. L. Port, A. A. Mikhail, and M. M. Patterson, *Behav. Neurosci.* **99,** 200 (1985).
13. D. G. Lavond and J. E. Steinmetz, *Behav. Brain Res.* **33,** 113 (1989).
14. H. S. Pennypacker, *Science* **144,** 1248 (1964).
15. D. G. Lavond and J. E. Steinmetz, *J. Neurosci. Methods* (1990).
16. A. B. Levey and I. Martin, *Psychol. Rev.* **75,** 398 (1968).
17. R. F. Thompson, *Science* **233,** 941 (1986).
18. E. R. Hilgard and D. G. Marquis, *Psychol. Monogr.* **47,** 186 (1936).
19. R. G. Finkbiner and D. S. Woodruff-Pak, *Neurosci. Abstr.* **17,** 698 (1991).
20. D. S. Woodruff-Pak and R. F. Thompson, *in* "Life-Span Development and Behavior'' Vol. 9 (P. B. Baltes, D. M. Featherman, and R. M. Lerner, eds.), Vol. 9, p. 1. Erlbaum, Hillsdale, New Jersey, 1988.
21. J. E. LeDoux, P. Cicchetti, A. Xoagoraris, and L. M. Romanski, *J. Neurosci.* **10,** 1062 (1990).
22. M. Davis, *Annu. Rev. Neurosci.* **15,** 353 (1992).
23. N. R. W. Selden, B. J. Everitt, L. E. Jarrard, and T. W. Robbins, *Neuroscience (Oxford)* **42,** 335 (1991).
24. J. Hitchcock and M. Davis, *Behav. Neurosci.* **100,** 11 (1986).

25. B. S. Kapp, R. C. Frysinger, M. Gallagher, and J. R. Haselton, *Physiol. Behav.* **29,** 1109 (1979).

26. J. E. LeDoux, *in* "Handbook of Physiology. The Nervous System V: Higher Functions," (F. Plum and V. Mountcastle, eds.), p. 419. American Physiological Society, Bethesda, Maryland, 1988.

# [29] Measurement of Song Learning Behavior in Birds

Douglas A. Nelson and Peter Marler

## Introduction

### Song Acquisition and Development

Birds learn song by at least three different processes. In the first, birds memorize songs heard from adults, and then use this memory to control their own vocal output via auditory feedback. This is the "sensorimotor" or "memory-based" process of song learning (1, 2). In the second, birds learn to select certain songs for retention in their repertoire (3, 4). This process of selective attrition has been termed "action-based" learning (1). Finally, "sensory" learning occurs when males (and perhaps females) learn to recognize the individual attributes of songs of territory neighbors or social companions without necessarily vocalizing themselves. Learning to recognize the songs of neighbors does not require an effect on the structure of a bird's own song, although it can do so, and song sharing or "dialects" may result. In this chapter we discuss methods for measuring sensorimotor learning.

The sensorimotor process consists of a sensory phase, during which songs are committed to memory, and a sensory–motor phase, during which the bird learns to match its own vocal output to the auditory memory previously acquired. If auditory feedback is denied by deafening, or if birds are not provided with appropriate song models, as in social isolates, songs with abnormal structure result.

Song development can be divided into three general stages: subsong, plastic song, and crystallized song (5). The first stage, subsong, consists of a quiet series of highly variable notes. The series may range from several seconds to a long, rambling sequence several minutes in duration. The subsong stage, with interruptions, may last from 1 or 2 weeks to 9 months, depending on the species, before merging into plastic song. The first evidence of imitations of specific tutor models becomes apparent in plastic song. Early in plastic song, individual notes and syllables are highly variable in structure, and song material acquired from different models is often combined in different orderings. In several species, plastic song includes imitated or invented material that is not incorporated into the final, crystallized repertoire. It is during the transition from plastic song to crystallized song that action-based learning occurs as an attrition process, with some song material lost from the production repertoire and the remainder sung with increased stereotypy.

## Dominant Themes in Song Learning Research

Two major themes pervade work on avian song learning. One theme has been to identify the physiological and sensory bases of the learning mechanisms. This theme has become increasingly important in recent years following the identification of discrete brain nuclei involved in the motor control and perception of song (6–8). The usual procedure here has been to apply an experimental treatment to one group of birds (e.g., lesion of a brain nucleus, nerve section, hormonal manipulation) and to compare their songs to a control group of birds. Methods are required to establish that the experimental perturbation has affected the outcome of the song learning process.

A second theme has involved identifying the temporal and acoustic selectivity of the learning process. Birds of many species are selective in what they will learn and when they will learn it (9, 10). Establishing the acoustic and temporal selectivity of the learning process takes advantage of the remarkable fidelity with which birds reproduce the songs they memorize. In a typical experiment, subjects are exposed to a variety of tutor songs at different times in development. Techniques are then required to compare the songs developed by each subject to the array of potential tutor stimuli. By determining which tutor song(s) most closely resembles the songs learned by subjects, inferences can be drawn about when song acquisition occurs.

This chapter first discusses issues in experimental design that bear on estimation of the temporal and acoustic selectivity of learning. Then we discuss methods for sampling song repertoires. This is a critical issue because many birds sing mature crystallized repertoires that contain multiple song types. Moreover, analysis of the plastic song repertoire, which may be larger than the crystallized song repertoire, sometimes provides a more accurate and complete estimate of what is actually learned, and how it is learned.

Next we describe current techniques for quantifying the effects of experimental manipulation on the structure of learned songs. Techniques fall into two major methods: (1) analysis of the acoustic structure of songs and (2) bioassays using the responses of male or female birds to broadcast of tape-recorded songs.

## Experimental Design

Many species of song birds will memorize tape-recorded songs broadcast from a loudspeaker. This ability, combined with the precision with which stimuli can be controlled, makes tape-tutoring the method of choice for experimentation on the sensory bases of song learning. The technique of presenting different tutor treatments associated with particular tutor song types has been termed "pattern type labeled tutoring" (11). An example is given in the next section. Although the use of live tutors has provided insight into the social factors that influence song learning, control of the stimulus is difficult with live tutors, but relatively easy with tape tutors. Digital computer methods for the storage and reproduction of sounds, which are replacing the functions of analog tape recorders, are capable of providing more precise control over the characteristics of a stimulus (12). Here we use "tape tutoring" to refer to both analog and digital methods for the manipulation, storage, and reproduction of tutor songs. Below we discuss several factors that influence song learning and that should be controlled when preparing tutor tapes, unless they are the independent variable(s) under study.

## Bout Structure

Birds that produce repertoires of multiple song types usually do not sing songs in random order. Instead, the serial pattern of delivery is often organized into bouts, which may assume two contrasting forms. When singing with immediate variety, birds cycle rapidly through their repertoires without immediate repetitions (e.g., ABCDEABCDE . . .). In contrast, in eventual variety, song types are repeated before changing to the next (FFFGGG . . . JJJ). The examples just given could be used as tutoring stimuli, with different temporal delivery patterns marked by unique song types (pattern type labeled tutoring). If birds learned song types F and J, for example, this would indicate they preferred learning songs presented with eventual variety. Song acquisition by male nightingales (*Luscinia megarhynchos*) is affected both by the bout structure and temporal spacing between songs (13, 14). In designing tutor tapes, the normal pattern of song delivery should be taken into account.

## Number of Songs Sufficient for Acquisition

The frequency with which a song stimulus is experienced may affect the probability of its being memorized. Song sparrows learned more song types when exposed for a 6-week period (2520 presentations) than in 1 week (420)(15). White-crowned sparrows learned more song types when they heard 1440 presentations over 40 days, rather than 800 repetitions over 10 days (16). Nightingales can memorize a song from as few as 10 repetitions (17). Increasing exposure from 25 to 50 to 100 repetitions did not increase acquisition success (18). In these studies, the number of presentations is confounded with the duration (in days) of presentation. It is not known how these factors may interact. At the moment, it appears that 1000+ repetitions, either alone or when presented over 2 weeks or more, maximize acquisition success in sparrows. More data are required to estimate the shape of song acquisition functions.

## Acoustic Selectivity

Some species of songbirds exhibit innate predispositions to memorize species-specific song. Digital methods for the synthesis of acoustic signals (19, 20) provide powerful tools for exploring the acoustic basis of selective learning. Tutor stimuli can be created by combining and rearranging parts of natural stimuli, or they can be created *de novo*.

## Temporal Selectivity

Sensitive phases for memorization of songs, during which heightened sensitivity for learning songs occurs relative to other times, are a common feature of the sensorimotor process. To estimate when learning is most likely to occur, subjects are presented with a sequence of tutor songs over a finite period of time; usually, the first year of life will be of most interest. Each song in the sequence is presented for a block of $n$ days, and the block duration limits the temporal resolution in estimating the duration of the sensitive phase. Across tutor blocks, presentation frequency, bout structure, and time of day during which tutoring occurs should be rigorously controlled. If birds require a certain number of song presentations for accurate imitation (see above), this may limit the temporal resolution of the analysis even further. To identify with confidence the particular tutor song(s) on which subjects base their imitations, the tutor songs must be sufficiently different from one another so that their imitations are not confusable. The tutor songs must also be equally attractive (i.e., conspecific songs in most subject species). Therefore, the acoustic selectivity that birds exhibit places an upper bound on the tutor songs available. The need for songs to be discriminable will limit the number of stimuli even further, and this will place an upper limit on the number of tutor blocks available. One way to increase temporal resolution is to restrict the analysis to a shorter period instead of the entire year.

## Sampling Repertoires

To assess whether and when song learning has occurred, a reliable estimate of the number of song types produced is necessary. Song repertoires of individual birds vary greatly among species. Individuals of many species sing a single song type in their adult, crystallized repertoires. Repertoires of a dozen song types are not uncommon, and they may range into the hundreds. Identifying song types is in itself a challenging problem in vocal taxonomy. Identifying song types in plastic song is even more difficult, and it is often more useful to estimate the repertoire of smallest, recombinable vocal units, namely, notes or syllables (21). Identification of types of vocal units has usually been done by visual inspection of sound spectrograms (see below), although quantitative techniques have also been applied (22).

To estimate repertoire size a catalog of vocal units (hereafter song types) is built. This is accomplished most easily by producing sound spectrograms via real-time spectrography (23). As song types are encountered, they are compared to types already in the catalog, and the catalog is updated as necessary. Three techniques have been used to estimate how large the catalog (repertoire) is. One technique involves intensive sampling of an individual's singing. This is practical for small repertoires but may involve a great deal of labor for larger repertoires.

A second technique estimates repertoire size by fitting an exponential curve to the number of song types encountered plotted against the number of song tokens sampled:

$$n - R \left(1 - e^{N/R}\right) \qquad (1)$$

where $n$ is the number of song types encountered in the sample, $N$ is the number of song tokens sampled, and $R$ is the estimated repertoire size (asymptote of the curve) (24). This method assumes that song types are distributed at random within a singing performance. This is not true for birds that sing with eventual variety (e.g., AAABBBCCC . . .) or in a predictable sequence (e.g., ABCABC . . .). $N$ must be adjusted, perhaps by counting the number of song bouts sampled (three and two bouts in the examples above) instead of the number of individual songs, to reflect dependencies in the bird's singing.

A third method includes a correction for sample coverage ($\Theta$), the estimated probability that the next sampled song belongs to a type already encountered in the sample (25):

$$R = n/\Theta \qquad (2)$$

where $\Theta = 1 - S/N$, and $S$ is the number of song types that occurred only once in the sample. This method has the advantage of providing an estimate of sample coverage, although similar information can be obtained from the exponential plot. Provided a "large" sample of songs is analyzed, these three methods yield similar results (25).

## Acoustic Analysis

Acoustic analysis involves comparing the structure of songs produced by experimental subjects to the songs of another group of birds (e.g., tutor birds, unmanipulated controls). Three basic techniques have been employed: (1) visual assessment of audiospectrograms by human judges, (2) measurement of acoustic variables, and (3) spectrogram cross-correlation. All three techniques make use of audiospectrograms, the standard method for graphically representing song structure. An audiospectrogram is produced by Fourier analysis [fast Fourier transform (FFT)], which plots the distribution of energy in a song as a function of frequency and time (Fig. 1). Spectrograms can be produced by special purpose machines (e.g., the Kay Sonagraph), but with the advent of powerful desktop computers they are increasingly produced with digital techniques (27, 28).

The user needs to understand the fundamental trade-off between frequency and temporal resolution inherent in Fourier analysis. There is a kind of uncertainty principle, such that an increase in temporal resolution results in a reciprocal decrease in frequency resolution (29, 30). A time-domain technique such as zero-crossing analysis does not have this liability and is faster to implement; however, it has not gained wide use, in part because its performance deteriorates drastically in the presence of noise, and it is cumbersome to use with signals that contain energy at more than one frequency at a point in time (e.g., harmonic sounds). Because bird songs often contain rapid amplitude and frequency modulation, most workers use a "wide-band" (300 Hz) analysis on the Kay Sonagraph, as this provides fine temporal resolution (3 msec) in a 8 kHz bandwidth. Digital spectrograms calculated with a 128-point discrete Fourier transform will provide temporal resolution of 5 msec in a 10 kHz bandwidth.

### Visual Assessment by Human Judges

Visual comparison of songs by human judges has been the most common method used in measuring song learning. Sound spectrograms are prepared of the songs of all animals in the study. To prevent observer bias, identities of the songs are coded. Several judges are independently asked to determine

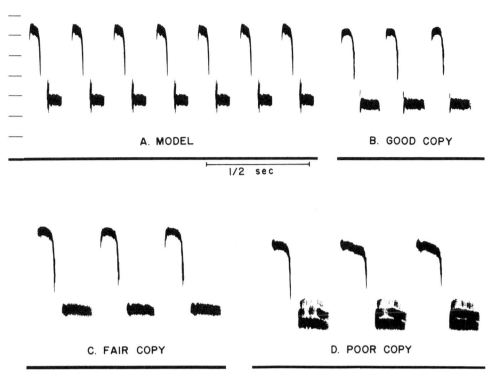

FIG. 1  Sound spectrograms from the trill portion of a song sparrow (*Melospiza melodia*) tutor song (A), and imitations of varying quality by three different subjects (B–D). [From Marler and Peters (26) with permission.]

how closely the songs in one set (e.g., tutor songs) resemble those in another set (subjects' songs). An ordinal scale representing the degree of similarity is often employed (Fig. 1). Agreement among judges can be expressed by the percentage of decisions in which they agreed or by Kendall's coefficient of concordance, which provides a statistical measure of agreement. The advantages of using human judges are that it is fast and it makes use of the most sophisticated pattern recognition device available: the human eye and brain. However, this technique is subjective and nonquantitative. Lack of a quantitative measure of similarity can impede further progress in analyzing the specific effects that an experimental procedure has on vocal learning.

## Acoustic Measurement

A more operational method for assessing the effects of vocal experience on development involves measurement of selected acoustic parameters. Audiospectrograms are the usual source of data, although temporal measurements can be made on the sound waveform, and some spectral measures can be obtained from single power spectra. To maximize precision, temporal measurements should be on "wide-band" (short FFT) spectrograms, whereas frequency measures should be derived from narrow-band (long FFT) spectrograms. Beecher (30) has suggested that analysis filters should be chosen to match the rate of nonperiodic FM in the sounds of interest. Although this is correct for a homogeneous sample of sounds, in practice the rate of FM can change rapidly within a song or even within a single note. Using several different filters to measure frequency characters would yield measurments made with varying precision. Measurement resolution for the analysis as a whole would be limited to that for the least precise measurement, thereby obviating any advantage gained by matching filters to FM rates.

Measurements are most easily made using digital signal analysis programs that permit procedures to be stated in operational terms or even automated. For IBM PC platforms, the SIGNAL program (31) is available, and for Macintosh platforms, MacRecorder (32) and Signalyze (33) are in use. The Kay DSP Sonagraph is a special-purpose analyzer.

The choice of measurement variables to include in an analysis is usually dictated by the nature of variation contained in the sample of songs. In other words, the user must select variables that describe the range of variation within the sample. In some cases, there may be an *a priori* basis for expecting that an experimental manipulation may affect a particular acoustic variable.

## Statistical Analysis

To state the obvious, the design of the experiment will determine the type of statistical procedure applied to the acoustic measurements. In general, bird songs are structurally complex, and multivariate statistical techniques are required to describe their complexity adequately. Using univariate tests on each variable is not only inappropriate, because variables are often not independent, but also will not be as sensitive at detecting differences as a multivariate procedure that includes variables simultaneously. Detailed discussion of such procedures is beyond the scope of this chapter (see Ref. 34).

To identify which of a number of potential tutors was likely to have influenced the development of a subject's song, one could employ a variety of distance metrics or correlation procedures. Nelson (4) used the Euclidean distance in an acoustic space of 14 standardized variables to compare songs of young field sparrows (*Spizella pusilla*) to the songs of potential tutors

(Fig. 2). Of the several songs in the repertoire of each young bird, the one that most closely resembled a tutor's song was preferentially retained in the repertoire.

Disadvantages of acoustic analysis are that it is time-consuming, the choice of variables to measure involves some subjectivity, and biological interpretation of acoustic differences may be difficult. The use of bioassays can address the last issue.

## Spectrogram Cross-correlation

A recently developed technique permits automated comparison of two digital spectrograms (35). In contrast to a comparison based on analysis of individual acoustic attributes, this technique compares the structure of entire spectrograms, thereby obviating problems involved in the selection of individual acoustic attributes. Imagine each spectrogram as a three-dimensional matrix or map, with frequency forming the rows (latitude), time forming the columns (longitude), and amplitude represented by elevation. Amplitude is first normalized within each spectrogram to remove differences due to absolute recording level. One matrix is then "slid" relative to the other in small time increments, and at each step the amplitude levels in corresponding cells are correlated. This generates a cross-correlation function, the peak of which represents the similarity between the two sounds (Fig. 3). Energy outside the frequency band occupied by the signals can be excluded from the analysis, thereby minimizing the effects of noise. Absolute frequency differences between the two sounds can also be eliminated prior to their comparison by digitally "rotating" one spectrogram up or down in frequency. This produces a cross-correlation influenced primarily by similarity in the shapes of spectrograms (frequency–time contours), thus achieving a judgment perhaps more similar to that based on visual comparisons, but in an objective and quantitative fashion.

This technique has several applications. In one, a subject's song, or part thereof, is compared to a library of tutor songs. The tutor song with the highest cross-correlation would be the likely model for the subject's song. In another, a developmental profile for a note can be constructed (35). Figure 4 shows two profiles for a note learned by a male swamp sparrow (*Melospiza georgiana*). The profiles are created by comparing notes sampled over a 16-week period from subsong to crystallized song to both (1) the bird's own crystallized song (the end point of the developmental process) and (2) the tutor note that the bird imitated in the first 6 weeks of life. The two profiles are similar into early plastic song, but then diverge, with the highest correlations

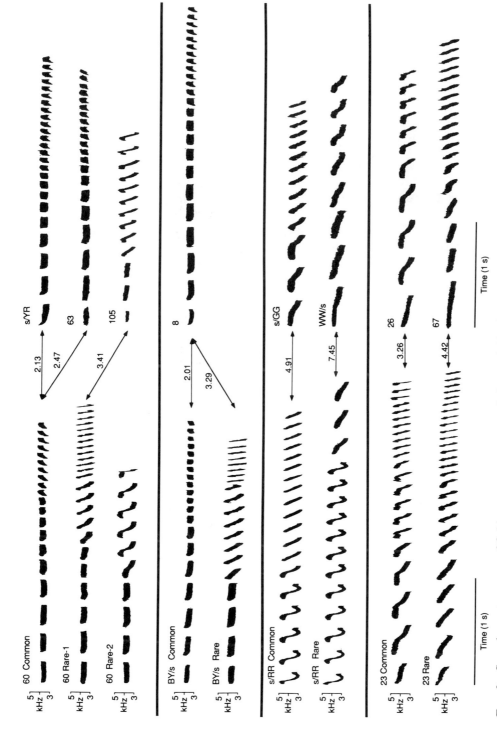

FIG. 2 Sound spectrograms of field sparrow songs recorded on neighboring territories. Numbers associated with arrows connecting pairs of songs are Euclidean distances calculated in a 14-dimensional space of acoustic variables. Small distances identify songs that are shared by territory neighbors. [From Nelson (4) with permission.]

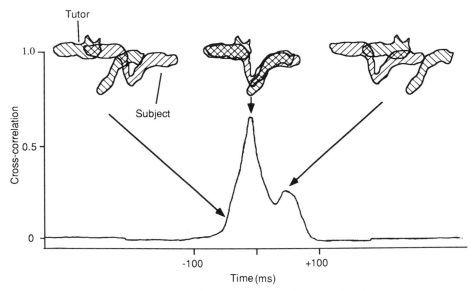

FIG. 3 Spectrogram cross-correlation technique. The two hatched areas represent the tutor note and the subject's imitation of that note. Three comparisons are shown, from left to right: tutor note leading the subject's note by 50 msec, tutor leading by 10 msec (yielding the maximal cross-correlation), and tutor lagging by 50 msec. Arrows indicate corresponding points on the cross-correlation function.

acheived with the bird's own crystallized song. Such profiles can be compared across treatment groups to gain insight into the course of song development.

A second application involves creating an average sound spectrogram based on a set of spectrograms. Figure 5 depicts average note spectrograms sampled over 9 weeks during development. Within each week the spectrograms are first aligned using the cross-correlation method before calculating the average. Notice how variable the note structure is in subsong (a blurry tracing), with variability then decreasing as development proceeds.

## Bioassays

A demonstration that the songs of tutored and untutored males differ acoustically is sufficient to show that tutoring affects the structure of learned songs. It does not follow, however, that the acoustic differences so produced function biologically, nor is it possible to rule out the possibility that biologically important effects were produced that are not detectable by conventional

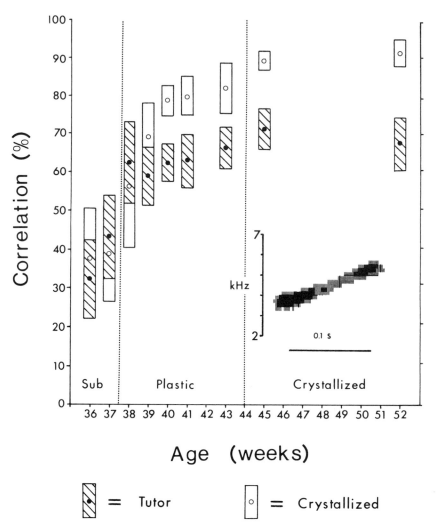

FIG. 4 Learning profiles for one male swamp sparrow note, generated by cross-correlating notes sampled at different points in development with either the tutor note the bird learned from or his own crystallized note. [From Clark *et al.* (35) with permission.]

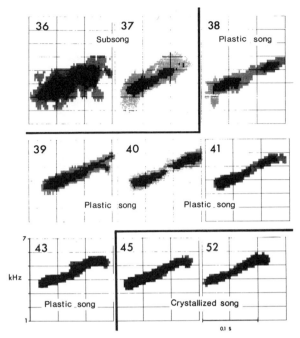

FIG. 5  Average sound spectrograms of the note depicted in Fig. 4 sampled at different times during development. [From Clark *et al.* (35) with permission.]

Fourier techniques. Because bird songs function as communication signals, one can use the natural, unconditioned responses that birds of either sex give to song stimulation as indices of whether learned aspects of song function in communication (Fig. 6) (36). Methods for testing male and female subjects differ.

## Male Subjects

Wild territorial male songbirds presented with tape playback of conspecific song typically respond with the same behaviors they show in a natural interaction with a live male intruding into the territory. Playback of heterospecific song usually evokes a weak response or none at all. The stimulus specificity of the response can be exploited to test whether learned aspects of song affect the potency of territorial responses. Only the most general aspects of song playback design can be discussed here (see Ref. 37). The Library of Natural Sounds at the Cornell University Laboratory of Ornithol-

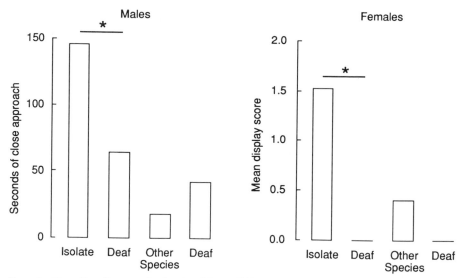

FIG. 6   Results of song playback of four different stimuli to male and female song sparrows. Playback of isolate song sparrow song to either males or females elicited stronger responses than did playback of deaf song sparrow song. Playback of deaf song sparrow song elicited responses similar to playback of other species song. Songs sung by isolated intact song sparrows retain sufficient species-typical structure to elicit territorial responses from males and copulation solicitation responses from females. Deaf song sparrow songs do not. [After Searcy and Marler (36).]

ogy and The Borror Laboratory of Bioacoustics at the Ohio State University are resources for information and equipment useful in field playback studies.

*Subjects and Stimuli*

There are often pronounced differences among males in their absolute level of responding. These differences arise in part from differences in the stage of the breeding cycle males are in. For this reason, within-subjects experimental designs will usually be the most sensitive at detecting treatment effects. Stimuli can be presented successively to each subject using a single loud-speaker, or simultaneously with two loudspeakers placed 15–20 m apart within the subject's territory. Each of $n$ independent subjects is exposed once to each treatment. Order of treatments should be randomized across subjects, using a Latin Square or other design as appropriate. To maintain independence of observations, avoid treating neighbors within a 2- or 3-day period. To avoid pseudoreplication of treatments, a variety of songs representing each treatment group should be used (38). For $n = 10$ subjects in a design comparing two treatments, ideally one would employ 10 different exemplars of each treatment. Practical considerations may reduce the number of stimuli available, but every effort should be made to use a representative sample of stimuli.

### Response Measures

In choosing response measures, there is no substitute for a thorough understanding of the natural behavior of the subject species. This can only be gained from naturalistic observation augmented by trial playbacks. Response measures should be easily observable and quantifiable. In single speaker designs, measures commonly used include approach distance to the speaker, numbers of songs, calls, or other displays, and numbers of flights (39). It is usually advantageous to record responses separately during and after stimulus playback, since responses may exhibit a ''ceiling effect'' during playback. In two-speaker designs, mean or median distance of approach to each loudspeaker is the only response variable available (40).

### Statistical Analysis

As with analysis of acoustic attributes of songs, multivariate analysis of response measures makes maximal use of the data and should be most powerful at detecting differences. A common technique has been to use principal components analysis (PCA) to reduce multiple response measures to a smaller set of independent principal components. Each component can then be used in repeated measures analysis of variance (ANOVA) or a nonparametric paired test to examine treatment effects (41).

## Female Subjects

In many species, it has proved exceedingly difficult to elicit from wild female songbirds noticeable, reliable responses to playback of song on their breeding territories. A technique (42) in which females of hole-nesting species are attracted to a loudspeaker placed atop an artificial nest box could be adapted to study female responses to learned aspects of song. Lacking a widely applicable assay for wild subjects, studies on female responses to song have been carried out in the laboratory. These studies have used the copulation solicitation posture evoked by song playback as the response measure.

### Subjects and Hormonal Implantation

Female brown-headed cowbirds (*Molothrus ater*) in breeding condition elicited by long days will produce the copulation solicitation posture in response to song (43). Females of all other species tested to date require treatment

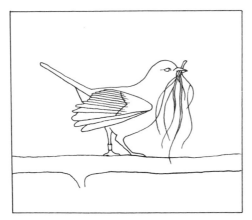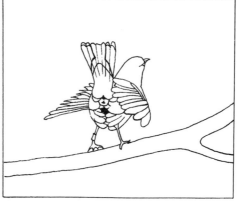

FIG. 7   Female copulation solicitation posture.

with estradiol (44) to elicit the response, and some species in addition require visual stimulation provided by a conspecific male (45). If testing is carried out in winter, subjects are placed on long days (16 hr light/8 hr dark) for 2 months prior to hormone implantation. 17$\beta$-Estradiol is administered via subcutaneous implantation of a hormone-packed length of Silastic medical grade tubing (1.96 mm outside diameter). The ends are sealed with Silastic adhesive (Dow Corning, Midland, MI). Based on published data for nine species, dosage is adjusted on the basis of body mass by the following formula:

$$\ln (\text{mm of estradiol}) = 0.041 (\text{body mass, g}) + 1.35 \qquad (3)$$

Subjects are housed individually in cages inside sound attenuation boxes for 6–10 days before testing begins. Disturbance to the birds should be minimized. Stimuli are broadcast from a loudspeaker within each sound attenuation box.

### Response

In the copulation solicitation display, the female arches her back as the tail is brought forward and the head back. The wings are moved away from the body and quivered (Fig. 7). Most studies have employed a subjective evaluation of responsiveness using an ordinal scale [0 = no response, 1 = incomplete display, 2 = full, short duration display (<1 sec), 3 = full, long display] (46). Recording of responses on videotape would be advantageous and would permit quantification of response magnitude and duration.

The response habituates rapidly, often after 2–5 presentations of one song type, so exposure to song should be minimized, unless the serial pattern of song delivery is an experimental variable. Most workers allow several hours to elapse between trials to the same subject. This technique has been used successfully to demonstrate species differences in relationships between male song repertoire sizes and sexual responsiveness of females (46, 47), as well as effects of lesions in the song system on the responsiveness of female canaries (*Serinus canarius*) to male song (48).

## Conclusion

Technical advances have opened up many new prospects in research on song learning in birds. Perhaps most important from a neurobiological point of view are computerized methods for synthesizing songs and songlike stimuli. It is now possible to explore the sensory domains of sound-responsive units rapidly and exhaustively by use of on-line synthesis programs. Such programs can generate wide ranges of complex acoustic stimuli that explore systematically the acoustic space surrounding biologically meaningful sound patterns such as songs and calls used in social communication (49, 50), with a degree of rigor and depth that was previously beyond our reach. There is an opportunity now for students of the auditory processing of complex communicative stimuli to make a unique contribution to advances in the neurobiological understanding of the kinds of neural circuitry that underly the detection of complex stimuli, as well as the modification of that circuitry as a consequence of learning and other forms of experience (51–53).

## Acknowledgments

This work was supported by a grant from the National Institutes of Health (MH 14651). We thank Linda Evans for preparing Fig. 7.

## References

1. P. Marler, *Trends Neurosci.* **14,** 199 (1991).
2. M. Konishi, *Z. Tierpsychol.* **22,** 770 (1965).
3. M. J. West and A. P. King, *Nature (London)* **334,** 244 (1988).
4. D. A. Nelson, *Behav. Ecol. Sociobiol.* **30,** 415 (1992).
5. P. Marler and S. Peters, *in* "Acoustic Communication in Birds" (D. E. Kroodsma and E. H. Miller, eds.), Vol. 2, p. 25. Academic Press, New York, 1982.

6. F. Nottebohm, T. M. Stokes, and C. M. Leonard, *J. Comp. Neurol.* **165,** 457 (1976).
7. M. Konishi, *Annu. Rev. Neurosci.* **8,** 125 (1985).
8. S. W. Bottjer and A. P. Arnold, *in* "Handbook of Behavioral Neurobiology" (E. M. Blass, ed.), Vol. 8, p. 129. Plenum, New York, 1986.
9. P. Marler and S. Peters, *in* "The Comparative Psychology of Audition: Perceiving Complex Sounds" (R. J. Dooling and S. H. Hulse, eds.), p. 243. Erlbaum, Hillsdale, New Jersey, 1989
10. P. Marler, *in* "Imprinting and Cortical Plasticity" (J. P. Rauschecker and P. Marler, eds.), p. 99. Wiley, New York, 1987.
11. H. Hultsch, R. Lange, and D. Todt, *Verh. Dtsch. Zool. Ges.* **77,** 249 (1984).
12. T. Dabelsteen and S. B. Pedersen, *Bioacoustics* **3,** 193 (1991).
13. H. Hultsch, *Anim. Behav.* **42,** 883 (1991).
14. H. Hultsch, *J. Comp. Physiol. A* **170,** 132 (1992).
15. P. Marler and S. Peters, *Ethology* **76,** 89 (1987).
16. P. Marler and D. A. Nelson, unpublished data (1993).
17. H. Hultsch and D. Todt, *J. Comp. Physiol. A.* **165,** 197 (1989).
18. H. Hultsch and D. Todt, *Anim. Behav.* **44,** 590 (1992).
19. S. R. Zoloth, R. J. Dooling, R. Miller, and S. S. Peters, *Z. Tierpsychol.* **54,** 151 (1980).
20. T. Dabelsteen and S. B. Pedersen, *Biol. Cybern.* **53,** 399 (1985).
21. P. Marler and S. Peters, *Dev. Psychobiol.* **15,** 369 (1982).
22. S. Nowicki and D. A. Nelson, *Ethology* **86,** 89 (1990).
23. C. D. Hopkins, M. Rossetto, and A. Lutjen, *Z. Tierpsychol.* **34,** 313 (1974).
24. J. L. Wildenthal, *Auk* **102,** 161 (1965).
25. D. E. Kroodsma and R. A. Canady, *Auk* **102,** 439 (1985).
26. P. Marler and S. Peters, *Ethology* **77,** 125 (1988).
27. P. K. Stoddard, *Bioacoustics* **2,** 217 (1990).
28. J. M. Williams and P. J. B. Slater, *Bioacoustics* **3,** 121 (1991).
29. P. Marler, *in* "Bird Vocalisations: Their Relations to Current Problems in Biology" (R. A. Hinde, ed.), p. 5. Cambridge Univ. Press, Cambridge, 1969.
30. M. Beecher, *Bioacoustics* **1,** 187 (1988).
31. "User's Guide." Engineering Design, Belmont, Massachusetts. 1991.
32. "MacRecorder User's Guide." Farallon Computing, Emeryville, California, 1990.
33. E. Keller, "Signalyze User's Guide." InfoSignal, Seattle, Washington, 1992.
34. P. W. Colgan (ed.), "Quantitative Ethology." Wiley, New York, 1978.
35. C. W. Clark, P. Marler, and K. Beeman, *Ethology* **76,** 101 (1987).
36. W. A. Searcy and P. Marler, *Dev. Psychobiol.* **20,** 509 (1987).
37. P. K. McGregor (ed.), "Playback and Studies of Animal Communication: Problems and Prospects." Plenum, New York, 1992.
38. D. E. Kroodsma, *Anim. Behav.* **37,** 600 (1989).
39. D. A. Nelson, *J. Comp. Psychol.* **103,** 171 (1989).
40. S. S. Peters, W. A. Searcy, and P. Marler, *Anim. Behav.* **28,** 393 (1980).

41. P. K. McGregor, *in* "Playback and Studies of Animal Communication: Problems and Prospects" (P. K. McGregor, ed.), p. 79. Plenum, New York, 1992.
42. D. Eriksson and L. Wallin, *Behav. Ecol. Sociobiol.* **19,** 297 (1986).
43. A. P. King and M. J. West, *Science* **195,** 1002 (1977).
44. W. A. Searcy and P. Marler, *Science* **213,** 926 (1981).
45. M. Baker, *Am. Nat.* **128,** 491 (1986).
46. W. A. Searcy and P. Marler, *Z. Tierpsychol.* **66,** 128 (1984).
47. C. K. Catchpole, J. Dittami, and B. Leisler, *Nature (London)* **312,** 564 (1984).
48. E. Brenowitz, *Science* **251,** 303 (1991).
49. D. A. Nelson, *Behavior* **106,** 158 (1988).
50. D. A. Nelson and P. Marler, *in* "Comparative Perception, Volume 2: Complex Signals" (W. C. Stebbins and M. A. Berkley, eds.), p. 443. Wiley, New York, 1990.
51. D. Margoliash, *J. Neurosci.* **3,** 1039 (1983).
52. H. Williams and F. Nottebohm, *Science* **229,** 279 (1985).
53. A. J. Doupe and M. Konishi, *Proc. Natl. Acad. Sci. U.S.A.* **88,** 11339 (1991).

# Index